Deterioration and Protection of Sustainable Biomaterials

ACS SYMPOSIUM SERIES 1158

Deterioration and Protection of Sustainable Biomaterials

Tor P. Schultz, Editor
Silvaware, Inc.
Starkville, Mississippi

Barry Goodell, Editor
Virginia Polytechnic Institute and State University (Virginia Tech)
Blacksburg, Virginia

Darrel D. Nicholas, Editor
Mississippi State University
Mississippi State, Mississippi

Sponsored by the
ACS Division of Cellulose and Renewable Materials

American Chemical Society, Washington, DC

Distributed in print by Oxford University Press

Library of Congress Cataloging-in-Publication Data

Deterioration and protection of sustainable biomaterials / Tor P. Schultz, editor, Silvaware, Inc., Starkville, Mississippi, Barry Goodell, editor, Virginia Polytechnic Institute and State University (Virginia Tech), Blacksburg, Virginia, Darrel D. Nicholas, editor, Mississippi State University Mississippi State, Mississippi ; sponsored by the ACS Division of Cellulose and Renewable Materials.
    pages cm. -- (ACS symposium series ; 1158)
  Includes bibliographical references and index.
  ISBN 978-0-8412-3004-0 (alk. paper)
  1. Biomedical materials--Congresses. 2. Biomedical materials--Deterioration--Congresses. 3. Sustainable engineering--Congresses. I. Schultz, Tor P., 1953- editor of compilation. II. Goodell, Barry, editor of compilation. III. Nicholas, Darrel D., editor of compilation. IV. American Chemical Society. Cellulose and Renewable Materials Division.
  R857.M3D48 2014
  610.28'4--dc23
                                    2014013674

The paper used in this publication meets the minimum requirements of American National Standard for Information Sciences—Permanence of Paper for Printed Library Materials, ANSI Z39.48n1984.

Copyright © 2014 American Chemical Society

Distributed in print by Oxford University Press

All Rights Reserved. Reprographic copying beyond that permitted by Sections 107 or 108 of the U.S. Copyright Act is allowed for internal use only, provided that a per-chapter fee of $40.25 plus $0.75 per page is paid to the Copyright Clearance Center, Inc., 222 Rosewood Drive, Danvers, MA 01923, USA. Republication or reproduction for sale of pages in this book is permitted only under license from ACS. Direct these and other permission requests to ACS Copyright Office, Publications Division, 1155 16th Street, N.W., Washington, DC 20036.

The citation of trade names and/or names of manufacturers in this publication is not to be construed as an endorsement or as approval by ACS of the commercial products or services referenced herein; nor should the mere reference herein to any drawing, specification, chemical process, or other data be regarded as a license or as a conveyance of any right or permission to the holder, reader, or any other person or corporation, to manufacture, reproduce, use, or sell any patented invention or copyrighted work that may in any way be related thereto. Registered names, trademarks, etc., used in this publication, even without specific indication thereof, are not to be considered unprotected by law.

PRINTED IN THE UNITED STATES OF AMERICA

Foreword

The ACS Symposium Series was first published in 1974 to provide a mechanism for publishing symposia quickly in book form. The purpose of the series is to publish timely, comprehensive books developed from the ACS sponsored symposia based on current scientific research. Occasionally, books are developed from symposia sponsored by other organizations when the topic is of keen interest to the chemistry audience.

Before agreeing to publish a book, the proposed table of contents is reviewed for appropriate and comprehensive coverage and for interest to the audience. Some papers may be excluded to better focus the book; others may be added to provide comprehensiveness. When appropriate, overview or introductory chapters are added. Drafts of chapters are peer-reviewed prior to final acceptance or rejection, and manuscripts are prepared in camera-ready format.

As a rule, only original research papers and original review papers are included in the volumes. Verbatim reproductions of previous published papers are not accepted.

ACS Books Department

Contents

Preface xi

Agents of Deterioration

1. **Current Understanding of Brown-Rot Fungal Biodegradation Mechanisms: A Review** 3
   Valdeir Arantes and Barry Goodell

2. **Fungal and Bacterial Biodegradation: White Rots, Brown Rots, Soft Rots, and Bacteria** 23
   Geoffrey Daniel

3. **Omics and the Future of Sustainable Biomaterials** 59
   Juliet D. Tang and Susan V. Diehl

4. **Genetic Identification of Fungi Involved in Wood Decay** 81
   Grant T. Kirker

5. **Evolution of Fungal Wood Decay** 93
   Daniel C. Eastwood

6. **Above Ground Deterioration of Wood and Wood-Based Materials** 113
   Grant Kirker and Jerrold Winandy

7. **Wood Deterioration: Ground Contact Hazards** 131
   Robin Wakeling and Paul Morris

8. **Thermal Degradation and Conversion of Plant Biomass into High Value Carbon Products** 147
   Xinfeng Xie and Barry Goodell

9. **Termites and Timber** 159
   Don Ewart and Laurie J. Cookson

Biocides

10. **Fungicides and Insecticides Used in Wood Preservation** 185
    Rod Stirling and Ali Temiz

11. Treatment Technologies: Past and Future 203
    Adam Taylor and Jeffrey J. Morrell

12. Copper-Based Wood Preservative Systems Used for Residential Applications in North America and Europe 217
    Stefan Schmitt, Jun Zhang, Stephen Shields, and Tor Schultz

13. Microdistribution of Copper in Southern Pine Treated with Particulate Wood Preservatives 227
    Philip D. Evans, Hiroshi Matsunaga, Holger Averdunk, Michael Turner, Ajay Limaye, Yutaka Kataoka, Makoto Kiguchi, and Tim J. Senden

14. Evaluating the Leaching of Biocides from Preservative-Treated Wood Products 239
    Stan T. Lebow

15. Discussion on Prior Commercial Wood Preservation Systems That Performed Less Well Than Expected 255
    Tor P. Schultz, Darrel D. Nicholas, and Patti Lebow

Nonbiocidal Modification

16. Processes and Properties of Thermally Modified Wood Manufactured in Europe 269
    H. Militz and M. Altgen

17. Wood Protection with Dimethyloldihydroxy-Ethyleneurea and Its Derivatives 287
    Yanjun Xie, Andreas Krause, and Holger Militz

18. Acetylation of Wood 301
    Roger M. Rowell and James P. Dickerson

Approval Processes

19. ICC-ES: The Alternate Path for Building Code Recognition 331
    Craig R. McIntyre

20. The Development of Consensus-Based Standards for Wood Preservatives/Protectants and Treated Wood Products 341
    Colin A. McCown

Global Trends

21. Wood Protection Trends in North America 351
    Tor P. Schultz, Darrel D. Nicholas, and Alan F. Preston

22. Preservation of Wood and Other Sustainable Biomaterials in China 363
   Jinzhen Cao and Xiao Jiang

Editors' Biographies 385

Indexes

Author Index 389

Subject Index 391

Preface

Wood and other structural lignocellulose biomaterials are renewable resources that provide sustainable products that require considerably less energy to manufacture into useable products than other alternatives produced from nonrenewable resources. However, these materials are readily biodegradable and as such must be protected if they are to be used in adverse environments. Consequently, their protection through chemical and nonchemical means plays a vital role in the satisfactory utilization of many products.

This publication represents the third ACS book by the three co-editors in a series addressing scientific and practical aspects of biodeterioration and protection of lignocellulose materials. The first book was published in 2003 and concentrated on basic wood deterioration mechanisms, decay detection and test methodology for evaluating wood preservatives, and development of new wood preservative systems. The second book was published in 2008, with major emphases on new developments in understanding basic wood biodeterioration, termite control methods, chemical and non-biocidal wood preservative systems, registration and approval systems for wood preservatives, and treated wood disposal issues.

The objective of this third book diverges to some extent from the prior texts in that we try to provide an overall view of our current understanding of the microbial and thermal degradation of plant biomass along with new developments in the rapidly changing field of wood protection. The latter is particularly important in light of dramatic changes in copper-based wood preservative systems that are used extensively to treat wood for residential construction, and in the commercial development of lignocellulose modification processes that protect bio-based materials without the addition of biocides. These changes, along with an update on new organic wood preservative systems, factors influencing wood biodeterioration above ground and in soil contact, wood treatment processes, registration and approval processes, applications of molecular biology in wood protection research, and the conversion of biomass into high value carbon products and worldwide trends in wood protection, are covered in this latest ACS book.

The individual chapters were authored by a world-class group of academic and industrial scientists in order to provide a state-of-the-art review and global perspective of this rapidly changing field and reviewed by internationally recognized scientists.

Darrel D. Nicholas, Ph.D.
Department of Sustainable Bioproducts
Mississippi State University
P.O. Box 9820
Mississippi State, MS 39762-9820

Barry Goodell, Ph.D.
Department of Sustainable Biomaterials
Virginia Polytechnic Institute and State University
216 ICTAS II Building (0917), 1075 Life Sciences Circle
Blacksburg, Virginia 24061

Tor P. Schultz, Ph.D.
Silvaware, Inc.
303 Mangrove Palm
Starkville, MS 39759

Agents of Deterioration

Chapter 1

Current Understanding of Brown-Rot Fungal Biodegradation Mechanisms: A Review

Valdeir Arantes[1] and Barry Goodell*,[2]

[1]University of British Columbia, 4035-2424 Main Mall,
V6T 1Z4, Vancouver BC, Canada
[2]Virginia Polytechnic Institute and State University (Virginia Tech),
216 ICTAS II Building (0917), 1075 Life Sciences Circle,
Blacksburg VA 24061, United States
*E-mail: goodell@vt.edu.

The biological decomposition of lignocellulosic materials, in particular woody biomass by wood-rotting Basidiomycetes, plays an essential role in carbon circle. Brown-rot fungi are perhaps the most important agents involved in the biodegradation of wood products and the dead wood in coniferous ecosystems. It has long been proposed that brown-rot attack is based on a two-step process. Over the last decades various pathways, mostly based on research with *Gloeophyllum* species, have been suggested as the potential biodegrative mechanism in brown-rot fungi. As research has advanced over time, some of these proposed pathways have been well supported, whereas others remain equivocal. This chapter provies an overview of the more widely reported pathways that are more likely to constitute the two-step biodegradative mechanism in brown-rot fungi and how this fundamental knowledge may contribute to the development of new environmentally benign organic wood preservatives.

© 2014 American Chemical Society

Introduction

Biodegradation of woody materials, the most abundant source of biomass on earth, is one of the most important processes in terrestrial forests and it is also a process that can severely impact the durability of wooden structures. Wood-decay fungi are some of the few organisms that can utilize or degrade nearly all biochemical forms of carbon in woody tissues facilitating the return of CO_2 to the atmosphere (1). Brown-rot basidiomycete fungi preferentially attack coniferous woods, and brown rot is also the most common and most destructive type of decay in structural wood products in the northern hemisphere (2).

It has been estimated that the equivalent of ten percent of the timber cut in the United States decays in service each year (3). Since the majority of timber used in construction in the northern hemisphere is from coniferous species, a large part of this destruction is due to the action of brown rot fungi.

Early or incipient stages of brown rot decay display little outward evidence of attack. However, extensive depolymerization of the cellulose within the wood cell wall occurs at very low weight loss, which causes the wood to rapidly lose strength in comparison with the rate of wood metabolism. When mass losses are between five and ten percent in brown rotted wood, in its hydrated state the incipiently decayed wood may only appear to be water stained. As decay progresses, the wood darkens, turning brown because of the preferential accumulation of modified lignin residues with removal of the holocellulose fraction. After drying, wood in advanced stages of degradation undergoes extreme shrinkage. In advanced decay stages, volume loss with drying in both longitudinal and transverse directions is much greater than in sound wood, and volume losses exceed thirty percent in some cases. The wood is friable and, in its weakened state, readily fractures across the grain due simply to drying stresses. The wood takes on a checked appearance with cubical pieces of the degraded wood prevalent.

Wood decay fungi and their degradative systems (and related biomimetic systems) have also attracted attention for their potential in bioremediation and bioconversion of recalcitrant wastes, bioconversion of lignocellulosic biomass into biofuels, biopulping of wood chips, biobleaching of cellulosic pulps, biosorption of dyes and heavy metals, paper deinking, lignin adhesive pre-treatment, and many other applications. The efficiency of fungal biotechnological applications and the development wood preservation methods will undoubtedly benefit from a better understanding of the chemical and biochemical mechanisms involved in brown rot decay of lignocellulosic polysaccharides and lignin.

Although it was postulated that low molecular weight systems must be involved in brown-rot since the 1960's (4), and previous hypotheses explored pathways for the action of proposed systems (5–8), the focus of research remained on enzymatic systems for these fungi until the mid-1990's.

In early stages of brown-rot decay, Cowling (4) first recognized that conventional enzymes were too large to penetrate into the intact wood cell wall. It is now well established that the initial stages of decay involve nonenzymatic action through the production of highly destructive hydroxyl radicals produced by extracellular Fenton chemistry (9–15). In addition, it has been shown that phenolic compounds such as 2,5-dimethoxy-1,4-bezoquinone and 4,5-dimethoxy-1,2-bezoquinone, produced by the fungi, and also biomimetic fungal phenolic compounds (*9, 16–18*) act as ferric iron chelators and sources of electrons for iron reduction, thus promoting mediated Fenton reactions.

Biology of Brown Rot Fungi and Physical Properties of the Degraded Wood

Brown-rot fungi, rather than a taxonomic group, constitute a physiological group of wood-rotting basidiomycete. They account for less than 10% of the taxonomic diversity of lignocellulose degrading basidiomycetes (*19–21*) but are prevalent in nature and represent the dominant wood decay fungi associated with northern coniferous forest ecosystems (*22*).

Brown-rot fungi are thought to have evolved from ancestral saprotrophic white-rot fungi (*23–25*). Comparative and functional genomics of wood decay fungi indicate that the evolution of brown-rot saprotrophy was accompanied by reductions and losses in key enzymes implicated in biomass breakdown in white rot, especially cellulases and lignin-modifying enzymes (*23, 25*). These studies are consistent with a recent genomewide analysis of cellulose- and hemicellulose-degrading enzymes in various white and brown rot fungi that revealed that relative to brown rot fungi, white rot fungi possess greater enzymatic diversity supporting lignocellulose attack (*26*).

Despite their loss of enzymatic systems, the brown-rot fungi are still capable of depolymerizing holocellulose and extensively modifying lignin. The brown-rot fungi have cast off the physiologically (energetically) expensive apparatus of lignocellulose degradation employed by white-rot fungi, and they have in turn acquired alternative lower-energy mechanisms to initiate fungal attack of wood (*23, 25*) to enhance the efficiency of their utilization of lignocellulse. It has also been hypothesized that some of the plant-derived modified lignin components may function in mediating/assisting in the production of hydroxyl radicals enhancing brown-rot fungi access to holocellulose and therefore digestible sugars (*27, 28*). In this way, fungal metabolites involved in mediating non-enzymatic processes would only need to be produced initially and in small amounts during early stages of decay, a pathway expected to be considerably less energetic than producing complex biomass deconstructing enzymes.

Decay by these fungi in the natural environment is typically initiated via the deposition of spores or mycelial fragments that are carried to wood or other lignocellulosic surfaces by the wind, water, or by insect or animal vectors; or because of mycelial growth into wood in direct soil contact. As with all wood inhabiting fungi, germination of spores and/or the initiation of fungal growth into the wood will not occur until appropriate conditions of moisture and temperature are met. In general, many types of brown rot decay can be initiated when: 1) The wood moisture content is above the fiber saturation point, but the cell lumen void space is not saturated, and 2) The temperature is between 10 and 45° C. Optimal temperatures for growth and degradative activity vary with fungal species (*3*) and there are many other factors that play a role in whether decay will initiate let alone grow into wood, and then begin metabolite production for initiation of the decay process. Of the millions to billions of spores that can be generated by a single fungus, only a few survive to actually initiate decay in lignocellulose materials.

The brown rot fungi initially invade wood when hyphae begin growing through the cell lumens and colonizing ray cells and axial parenchyma where stored carbohydrate is accessible as a ready energy source for the fungus. Once established in the parenchyma, the fungal mycelia penetrate through pit membranes to access tracheid lumens where they can further proliferate (*29*). These fungi develop the capacity to penetrate the wood cell walls via bore hole production, but for some brown rot fungi penetration via pit membranes remains the predominant mode of passage (*29*). In other species there is no preference for pit penetration as opposed to bore hole production (*19*). Formation of bore holes is initiated via the production of smaller hyphae (bore hyphae) with enlargement of the bore hole occurring as the decay process proceeds. The mechanism for bore hole production is not precisely known but is thought to include a combination of mechanical, chemical, and enzymatic processes.

As decay progresses, the hyphae of brown-rot fungi become established and proliferate in the lumen of wood cells (*19, 21, 30*). The hyphae secrete a glucan layer which coats the wood cell walls, and this 'hyphal sheath' helps to bind the hyphae to the S3 layer of the wood cell wall (*31*). While the S3 layer may remain relatively undegraded until late in decay (*31, 32*), the S2 layer of the wood cell wall is intensely degraded. Degradation of the S2 layer is not localized near the hyphae of fungi, which indicates that the degradation reagents produced by brown rot fungi are capable of diffusing through the S3 into the S2 layer of the wood cell wall. The S2 layer of wood cell walls undergoing brown rot attack is depolymerized preferentially to other layers. This layer has a reduced lignin content and lignin density compared to the S1 or S3 layers (*33*), which may help explain the preferential and early depolymerization of the cellulose there. In advanced stages of decay, an intact residual S3 layer is often observed within the cell wall while at the same time much of the S2 layer has been severely degraded and metabolized (*31, 32*).

Brown rot degradation is considered to be more problematic in structural wood products than white rot because of the way these fungi attack the wood cell wall and promote rapid strength losses. Although selective delignification of the cell wall has been observed in some white rot fungi (*34*), white rot attack of the wood cell wall typically occurs via thinning from the lumen outward. This is thought to occur because enzymes secreted by the fungus can act only on exposed wood layer surfaces. They are confined to initiating reactions only at the wall layer surface because spatial considerations and diffusion do not permit access into interior regions of the cell wall. Since all known enzymes exceed the micropore size of the intact structure of the wood matrix, in the absence of a diffusible low-molecular weight degradation system, enzymatic action is at least initially confined to thinning or erosion of the wood cell wall starting at the S3 layer surface. Considering the strength properties of a wood cell, assuming a cylindrically shaped model approximates cells such as tracheids or fibers, the interior thinning of a cylindrical wall has only a limited effect on strength. (The mechanical properties in bending mode of a hollow pipe vs. a solid rod of the same size are similar). This explains why white rot fungi cause only a gradual loss of wood strength and stiffness of the material as decay progresses. Conversely, brown rot fungi employ a low molecular weight decay system and rapidly depolymerize the cellulosic fraction of the cell wall. Following, or concurrent with, depolymerization of cell wall components using this system, enzymatic degradation also occurs. The rapid decrease in cellulose DP throughout the wood cell wall caused by the low molecular weight system is perhaps the key defining feature of brown rot attack at the chemical level. Although the brown rot fungi metabolize only limited amounts of the cellulose breakdown products in the initial stages of degradation, the extensive depolymerization of the cellulose in the wood cell wall dramatically reduces the strength of the wood.

Mechanical property loss caused by brown rot fungi have been reported to be as much as 70% of modulus of elasticity (MOE) and modulus of rupture (MOR) (*21*) in early, or incipient, decay stages. In other work (*35*) with *G. trabeum* brown rotted wood, loss of mass from the wood was undetectable until 40% loss in MOR had occurred. As discussed above, extensive cellulose depolymerization with limited glucose metabolism has been well documented as the cause of this mechanical property loss in early stages of degradation. However, Winandy and Morrell have shown that hemicellulose losses in early stages of brown rot degradation also correlate with mechanical property loss in wood (*36*).

Brown-Rot Biodegradative Mechanism

Fenton-Based Free Radicals in Brown-Rot Fungal Attack

Fenton chemistry was first proposed to be involved in wood biodegradation in the 1960's, when Halliwell (*37*) observed that cellulose could be degraded by Fenton-based free radicals, suggesting a possible role for Fenton reactions in the biodegradation of lignocellulose. Subsequently, Koenigs (*38–40*) demonstrated that Fenton chemistry oxidatively degraded the cellulosic fraction of softwood. These works also showed that wood decay fungi produce extracellular hydrogen peroxide, and demonstrated that wood contains enough iron to make the oxidant hypothesis possible (*38–40*). Later work by Highley (*41*) demonstrated that the changes observed in brown-rotted cellulose (increased reducing capacity and increased alkali solubility) resembled changes in cellulose caused by treatments with Fenton reagent. It was also observed that small amounts of carboxyl groups were generated in cotton cellulose depolymerized by the brown-rot fungus *Postia (Poria) placenta* or by Fenton reagent.

The aforementioned findings coupled with more recent evidence — reviewed elsewhere (*10*) — show that plant cell wall deconstructing enzymes are too large to penetrate into the intact non-modified wood cell wall. Earlier works (*42, 43*), supported the hypothesis of the involvement of smaller agents in the initiation of brown-rot decay. Support for a nonenzymatic oxidative mechanism involving mediated Fenton reactions for hydroxyl radical production in brown-rot decay has been provided by several research groups as discussed below.

Various studies have demonstrated the generation of hydroxyl radicals in liquid culture medium at both early (10 days) and more advanced (4–5 weeks) stages of wood colonization by brown-rot fungi (*31*, *44*). It has also been observed that as the biodegradation time increases, the level of OH radicals increases as well (*44*). More recently, genome, transcriptome, and secretome analysis of brown-rot fungi has supported the biodegradative role of Fenton chemistry in lignocellulose conversion (*23*, *25*).

Non-Enzymatic Brown-Rot Pathways

To attack plant cell wall constituents via Fenton chemistry directly or indirectly, brown-rot fungi require the presence of specific mechanisms to solubilize ferric iron from iron oxy(hydr)oxides present in plant tissues, and also to reduce iron to its ferrous state. Additional mechanisms are needed to produce H_2O_2.

The hydroxyl radical is the most powerful non-specific oxidant in biological systems, with an extremely short half-life (10-9s). In order to function optimally, Fenton reactions must occur immediately adjacent to the target for oxidative action within the secondary plant cell wall because of spatial diffusion limitations (•OH would be unable to diffuse from the environment surrounding the fungal hyphae into the wood cell wall because of the extremely short half-life) (*9*). Therefore, the Fenton system must be activated at a distance from the fungal hyphae to avoid the generation of highly reactive species in close proximity to the fungus, that could attack and damage fungal hyphae. Further, •OH must be generated very close to the cellulose or lignin for radical to be effective in acting on these substrate molecules. Thus, a mechanism must exist to generate or activate Fenton substrates withing the lignocellulose cell wall, and within close proximity to the target holocellulose or lignin molecules.

Despite the strong support for the biochemical Fenton reaction in brown-rot fungi, and continuing research on the production of species involved in Fenton chemistry, further research is needed to pinpoint the source(s) of Fe^{2+} and H_2O_2 in the brown-rot system. Further, because brown rot fungi are undergoing repeated systematic reclassification as new genomic databases for these fungi is generated, establishing extracellular Fenton reactions as a universal characteristic of brown-rot fungi may not be possible until more fungal genomes have been explored, and classification has stabilized. A summary of some most probable pathways leading to Fe^{2+} and H_2O_2 formation during the brown-rot decay process is provided below.

Location and solubilisation of iron: In plant tissues, iron can be found as insoluble Fe(oxydydr)oxide complexes in the wood cell lumen or bound to wood components in the wood cell wall (*9*, *28*), which must be solubilized and reduced to ferrous iron to participate in Fenton ractions.

Oxalic acid, is often accumulated extracellularly by brown-rot fungi (*44*, *45*), and it is capable of binding and solubilizing iron from Fe(oxyhydr)oxide complexes in the wood lumen region (*9*). pH conditions can strongly influence the dissolution of iron oxides (*46*). Protons facilitate the dissolution process by protonating the OH binding groups, thereby contributing to a weakening of the

Fe-O bond. Non-reductive dissolution is a simple adsorption process, where ferric ions are transferred to the bulk solution by a ligand with a high affinity for iron (46).

It has been proposed that, in the incipient stage of decay, iron is chelated by oxalic acid in the low pH environment of the brown-rot fungal hyphae within the wood cell lumen area (9) as a result of the oxalic acid production at high concentration in this area. Experimental evidence supports this hypothesis with regard to the solubilisation of ferric iron present in wood by oxalic acid (47). The authors showed that at a pH of around 2, in the presence of high oxalate:Fe molar ratios (physiological conditions typically found near fungal hyphae in wood undergoing brown-rot decay), oxalic acid can bind iron forming soluble oxalate:Fe complexes. Under very acidic conditions such as occur close to the hyphae in the lumen, fungal Fe^{3+}-reductants (phenolate and hydroquinone compounds reviewed below) have been shown to be weaker iron chelators compared to oxalate, due to the predominant formation of stable Fe-oxalate complexes ($[Fe(C_2O_4)^{2-}]$ and $[Fe(C_2O_4)_3^{3-}]$. Sequestration of iron in this microsite location would prevent low pH iron reduction, and damage to the hyphae from Fenton generated ·OH (9, 47, 48). As the newly formed Fe-oxalate complexes diffuse from the lumen into the cell wall, the complex diffuses through a pH gradient as proposed by Hyde and Wood (6) or pH differential proposed by Goodell (2). The wood cell wall is a highly buffered environment and would retain its natural pH (approx. pH 5.5), and in addition lower oxalic acid concentrations would be present at a distance from the hyphae. Once inside the wood cell wall, the inherently high buffering capacity of the lignocellulose matrix supports maintenance of the naturally greater pH of the wood against any pH change (49). Under these conditions (at a distance from the hyphae) of pH 3.6 or higher, and at a low oxalate:Fe molar ratio, oxalate has a decreased affinity for iron and this enables the temporary transfer of Fe^{3+} (from Fe-oxalate complexes, mostly $[Fe(C_2O_4)^+]$ to the cellulosic fraction of the wood (47). As described in more detail below and shown in Figure 1, in this pH environment (pH 3.6 or higher) iron would then be sequestered directly from Fe-oxalate complexes or from cellulose (iron originally bound or previously sequestered from Fe-oxalate complexes) by Fe^{3+}-chelating/reducing agents, which possess a greater affinity for Fe^{3+} than that of the cellulosic fraction in the cell wall or oxalic acid, as was demonstrated with fungal biomimetic reductants. This proposed pathway provides a reasonable mechanism for accumulating Fe^{3+} and safely reducing it at a distance from the fungal hyphae (9, 47).

Hydrogen peroxide: There are two proposed sources of the H_2O_2 associated with mechanisms involved in brown-rot fungal decay, reduction of molecular oxygen and oxidation of methanol. Several pathways have been proposed for the extracellular reduction of O_2 to H_2O_2 (8, 17, 50, 51). Since they are coupled with the extracellular reduction of Fe^{3+}, formation of H_2O_2 via reduction of O_2 is discussed in the next section.

During brown-rot fungal attack, methanol is primarily generated through demethylation of the lignin substructures (52). Methanol has been found to be a preferred physiological substrate of the alcohol oxidase produced by brown-rot fungi (50, 53). Martinez et al. (25) observed the cellulose-induced expression of methanol oxidases in the brown-rot fungus *Postia placenta* (25), and Daniel et al.

(*50*) observed alcohol oxidase localized on the brown-rot fungus *Gloeophyllum trabeum* hyphae and extracellular slime and within the secondary cell walls of degraded wood fibers in liquid culture. Since methanol is unlikely to be of nutritional value for brown-rot fungi, it has been suggested that alcohol oxidase and methanol are more likely to serve as an important extracellular source of H_2O_2 during brown-rot decay (*50*). A controversial aspect of the proposed role of methanol in wood decay is the instability of alcohol oxidase at the acidic pH conditions that are characteristic of brown-rot decay and the limited permeability of enzymes into the S2 cell wall during incipient brown-rot decay (*43*). In addition, it cannot be ruled out that the oxidation of methanol by alcohol oxidase may, instead, be required to control the concentration of methanol in the fungal system because of its toxicity to the fungi. It is likely that the buffered pH of the extracellular slime layer permits activity of alcohol oxidase thus permitting a dual function of protecting the fungal from methanol toxicity and also providing a source of H_2O_2 which can then diffuse into the wood cell wall to react with reduced iron.

Iron-reducing agents: there have been various mechanisms proposed to explain the reduction of Fe^{3+} to Fe^{2+} during brown-rot decay. However, over time, some of these hypotheses have been well supported, whereas others remain equivocal. The more widely supported mechanisms for iron reduction involve extracellular low molecular weight fungal aromatic compounds. A simplified generic mechanism involving low molecular weight Fe^{3+}-reductants for in situ generation of Fe^{2+} and H_2O_2, and degradation of the major plant cell wall constituents by brown-rot fungi via •OH-producing Fenton reactions is shown in Figure 1. Lignin degradation products may function similarly to the fungal aromatic compounds after initiation of reactions by the fungal compounds, but more research is needed to confirm this hypothesis. The fungal aromatic compounds bear electron-donating substituents (such as –OH, –OCH$_3$), that auto-oxidize rapidly in the presence of most Fe^{3+} salts, thus generating Fe^{2+} (*27*). Several of these low molecular weight Fe^{3+}-chelating and -reducing compounds (herein referred to as fungal Fe^{3+}-reductants) of phenolate and hydroquinone origin have been shown to be produced by brown-rot fungi (*9, 48, 54–56*). These compounds can function to mediate the Fenton reaction by reducing, within the cell wall, Fe^{3+} to Fe^{2+} (*9, 48, 55, 56*).

The fungal Fe^{3+}-reductants, unlike enzymes, are of relatively low molecular weight and can penetrate through the wood cell wall matrix (*9*), and have been shown in immunolabeling studies to be present throughout the S2 layer of the brown-rot degraded cell wall (*57*). The low molecular weight 2,5-dimethoxyhydroquinone (2,5-DMHQ) has been implicated in non-enzymatic fungal depolymerization of lignocelluose components and, in this context, was first isolated from *Gloeophyllum* trabeum by Paszczynski et al. (*14*). It was subsequently found in *Postia placenta* (*58*), in various cultures of *Gloeophyllum* species (*59–61*), and *Serpula lacrymans* (*55*). Fungal Fe^{3+}-reductants such as 2,5-DMHQ, reduce Fe^{3+} to Fe^{2+} with the concurrent generation of a semiquinone radical. In addition, the semiquinone radical is capable of reducing O_2 to give •OOH, which can yield H_2O_2 by a dismutation reaction, or can reduce additional Fe^{3+} to Fe^{2+}, forming a quinone derived from the original Fe^{3+}-reductant (*17,*

60). It is also possible that in the simultaneous presence of •OOH and Fe^{2+}, the Fe^{2+}/Fe^{3+} couple equilibrates with the •OOH/O_2 couple, while Fe^{2+} reduces •OOH to yield H_2O_2 (*62, 63*). This mechanism of Fe^{3+} reduction and hydroxyl radical formation has been referred to as the chelator-mediated Fenton (CMF) reaction (*9, 48, 64*), and also subsequently by other terms including hydroquinone-redox cycling, or dihydroxybenzene-driven Fenton reaction (*42, 59*). Generation of hydroxyl radical via this mechanism has been demonstrated, and a key part of the mechanism involves the generation of hydroxyl radicals at a distance from the fungal hyphae (*9, 27, 65, 66*). This mechanism is proposed to occur via chemistry that is dictated by a pH and oxalate concentration differential which occurs between the fungal hyphae and the interior of the wood cell wall (*9, 22, 65, 66*) and other research supports that such a pH differential, or gradient, exists in brown rot fungi (*67*). The hypothesis put forth by Korripally et al. (*68*) that 2,5 dimethoxyhydroquinone is the sole Fe^{3+}-reductant produced across three brown rot fungal orders (Gloeophyllales, Polyporales and the Boletales) is unlikely given that other reducing chelators have been isolated by several other groups (*9, 40*). Further, Korripally et al. (*68*) have proposed a mechanism where 2,5-DMHQ, will generate a complete Fenton system without the presence of hydrogen peroxide, and without a pH. This hypothesis implies that the addition of 2,5-DMHQ to moist wood would initiate depolymerization (mimicking brown rot fungal degradation of wood) without the addition of additional chemical components. This does not occur. Further, the Korripally et al. (*68*) hypothesis would require that hydroxyl radicals be generated by 2,5-DMHQ in the absence of a oxalate concentration differential (*69*). If this were the case, then •OH-producing Fenton reactions would occur in the immediate vicinity of the fungal hyphae which would be deleterious to the fungus and would be inconsistent with prior research findings (*6, 69*).

As outlined above, during brown-rot degradation, lignin is extensively demethylated (*27, 70, 71*) with some of the degradation products proposed to have iron-reducing capabilities (*27, 64*). In support of this hypothesis is the close relationship between lignin demethylation and holocellulose loss, which occur simultaneously, that is, demethylation is proportional to holocellulose loss during brown-rot degradation of wood (*27*). This suggests that the two processes (holocellulose breakdown and lignin demethylation) may be mechanistically linked as part of the mediated Fenton-chemistry driven degradation of wood by brown-rot fungi (*27*). In addition, brown-rot degraded wood has also been shown to have substantially greater iron-reducing capability than extracts from wood colonized by white-rot or non-decay fungi (*64*).

A single mole of fungal or biomimetic Fe^{3+}-reductants can reduce multiple moles of Fe^{3+} (*9, 11, 66*). Although this non-stoichiometric reduction has not been well elucidated, it has been suggested that these Fe^{3+}-reductants could be regenerated (*13*) by a quinone oxidoreductase (*72*) or mineralized (*9, 73*). Pratch et al. (*73*) have shown that Fe^{3+}-reductants can be partially or completely mineralized to CO_2 while reducing ferric iron continuously. Consequently, fungal compounds may undergo a similar pathway of mineralization leading to the non-stoichiometric reduction of iron. On the other hand, it has been shown that an intracellular NADH:quinone oxidoreductase is produced in the brown-rot

Gloeophyllum trabeum (*74*) and is able to regenerate the fungal Fe^{3+}-reductant (*72, 75*). Recently, genes encoding a quinone reductase were found in *P. placenta* (*25*). Since the brown-rot fungal mediated Fenton reaction would necessarily be required to occur at a distance from the hyphae, diffusion of the oxidized Fe^{3+}-reductant from the S2 layer of the wood cell wall to the lumen area for transfer to the fungal hyphae would need to occur for regeneration by the quinone reductase. At this time, this diffusion mechanism has yet to be tested.

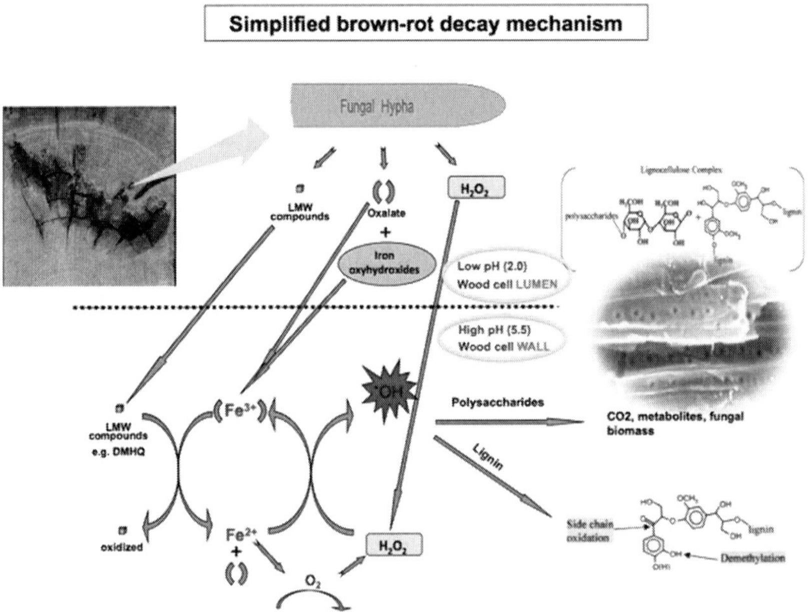

Figure 1. Simplified mechanism for in situ generation of Fe^{2+} and H_2O_2, and degradation of major plant cell wall macrocomponents by brown rot fungi via •OH-producing Fenton reactions. In brief, brown-rot fungal hyphae growing in the lumen area produce oxalic acid, low molecular weigh (LMW) iron reducing compounds, and hydrogen peroxide (H_2O_2). Oxalic acid binds Fe^{3+}, and the complex diffuses into the cell wall. H_2O_2 and LMW compounds also diffuse into cell wall, where LMW compound sequesters Fe^{3+} from the Fe–oxalate complexes and reduces it to Fe^{2+}. Fe^{2+} then reacts with H_2O_2 (Fenton reaction) and generates hydroxyl radicals (•OH). Upon attack of OH radicals, lignocellulose matrix is disrupted. In the incipient stage of decay, the major changes are cleavage of glycoside bonds in cellulose and hemicellulose and lignin demethylation, depolymerization, and repolymerization. (Photos courtesy of Barry Goodell.)

An alternative mechanism for iron reduction involving cellulose dehydrogenase (CDH) was proposed by Hyde and Wood (76). A potential problem associated with this pathway is that, to date, CDH has not been found ubiquitously in brown-rot fungi. Another suggested alternative involves extracellular low molecular weight glycopeptides, which have been identified in cultures of brown-rot fungi and also proposed to reduce Fe^{3+}. Recently, genes that apparently encode these peptides were found to be present in, and expressed by, *P. placenta* (25). These fungal glycopeptides have been reported to be NADH-dependent (12, 77), however NADH is physiologically unlikely to be found extracellularly in wood undergoing brown-rot decay.

The Chemistry of Polysaccharide and Lignin Biodegradation

Polysaccharide Biodegradation

Holocellulose is a major carbon/energy source for the growth of many fungi. Cellulolytic systems in brown-rot fungi differ significantly from that of other well-characterized groups of cellulose-degrading fungi such as the ascomycete *Trichoderma reesei* (syn. *Hypocrea jecorina*, an industrially relevant cellulase producer) and the white-rot basidiomycete fungus *Phanerochaete chrysosporium* (a model white rot fungus), which are thought to efficiently degrade cellulose via a synergistically acting system involving a set of hydrolytic enzymes composed of cellobiohydrolases, endoglucanases, and beta-glucosidases (78, 79).

As reviewed in the introduction, brown-rot decay is characterized by a rapid and extensive holocellulose depolymerization, with the accumulation of partially degraded sugars at initial decay stages when weight loss is minimal (4, 78, 80). Depolymerization occurs in the S2 layer of the wood cell wall at a distance from the hyphae. The cellulose that remains has an average DP of 150-200 and the overall percent crystallinity increases (80–82), presumably through degradative loss of the amorphous regions of the cellulose microfibrils (83) and preferential initial removal of hemicelluloses (59, 82, 84–88), with a near-complete removal of hemicelluloses occurring at approximately 20% weight loss (35), likely opening up the cell wall structure, thus increasing the accessibility of the cellulose.

Brown-rot fungi are known to produce many hemicellulases but, with the exception of *Coniophora puteana* (89), they have an incomplete cellulase system and lack cellobiohydrolases (26, 90, 91). In some species, it has been reported that the absence of cellobiohydrolase is compensated for by the production of processive endoglucanases to provide the functionality similar to a cellobiohydrolase (75). Degradation of native or insoluble cellulose has not been observed with isolated brown-rot fungal enzymes (41, 92), and widespread exoglucanases and CDH activities in brown-rot fungi has yet to be shown. These observations suggest that, brown-rot hydrolytic enzymes, mainly endoglucanases and hemicellulases, are likely to only contribute to the overall holocellulose breakdown after free radical generation and initial attack on the cell wall occurs. This is in agreement with Rättö et al. (93) who found that endoglucanases from the brown-rot fungus *P. placenta* and *T. reesei* as well as a commercial *T.*

reesei cellulase cocktail displayed higher hydrolysis yields after Fenton reaction treatment of spruce sawdust. The authors also found that the oxidation appears to modify the residual cellulose to a form that is more hydrolysable with the brown-rot endoglucanases (*93*). The decay pattern found by Fackler et al. (*84*), while characterizing brown-rotted spruce wood cell walls using FT-IR imaging microscopy, also supports the hypothesis that a non-specific primary attack of polysaccharides leads to the accumulation of decomposition products that then become accessible to enzymatic hydrolysis. In addition, it has recently been shown that the response of the brown rot fungus *Postia placenta* during incipient decay of modified wood (acetylated, DMDHEU-treated and thermally modified wood) up-regulates the expression of the oxidative machinery involved in polysaccharide degradation. The levels of expression of the genese investigated in the modified wood were equal to, or lower than, those in untreated wood (*94*).

Lignin Modification

With the advancement of analytical techniques, our view of brown-rot attack of lignin has changed over the past 10–15 years. Barr and Aust (*95*) and Gierer (*96*) discussed depolymerization and repolymerization of lignin in the presence of hydroxyl radicals, and Goodell et al. (*9*) suggested a similar mechanism existed in brown-rot when the chelator-mediated Fenton reaction was active. More recent research suggests that brown-rot fungi have a greater ligninolytic capability (*65, 69–71*) than previously thought (*78, 96*). Although brown-rot fungi do not remove lignin to an appreciable extent (*4, 97*), and the residual modified lignin retains most of its original aromatic residues (*70*) and several of the major types of lignin linkages (*71, 97*), brown-rotted softwood lignin (*27, 70*) and hardwood lignin (*97*) residues have been shown to be highly modified. These brown-rot alterations to lignin moieties may play a crucial role in brown-rot decay.

Brown-rotted lignins have been found to undergo extensive oxidative demethylation (*27, 71, 98, 99*), as well as significant side chain oxidation (*71, 97, 99*), depolymerization (*96*) and potentially repolymerization (*65, 69, 70, 81, 83*). Other significant alterations are limited aromatic ring cleavage (*99*), hydroxylation of aromatic rings (*99*); C$_\beta$-ether cleavage, partial side chains hydroxylation (*97*), and formation of new aryl-O-aryl ether (*71*), and aryl-aryl or side-chain (*70*) linkages. In spite of all these alterations, brown-rotted lignin is still polymeric (*96, 98, 100*), which may explain why earlier studies concluded that brown-rot fungi only slightly modified lignin without degrading it significantly (*78, 96, 98, 100*).

Unlike the white-rot fungi, which cleave the encrusting lignin to access the underlying polysaccharides through the action of lignin-degrading enzymes (e.g., laccase, managanese-peroxidase, lignin-peroxidase), brown-rot fungi do not typically produce such enzymes (*101, 102*). In addition, ultrastructural changes characteristic of brown-rotted lignin have also been observed within the S2 layer of the secondary cell wall (*78*), where ligninases are not expected to penetrate and attack due to spatial restrictions because of the mass of those enzymes, and the relative impermeability of the S3 layer.

Oxidative alteration of lignin side-chain linkages is characterized by the depletion of the major intermonomer linkages arylglycerol-β-aryl ether (β-O-4) (*70, 71*) and pinoresinol (β-β') (*68*), and C_α-C_β (*70, 71, 97*) during brown-rot decay. Recently, the cleavage of the β-O-4 linkage, in addition to demethylation, was observed in spruce treated with biomimetic brown-rot Fenton reactions (*65, 69*). This supports the hypothesis by Yelle et al. (*70*) suggesting that lignin attack by hydroxyl radicals during brown-rot decay should result in simultaneous cleavage of both methoxyl groups and intermonomer ether linkages. Another expected reaction associated with hydroxyl radical attack of the lignin side-chain is the oxidative cleavage of C_α-C_β, yielding benzoic acid residues and benzaldehydes (*78, 103, 104*). Over the last decade, many studies have detected an increase in the content of aryl carbonyl (benzoic acid and benzaldehyde) in brown-rotted lignin compared to undegraded lignin (*70, 71, 97, 105–107*), suggesting the ability of the brown-rot fungi degradative system to cleave the C_α-C_β linkage.

It has also been proposed that brown-rotted lignin has been re-polymerized (*65, 69–71, 108*). The high concentration of phenolic containing lignin fragments resulting from lignin demethylation, combined with the low or no fluid flow that would permit extensive extraction of the soluble demethylated lignin fragments out of the cell wall, and in combination with other micro-environmental conditions in brown-rotted cell wall and in the presence of reactive hydroxyl radicals, would provide appropriate conditions for new oxidation reactions to take place (*69*). This, in turn, would lead to highly unstable radical cation intermediates (phenoxy radicals) that can undergo polymerization by radical coupling reactions to produce a repolymerized lignin that the brown-rot fungi are evidently unable to significantly decompose as evidenced by the persistence of residual brown-rotted lignin observed in laboratory experiments. This would also be in agreement with the observation that brown-rotted lignin residues can persist in soil environments for decades to centuries.

Lignin repolymerization during brown-rot is in agreement with the finding by Filley et al. (*27*) that after a progressive increase in lignin demethylation during spruce biodegradation by *G. trabeum* and *P. placenta*, demethylation values dropped significantly, suggesting condensation of phenolics or the possible metabolism of the demethylated dihydroxy molecules (*27*). Lignin repolymerization is also consistent with the formation of new aryl-O-aryl ether linkages (*71*) in brown-rotted lignin, as well as the limited lignin removal/loss during brown-rot decay despite extensive degradative oxidation of several of the lignin moieties (*27, 71*).

These observations indicate that brown-rot ligninolysis via Fenton chemistry rearranges the lignin rather than degrading its substructures, a pathway that is likely to play a bigger role in the overall importance of brown-rot decay than is yet widely recognized, potentially, by facilitating the access of hydrolytic enzymes to polysaccharides. Therefore, in brown rot, although lignin is a barrier to carbohydrate metabolism by most microorganisms, the brown-rotted lignin matrix ultimately may assist in the carbohydrate break down necessary to support brown-rot fungal growth after the initial depolymerization of the wood cell wall.

Understanding of Brown-Rot Fungal Biodegradation Mechanisms Can Aid in the Development of New Environmentally Benign Organic Biocides for Wood, Bamboo, and Other Sustainable Biomaterials Protection

Many solid and composite lignocellulose products are treated to prevent biodegradation by lignocellulose-attacking organisms, such as brown-rot fungi, which represent the dominant wood decay fungi associated with northern coniferous forest ecosystems (*22*).

Over the last decade, there has been increasing interest in the development of alternative biocidal protection systems based on natural environmentally compatible compounds with the objective of replacing more toxic chemical preservatives currently still in use. Though progress in replacing traditional wood protection systems that employ toxic chemicals which can adversly affect human health and the environment has been slow, several studies have reported promising results on the potential of wood protection methods/processes based on natural products such as heartwood extractives alone or in combination with other biocides to control wood decaying fungi (*109–111*).

Some heartwood extractives have been shown to have limited fungicidal activity but have excellent antioxidant (free radical scavengers) and metal-chelating properties (*109*). These antioxidant activities and metal-chelating capability of heartwood extractives have been postulated to be key properites responsible for the natural resistance of heartwood of certain trees to decay fungi and/or insects (*112*). This is in agreement with our current understanding of brown-rot fungal biodegradation mechanisms, which is based on the oxidative attack of free radicals (i.e. hydroxyl radicals) generated via reaction of transition metals (i.e., iron) with hydrogen peroxide (Fenton reaction) as discussed earlier. This is also in agreement with recent studies showing that the response of a brown rot fungus during incipient decay of treated wood is to up-regulate the expression of the oxidative machinery involved in polysaccharide degradation (*94*).

Currrently, there is growing evidence that besides brown-rot fungi, other microorganisms, and even species in the Animal Kingdom, also make use of a non-enzymatic oxidative free radical attack in incipient decay and/or degradation processes. Therefore, it is reasonable to expect that the antioxidant properties of the heartwood extractives may also prevent free radical formation and, therefore, inhibit the degradative mechanisms of a wide range of organisms.

Future work on the fundamental knowledge of brown-rot fungal biodegradation mechanisms should help to provide guidance in the developing more efficient and cost-effective extractive-based wood protection methods/processes.

References

1. Bagley, S. T; Richter, D. L. In *Industrial Applications*; Osiewacz, H. D., Ed.; Springer: Berlin, 2002; Vol. 10, pp 327–341.
2. Goodell, B. *Wood Deterioration and Preservation: Advances in Our Changing World*; Goodell, B., Nicholas, D., Schultz, T., Eds.; ACS Symposium Series 845; American Chemical Society: Washington, DC, 2003; pp 97–118.
3. Zabel, R. A.; Morrell, J. J. *Wood Microbiology: Decay and its Prevention*; Academic Press, Inc.: San Diego, CA, 1992; pp 476.
4. Cowling, E. B. *USDA Tech. Bull.* **1961**, *1258*, 79.
5. Schmidt, C. J.; Whitten, B. K.; Nicholas, D. D. *Proc. Am. Wood-Preserv. Assoc.* **1981**, *77*, 157–164.
6. Hyde, S. M.; Wood, P. M. *Int. Res. Group Wood Preserv.* **1995** Doc. 95-10104.
7. Tanaka, H.; Fuse, G.; Enoki, A. *Mokuzai Gakkaishi* **1991**, *37*, 986–988.
8. Hirano, T.; Tanaka, H.; Enoki, A. *Mokuzai Gakkaishi* **1995**, *41*, 334–341.
9. Goodell, B.; Jellison, J.; Liu, J.; Daniel, G.; Paszczynski, A.; Fekete, F.; Krishnamurthy, S.; Jun, L.; Xu, G. *J. Biotechnol.* **1997**, *53*, 133–162.
10. Arantes, V.; Milagres, A. M. F. *Quim Nova* **2009**, *30*, 1586–1595.
11. Arantes, V; Milagres, A. M. F. *J. Chem. Technol. Biotechnol.* **2006**, *81*, 413–419.
12. Jensen, J. R. K. A.; Ryan, Z. C.; Wymelenberg, A. V.; Cullen, D.; Hammel, K. *Appl. Environ. Microbiol.* **2002**, *68*, 2699–2703.
13. Hammel, K. E.; Kapich, A. N.; Jensen, K. A., Jr; Ryan, Z. C. *Enzyme Microb. Technol.* **2002**, *30*, 445–453.
14. Karripally, P.; Timokhin, V. I.; Houtman, C. J.; Mozuch, M. D.; Hammel, K. E. *Appl. Environ. Microbiol.* **2013**, *79*, 2377–2383.
15. Martinez, D.; Challacombe, J.; Morgenstern, I.; Hibbett, D.; Schmoll, M.; Kubicek, C. P.; Ferreira, P.; Ruiz-Duenas, F. J.; Martinez, A. T.; Kersten, P.; et al. *Proc. Natl. Acad. Sci. U.S.A.* **2009**, *106*, 1954–1959.
16. Enoki, A.; Itakura, S.; Tanaka, H. *Biotechnology* **1997**, *53*, 265–272.
17. Kerem, Z.; Jensen, K. A.; Hammel, K. E. *FEBS Lett.* **1999**, *446*, 49–54.
18. Paszczynski, A.; Crawford, R.; Funk, D.; Goodell, B. *Appl. Environ. Microbiol.* **1999**, *65*, 674–679.
19. Liese, W. *Annu. Rev. Phytopathol.* **1970**, *8*, 231–257.
20. Gilbertson;, R. L. *Mycologia* **1980**, *72*, 1–49.
21. Wilcox, W. W.; Parameswaran, N.; Liese, W. *Holzforschung* **1974**, *28*, 211–217.
22. McFee, W. W.; Stone, E. L. *Soil Sci. Soc. Am. J.* **1966**, *29*, 432–436.
23. Eastwood, D. C.; Floudas, D.; Binder, M.; Majcherczyk, A.; Schneider, P.; Aerts, A.; Asiegbu, F. O.; Baker, S. E.; Barry, K.; Bendiksby, M.; Blumentritt, M.; Coutinho, P. M.; Cullen, D.; de Vries, R. P.; Gathman, A.; Goodell, B.; Henrissat, B.; Ihrmark, K.; Kauserud, H.; Kohler, A.; LaButti, K.; Lapidus, A.; Lavin, J. L.; Lee, Y-H.; Lindquist, E.; Lilly, W.; Lucas, S.; Morin, E.; Murat, C.; Oguiza, J. A.; Park, J.; Pisabarro, A. G.; Riley, R.; Rosling, A.; Salamov, A.; Schmidt, O.; Schmutz, J.; Skrede, I.;

Stenlid, J.; Wiebenga, A.; Xie, X.; Kuües, U.; Hibbett, D. S.; Hoffmeister, D.; Högberg, N.; Martin, F.; Grigoriev, I. V.; Watkinson, S. C. *Science* **2011**, *333*, 762–765.
24. Hibbett, D. S.; Donoghue, M. J. *Syst. Biol.* **2001**, *50*, 215–242.
25. Martinez, D.; Challacombe, J.; Morgenstern, I.; Hibbett, D.; Schmoll, M.; Kubicek, C. P.; Ferreira, P.; Ruiz-Duenas, F. J.; Martinez, A. T.; Kersten, P.; Hammel, K. E.; Vanden Wymelenberg, A.; Gaskell, J.; Lindquist, E.; Sabat, G.; Bondurant, S. S.; Larrondo, L. F.; Canessa, P.; Vicuna, R.; Yadav, J.; Doddapaneni, H.; Subramanian, V.; Pisabarro, A. G.; Lavin, J. L.; Oguiza, J. A.; Master, E.; Henrissat, B.; Coutinho, P. M.; Harris, P.; Magnuson, J. K.; Baker, S. E.; Bruno, K.; Kenealy, W.; Hoegger, P. J.; Kues, U.; Ramaiya, P.; Lucas, S.; Salamov, A.; Shapiro, H.; Tu, H.; Chee, C. L.; Misra, M.; Xie, G.; Teter, S.; Yaver, D.; James, T.; Mokrejs, M.; Pospisek, M.; Grigoriev, I. V.; Brettin, T.; Rokhsar, D.; Berka, R.; Cullen, D. *Proc. Natl. Acad. Sci. U.S.A.* **2009**, *106*, 1954–1959.
26. Hori, C.; Gaskell, J.; Igarashi, K.; Samejima, M.; Hibbett, D. S.; Henrissat, B.; Cullen, D. *Mycologia* **2013**, *105*, 1412–1427.
27. Filley, T. R.; Cody, G. D.; Goodell, B.; Jellison, J.; Noser, C.; Ostrofsky, A. *Org. Geochem.* **2002**, *33*, 111–124.
28. Xu, G.; Goodell, B. *J. Biotechnol.* **2001**, *87*, 43–57.
29. Daniel, G. *FEMS Microbiol. Rev.* **1994**, *13*, 199–233.
30. Kuo, M.-L.; Stokke, D. D.; McNabb, H. S. *Wood Fiber Sci.* **1998**, *20*, 405–414.
31. Illman, B.; Highley, T. L. In *Biodeterioration Research*; Llewellyn, G. C., Ed.; Perseus Books, Basic Books: New York, 1989; Vol 2.
32. Kim, Y. S.; Goodell, B.; Jellison, J. *Holzforschung* **1991**, *45*, 389–393.
33. Sachs, I. B.; Clark, I. T.; Pew, J. C. *J. Polym. Sci.* **1963**, *C-2*, 203–212.
34. Blanchette, R. A.; Burnes, T. A.; Eerdmans, M. M.; Akhtar, M. *Holzforschung* **1992**, *46*, 106–115.
35. Curling, S.; Clausen, C. A.; Winandy, J. E. *Int. Res. Group Wood Preserv.* **2001**Doc. 01-20219.
36. Winandy, J.; Morrell, J. *Wood Fiber Sci.* **1993**, *25*, 278–288.
37. Halliwell, G. *Biochem. J.* **1965**, *95*, 35–40.
38. Koenigs, J. W. *Mater. Org.* **1972**, *7*, 133–147.
39. Koenigs, J. W. *Wood Fiber* **1974**, *6*, 66–80.
40. Koenigs, J. W. Cellulose as a Chemical and Energy Resource. In *Symposium 5, Biotechnology and Bioengineering*; Wilke, C. R., Ed.; Wiley: New York, 1975; Vol. 5, pp 151–159.
41. Highley, T. L. *Mater. Org.* **1977**, *12*, 161–174.
42. Blanchette, R. A.; Krueger, E. W.; Haight, J. E.; Akhtar, M.; Akin, D. E. *J. Biotechnol.* **1997**, *53*, 203–213.
43. Flournoy, D. S.; Kirk, T. K.; Highley, T. L. *Holzforschung* **1991**, *45*, 383–388.
44. Dutton, M. V.; Evans, C. S.; Atkey, P. T.; Wood, D. A. *Appl. Microbiol. Biotechnol.* **1993**, *39*, 5–10.
45. Takao, S. *Appl. Microbiol.* **1965**, *13*, 732–737.

46. Lee, S. O.; Tran, T.; Park, Y. Y.; Kim, S. J.; Kim, M. J. *Intern. J. Miner. Process* **2006**, *80*, 144–152.
47. Arantes, V.; Qian, Y.; Milagres, A. M. F.; Jellison, J.; Goodell, B. *Int. Biodeterior. Biodegrad.* **2009**, *63*, 478–483.
48. Goodell, B. In *Wood Deterioration and Preservation: Advances in Our Changing World*; Goodell, B., Nicholas, D. D., Schultz, T. P., Eds.; ACS Symposium Series 845; American Chemical Society: Washington, DC, 2003; pp 97–118.
49. Goodell, B.; Qian, Y.; Jellison, J. In *Development of Commercial Wood Preservatives: Efficacy, Environmental, and Health Issues*; Schultz, T., Nicholas, D., Militz, H., Freeman, M. H., Goodell, B., Eds.; ACS Symposium Series 982; American Chemical Society: Washington, DC, 2008; pp 9–31.
50. Daniel, G.; Volc, J.; Filonova, L.; Plíhal, O.; Kubátová, H. P. *Appl. Environ. Microbiol.* **2007**, *73*, 6241–6253.
51. Hyde, S. M.; Wood, P. M. *Microbiology* **1997**, *143*, 259–266.
52. Niemenmaa, O.; Uusi-Rauva, A.; Hatakka, A. *Biodegradation* **2007**, *19*, 555–565.
53. Vanden Wymelenberg, A.; Gaskell, J.; Mozuch, M.; Sabat, G.; Ralph, J.; Skyba, O.; Mansfield, S. D.; Blanchette, R. A.; Martinez, D.; Grigoriev, I.; Kersten, P. J.; Cullen, D. *Appl. Environ. Microbiol.* **2010**, *76*, 3599–3610.
54. Hastrup, A. C. S.; Jensen, T. Ø.; Jensen, B. *Holzforschung* **2012**, *67*, 99–106.
55. Shimokawa, T.; Nakamura, M.; Hayashi, N.; Ishihara, M. *Holzforschung* **2004**, *58*, 305–310.
56. Suzuki, M. R.; Hung, C. G.; Houtman, C. J.; Dalebroux, Z. D.; Hammel, K. E. *Environ. Microbiol.* **2006**, *8*, 2214–2223.
57. Jellison, J.; Connolly, J. H.; Goodell, B.; Doyle, B.; Illman, B.; Fekete, F.; Ostrofsky, A. *Int. Biodeterior. Biodegrad.* **1997**, *39*, 165–179.
58. Wei, D.; Houtman, C. J.; Kapich, A. N.; Hunt, C. G.; Cullen, D.; Hammel, K. E. *Appl. Environ. Microbiol.* **2010**, *76*, 2091–2097.
59. Suzuki, M. R.; Hung, C. G.; Houtman, C. J.; Dalebroux, Z. D.; Hammel, K. E. *Environ. Microbiol.* **2006**, *8*, 2214–2223.
60. Newcombe, D.; Paszczynski, A.; Gajewska, W.; Kroger, M.; Feis, G.; Crawford, R. *Enzyme Microb. Technol.* **2002**, *30*, 506–517.
61. Shimokawa, T.; Nakamura, M.; Hayashi, N.; Ishihara, M. *Holzforschung* **2004**, *58*, 305–310.
62. Buettner, G. R. *Arch. Biochem. Biophys.* **1993**, *300*, 535–543.
63. Halliwell, B.; Gutteridge, J. M. C. *Free Radicals in Biology and Medicine*; Oxford University Press: Oxford, 1999.
64. Goodell, B.; Daniel, G.; Jellison, J.; Qian, Y. *Holzforschung* **2006**, *60*, 630–636.
65. Arantes, V.; Qian, Y.; Kelley, S. S.; Milagres, A. M. F.; Filley, T. R.; Jellison, J.; Goodell, B. *J. Biol. Inorg. Chem.* **2009**, *8*, 1253–1263.
66. Goodell, B.; Qian, Y.; Jellison, J.; Richard, M.; Qi, W. In *Progress in Biotechnology, Biotechnology in the Pulp and Paper Industry*; Viikari, L., Lantto, R., Eds.; Elsevier: Amsterdam, 2002; Vol. 21; pp 37–78.
67. Hyde, S. M.; Wood, P. M. *Microbiology* **1997**, *143*, 259–266.

68. Korripally, P.; Timokhin, V. I.; Houtman, C. J.; Mozuch, M. D.; Hammel, K. E. *Appl. Environ. Microbiol.* **2013**, *79*, 2377–2383.
69. Arantes, V.; Milagres, A. M. F.; Filley, T. R.; Goodell, B. *J. Ind. Microbiol. Biotechnol.* **2011**, *38*, 541–555.
70. Yelle, D. J.; Ralph, J.; Lu, F.; Hammel, K. E. *Environ. Microbiol.* **2008**, *10*, 1844–1849.
71. Martinez, A. T.; Rencoret, J.; Nieto, L.; Jiménez-Barbero, J.; Gutiérrez, A.; del Río, J. C. *Environ. Microbiol.* **2011**, *13*, 96–107.
72. Jensen, J. R. K. A.; Ryan, Z. C.; Wymelenberg, A. V.; Cullen, D.; Hammel, K. *Appl. Environ. Microbiol.* **2002**, *68*, 2699–2703.
73. Pratch, J.; Boenigk, J.; Isenbeck-Schroter, M.; Keppler, F.; Scholer, F. H. *Chemosphere* **2001**, *44*, 613–619.
74. Qi, W.; Jellison, J. *Int. Biodeterior. Biodegrad.* **2004**, *54*, 53–60.
75. Cohen, R.; Suzuki, M. R.; Hammel, K. E. *Appl. Environ. Microbiol.* **2004**, *70*, 324–331.
76. Hyde, S. M.; Wood, P. M. *Microbiology* **1997**, *143*, 259–266.
77. Enoki, A.; Hirano, T.; Tanaka, H. *Mater. Org.* **1992**, *27*, 247–261.
78. *Microbial and Enzymatic Degradation of Wood and Wood Components*; Eriksson, K. E. L., Blanchette, R. A., Ander, P., Eds.; Springer: Berlin, 1990; pp 1–72.
79. Wood, T. M.; McCrae, S. I.; Bhat, K. M. *Biochem. J.* **1989**, *260*, 37–43.
80. Kirk, K; Ibach, R; Mozuch, M. D.; Conner, A. H.; Highley, T. L. *Holzforschung* **1991**, *45*, 239–244.
81. Highley, T. L.; Dashek, W. V. In *Forest Products Biotechnology*; Palfreyman, J. W., Ed.; Taylor and Francis: London, 1998; pp 15–36.
82. Howell, C. A.; Hastrup, C.; Goodell, B.; Jellison, J. *Int. Biodeterior. Biodegrad.* **2009**, *63*, 414–419.
83. Kleman-Leyer, K.; Agosin, E.; Conner, A. H.; Kirk, K. *Appl. Environ. Microbiol.* **1992**, *58*, 1266–1270.
84. Fackler, K.; Stevanic, J. S.; Ters, T.; Hinterstoisser, B.; Schwanninger, M.; Salmén, L. *Enzyme Microb. Technol.* **2010**, *47*, 257–267.
85. Fackler, K.; Schwanninger, M. *Appl. Microbiol. Biotechnol.* **2012**, *96*, 587–99.
86. Highley, T. L. *Mater. Org.* **1987**, *21*, 39–45.
87. Monrroy, M.; Ortega, I.; Ramírez, M.; Baeza, J.; Freer, J. *Enzyme Microb. Technol.* **2011**, *49*, 472–477.
88. Schilling, J. S.; Tewalt, J. P.; Duncan, S. M. *Appl. Microbiol. Biotechnol.* **2009**, *84*, 465–475.
89. Schmidhalter, D. R.; Canevascini, G. *Arch. Biochem. Biophys.* **1993**, *300*, 551–558.
90. Martinez, D.; Challacombe, J.; Morgenstern, I.; Hibbett, D.; Schmoll, M.; Kubicek, C. P.; Ferreira, P.; Ruiz-Duenas, F. J.; Martinez, A. T.; Kersten, P.; Hammel, K. E.; Vanden Wymelenberg, A.; Gaskell, J.; Lindquist, E.; Sabat, G.; Bondurant, S. S.; Larrondo, L. F.; Canessa, P.; Vicuna, R.; Yadav, J.; Doddapaneni, H.; Subramanian, V.; Pisabarro, A. G.; Lavin, J. L.; Oguiza, J. A.; Master, E.; Henrissat, B.; Coutinho, P. M.; Harris, P.; Magnuson, J. K.; Baker, S. E.; Bruno, K.; Kenealy, W.; Hoegger, P. J.;

Kues, U.; Ramaiya, P.; Lucas, S.; Salamov, A.; Shapiro, H.; Tu, H.; Chee, C. L.; Misra, M.; Xie, G.; Teter, S.; Yaver, D.; James, T.; Mokrejs, M.; Pospisek, M.; Grigoriev, I. V.; Brettin, T.; Rokhsar, D.; Berka, R.; Cullen, D. *Proc. Natl. Acad. Sci. U.S.A.* **2009**, *106*, 1954–1959.
91. Nilsson, T. *Mater. Org.* **1974**, *9*, 173–198.
92. Murmannis, L.; Highley, T. L.; Palmer, J. G. *Wood Sci. Technol.* **1988**, *22*, 59–66.
93. Rättö, M.; Ritschkoff, A. C.; Viikari, L. *Appl. Microbiol. Biotechnol.* **1997**, *48*, 53–57.
94. Ringman, R.; Pilgård, A.; Richter, K. *Int. Biodeterior. Biodegrad.* **2014**, *86*, 86–91.
95. Barr, D. P.; S. D. Aust, S. D. *Environ. Sci. Technol.* **1994**, *28*, 78A–87A.
96. Agosin, E.; Jarpa, S.; Rojas, E.; Espejo, E. *Enzyme Microb. Technol.* **1989**, *11*, 511–517.
97. Koenig, A. B.; Sleighter, R. L.; Salmon, E.; Hatcher, P. G. *J. Wood Chem. Technol.* **2010**, *30*, 61–85.
98. Kirk, T. K. *Holzforschung* **1975**, *29*, 99–107.
99. Kirk, T. K.; Adler, E. *Acta Chem. Scand.* **1970**, *24*, 3379–3390.
100. Jin, L.; Schultz, T. P.; Nicholas, D. D. *Holzforschung* **1990**, *44*, 133–138.
101. Machuca, A.; Ferraz, A. *Enzyme Microb. Technol.* **2001**, *29*, 386–391.
102. Milagres, A. M. F.; Sales, R. *Enzyme Microb. Technol.* **2001**, *28*, 522–526.
103. Howard, J. A. In *Free Radical*; Kotchi, J. K., Ed.; Wiley: New York, 1973, pp 3–62.
104. Snook, M. E.; Hamilton, G. A. *J. Am. Chem. Soc.* **1974**, *96*, 860–869.
105. Davis, M. F.; Schroeder, H. A.; Maciel, G. E. *Holzforschung* **1994**, *48*, 301–307.
106. Martínez, A. T.; González, A. E.; Valmaseda, M.; Dale, B. E.; Lambregts, M. J.; Haw, J. F. *Holzforschung* **1991**, *45*, 49–54.
107. Sun, Q. N.; Qin, T. F.; Li, G. Y. *Int. J. Polym. Anal. Chem.* **2009**, *14*, 19–33.
108. Harvey, P. J.; Schoemaker, H. E.; Palmer, J. M. *FEBS Lett.* **1986**, *195*, 242–246.
109. Schultz, T. P.; Nicholas, D. D. *Phytochemistry* **2000**, *54*, 47–52.
110. Goodell, B.; Jellison, J.; Liu, J.; Krishnamurthy, S. U.S. Patent 6,046,375.
111. Sen, S.; Tascioglu, C.; Tirak, K. *Int. Biodeterior. Biodegrad.* **2009**, *63*, 135–141.
112. Schultz, T. P.; Nicholas, D. D.; Henry, W. P.; Pittman, C. U.; Wipf, D. O.; Goodell, B. *Wood Fiber Sci.* **2005**, *37*, 175–184.

Chapter 2

Fungal and Bacterial Biodegradation: White Rots, Brown Rots, Soft Rots, and Bacteria

Geoffrey Daniel*

Department of Forest Products/Wood Science,
Swedish University of Agricultural Sciences,
Box 7008, Uppsala, Sweden
*E-mail: geoffrey.daniel@slu.se.

Wood (lignocellulose) is colonized and degraded by a wide range of fungi and bacteria in aerobic environments. These microorganisms can cause very different types of decay (e.g. white rot, brown rot, soft rot, bacteria- tunneling and erosion) depending on the organisms involved and their inherent biochemical capabilities, type of wood substrate (native/modified), environmental situation and interactive competition. This review outlines some of the principle aspects concerning our understanding of morphological and biochemical aspects of white rot and bacterial decay with lesser emphasis on brown- and soft rot decay.

© 2014 American Chemical Society

White Rot Decay of Wood and Lignocellulose: Our Current Knowledge

Research on white rot decay of wood is in many respects advance to all other major types of decay. Developments to understand white rot and their mechanisms of decay have been driven by the biotechnological potential of these fungi and the specificity of the action of the enzymes so far discovered. In particular, the possibility for using white rot fungi and their enzymes in biopulping, xylanase bleaching in kraft pulping, energy savings during refining with cellulases, pitch removal with lipases, slime removal with enzymes, as well as fibre modification in the pulp and paper and related forest industries has financed much of the research with studies stimulated not only for environmental, but also for energy (e.g. in mechanical pulping) savings (e.g. (*1–5*)). White rot fungi have also been evaluated in bioremediation and detoxification of aromatics (*5–9*) with recent work directed to ecofriendly wood protection with laccase from the white rot fungus *Trametes versicolor* in laccase-catalysed iodination of wood surfaces (*10*). The potential of white rot fungi (e.g. *Physisporinus vitreus*) has also been assessed more recently for the "bioincising" of wood to increase the permeability of refractory wood species (i.e. Norway spruce) to improve penetration of wood preservatives and also improve acoustic properties of wood (mycowood) for musical instruments (*11–13*). Despite intensive efforts over many years, few of the major approaches using whole fungi in processes (e.g. biopulping) have turned out to be industrial viable for various reasons while the use of novel and recombinant enzymes to modify wood fibres in industrial processes (e.g. bleaching, pitch removal) seems to have greater potential (*5, 10*). In native terrestrial situations, white rot fungi are found colonizing hardwoods causing decay often referred to as "white pocket rot", "stringy rot" or "spongy rot". Apart from a few species, white rot fungi generally show poor tolerance to conventional metal wood preservatives at least under laboratory test conditions (*14*). Nevertheless, white rot fungi have been shown as degraders of wood in marine pilings in the sea (*15*) such as *Syncarpia* spp. (Cookson, Aust. pers. Comm.), test stakes treated with tributyl-tin oxide and AAC (*16*) and from cooling towers under in-service conditions (*17–19*). A limited number of marine basidiomycete fungi causing white rot decay of hardwoods have also been reported (*20–22*) although their importance in marine environments appears to be limited.

White rot fungi belonging to the Basidiomycetes are characterized like no other taxonomic group of fungi by their unique capacity to biomineralize not only polysaccharides in lignocellulose but also lignin. Under aerobic conditions these fungi can completely mineralize lignin and wood polysaccharides (cellulose, hemicelluloses) to CO_2 and H_2O. The process is characterized by the involvement of a diverse range of physiological and biocatalytic activities that occur while fungal hyphae colonize and degrade the wood structure (e.g. (*23–28*)). A wide range of different morphological decay patterns may be produced in wood with attack varying with fungal species, their physiological status as well as decay capacity (*29–31*) and wood type. Like all fungi, the principle path for white rot hyphal colonization into the wood structure is via the ray canals that provide, not only rapid entrance into the radial structure of the wood, but also provision of easily accessible nutrients stored in parenchyma cells. Thereafter, the hyphae can ramify throughout the wood structure with growth through native pits in the cellular elements or by the development of specialized bore hyphae that penetrate through the wood cell walls. In the majority of cases, the hyphae develop and are maintained within the cell lumena of wood cells during decay, although reports of a specialized type of attack with hyphae growing within middle lamella regions exist (*32*). Colonization of soft- and hardwoods is in principle similar although in the latter, the vessels provide a more rapid means of longitudinal penetration through the wood structure (*24*).

Two principle morphological decay types based on microscopic observations of decay patterns and described as early as 1863 by Schacht (*33*) are known. The two decay types: "simultaneous" and "selective" (preferential) reflect the rate and morphological appearance of the wood (wood fibres) during which the wood cell wall components are degraded (Figures 1-4). Some white rot fungi are able to cause both simultaneous and preferential attack often in the same wood material (e.g. *Heterobasidium annosum,* (*24*)), this double response depending on the fungal genetic condition and local environmental conditions of the wood. With simultaneous white rot (e.g. *Phanerochaete chrysosporium, Phlebia radiata, Trametes versicolor* wild types) all the main wood components (cellulose, hemicelluloses, lignin) are more or less degraded simultaneously from the cell lumen outwards. This has been proven by both gross chemical analyses of degraded wood samples and through microscopical observations (e.g. (*23, 30, 31*)). Depending on the overall shape of the wood fibres the progress of decay may differ, and in softwoods decay often progresses more rapidly along the thinner tangential walls than the thicker radial cell walls (Figure 1) (*24, 30, 34*). As attack continues outwards, even the lignin rich middle lamellae regions between fibres are degraded, with decay frequently progressing into the cell walls of adjacent fibres (Figures 1-2) (*35*). The middle lamella cell corners are the last regions of the cell wall structure to be degraded and sometimes these remain intact even at very advanced stages of decay when the entire wood structure has been almost totally lost (Figures 1-2). The most characteristic feature of this type of decay at the cellular level is distinguished by the development of a thin advancing zone in which lignocellulose mineralization takes place. This thin zone is easily recognized by a stronger staining with safranin during light microscopy and recognized at the ultrastructural level by as an electron-lucent layer in the lumen periphery of the secondary cell wall in which the cellulose microfibrils often become evident (*24, 30, 34, 36*). The thin zones of decay are evident on all cell wall layers even the lignin rich middle lamella cell corner regions remaining in advanced stages of attack where a difference in density and staining are apparent (Figures 1-2) (*30, 35*).

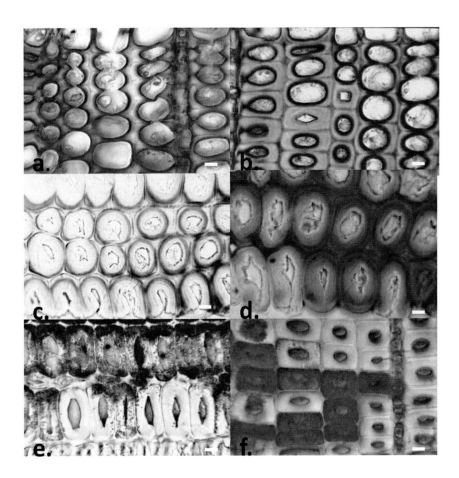

Figure 1. Aspects of simultaneous (a, b) and preferential white rot (c, d) and tunneling (e) and erosion (f) bacteria decay of pine and spruce (e). a) Advanced cell wall thinning of pine leads to early rupture of the tangential cell walls as these are thinner than the radial cell walls; b) Phlebia radiata cell wall thinning with dark brown zones indicating sites of simultaneous attack of all cell wall components and presence of peroxidases; (c, d) Preferential decay by Phlebia radiata Cel 26, after prussian blue staining (c) the rings indicating "in time and space" sites of lignin and hemicellulose attack; d) As for c) but after permanganate staining showing lignin removal; (e) Advanced tunneling bacteria attack developing across all cell wall layers including middle lamellae; f) Advanced bacterial erosion decay of secondary cell walls leaving middle lamellae. Bars: a, b, c, d, e, f, 5.0 μm. (see color insert)

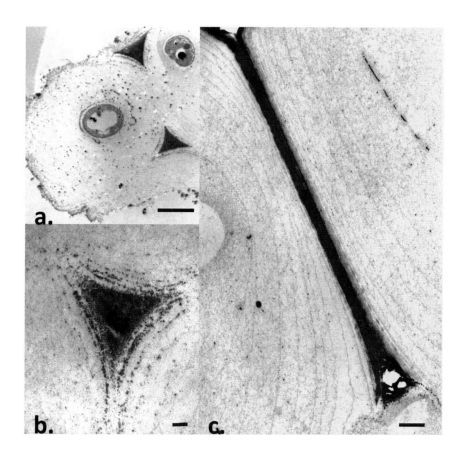

Figure 2. TEM micrographs showing aspects of advanced simultaneous white rot of birch by Phlebia radiata and involvement of extracellular slime. a, b, c) Remaining cell corner (a, b), hyphae (a) and middle lamellae (c) regions encapsulated in extracellular slime. The slime unites the cell wall regions with the remote fungal hyphae and shows a characteristic concentric arrangement in the previous location of the wood secondary cell wall. Bars: a, 3.0 μm; b, c 1.0 μm.

With "selective" or "preferential" white rot, the lignin and hemicelluloses are preferentially degraded leaving modified and separated fibres (i.e. defibrated) composed primarily of modified cellulose.

This type of white rot decay has received considerable attention and studied using both native- and prepared cellulase-less mutants (i.e. fungi with impaired or very reduced cellulose activity) (e.g. *Phlebia radiata* Cel 26 (*37*), *Cerioporiopsis subvermispora, Dichomitus squalens*) (*3, 24, 30, 38–40*) in order to unravel the events of decay. Morphologically, stages of attack are best recognized with observations of latewood fibres from softwoods where the zones of decay are readily recognized as rings progressing outwards across the wood secondary cell walls (Figures 1, 3, 4) (*30, 35*). The outer perimeter of the zones in the wall therefore reflect the sites in "time and in space" where biomineralization of lignin and hemicelluloses take place (Figures 1, 3, 4). Essentially the thick decay zones could represent an accentuated form of simultaneous decay (Figure 1), in which the majority of the cellulose remains albeit modified. Early work with chemical marking of lignin using bromination and staining with potassium permanganate in conjunction with TEM, in addition to UV-examination have indicated attack of lignin in the outer perimeter of the zones as decay progresses across wood cell walls (*30, 37, 39, 41, 42*). Like simultaneous white rot, the decay zones can progress across middle lamella regions of adjacent cells (Figures 1, 3). When this happens, the fibres can easily separate from one another when subjected to mechanical disturbance. The characteristic nature of this type of wood and cell wall attack formed the scientific basis for subsequent studies concerning biopulping where it was hoped that pretreatment of wood chips with white rot fungi could save both energy (e.g. during refining) and use of hazardous chemicals (e.g. during bleaching) thereby improving efficiency.

While the microscopical events of wood decay by white rot fungi are quite well known and relatively easy to demonstrate with fungal monocultures cultures under laboratory conditions, understanding the biochemistry of decay is much more difficult especially when trying to correlate events *in-situ* with *in-vitro* experiments with isolated enzymes and their different cofactors. The same is also true for brown rot fungi (see later chapters). Several fungi can be shown to degrade lignin model compounds or cellulose *in-vitro*, but are unable to degrade solid wood emphasizing the important role played by the wood substrate itself. The importance of lignin chemical structure (guiaicyl vs syringyl) and concentration during white rot delignification of wood has been shown by the selective removal of lignin from vessels (i.e. guiacyl lignified) in hardwoods by some species of white fungi rot (e.g. *Phlebia tremellosa*) but not others (e.g. *Dichomitus squalens*) (*39*). It has been well established that syringyl lignin is more easily mineralized than guiaicyl lignin at least under *in-vitro* conditions by white rot fungi (*43, 44*).

Another morphological type of white rot whereby middle lamella regions remain after attack has been described for a number of higher Ascomycetes like *Xylaria* and *Daldinia spp.* (*30, 45*).

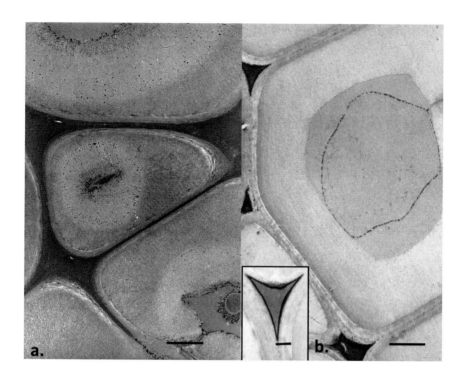

Figure 3. TEM micrographs showing preferential white rot decay of pine and birch secondary cell walls. a) Cross section of pine latewood tracheid showing the characteristic zones (electron lucent regions) produced in the S2 wall during preferential removal of lignin and hemicelluloses by luminal hyphae; b) Advanced preferential decay with majority of the lignin and hemicelluloses removed from the secondary (S1, S2) cell walls with only attacked cell wall corners remaining. Note absence of lumen hyphae in the highly degraded birch cells. Inset shows higher magnification of the middle cell wall corners with characteristic thin zones of attack in peripheral regions. Bars: a, 2.0 µm; b, 3.5 µm; Inset, 350 nm.

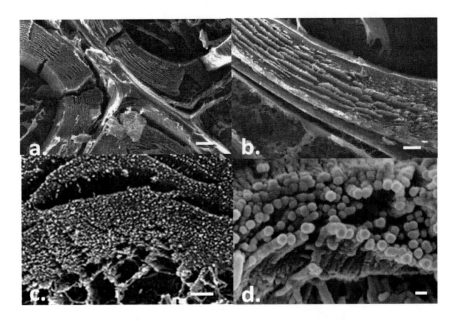

Figure 4. Cryo-FE-SEM micrographs showing preferential white rot decay of birch secondary cell walls. a, b) Cross-sections showing characteristic concentric fracture planes produced in the secondary S2 wall layer after preferential removal of lignin and hemicelluloses by lumenal hyphae; c, d) Remaining macrofibrillar structure comprising the S2 cell wall after lignin and hemicellulose removal during preferential white rot giving an impression of advanced decay and possibilities for enzyme/radical penetration. Bars: a, 2.0 μm; b, 1.0 μm; c, d, 100 nm. (Images c, d reproduced from reference (24). Copyright 2003 American Chemical Society).

Biochemical Aspects

A wide range of enzymes capable of mineralizing wood polysaccharides and lignin have now been isolated from a diverse array of white rot fungi and in many cases recombinant proteins produced (27). Major white rot species where diverse enzymes have been isolated and characterized include: *Phlebia radiata* (e.g. (46)), *Phanerochaete chrysosporium* (47, 48) and *Cerioporiopsis subvermispora*. Initial studies on enzymes involved *in-vitro* culturing of fungi in liquid cultures, establishment of optimal conditions for growth, up-scaling, and the isolation and purification of extracellular proteins by various chromatography techniques (e.g. electrophoresis, FPLC etc). The extracellular proteins have thereafter been physically and biochemically characterized (e.g. substrate specificity, activity, sugar content, size, charge etc) and frequently antibodies (initially polyclonal, later monoclonal) produced. Correlated immunocytochemical studies on degraded wood have provided a means of showing the indirect involvement and presence of enzymes *in-situ* at sites of attack during actual wood degradation.

The Enzymes System Involved

In order to degrade wood, fungi may produce a wide variety of polysaccharide and lignin degrading enzymes. However, the ability to produce abundant enzymes such as cellulase does not imply an ability to degrade wood. The classical example being the mould fungus *Trichoderma reesei* (anamorph of *Hypocrea jecorina*) (*49*) which has been frequently classified as a soft rot fungus although its effect on lignified wood cell walls is very limited. *T. reesei* has been used industrially for decades to produce cellulases but when grown on solid wood particularly softwoods it can only slightly increase permeability through the attack of non-lignified ray cells and pit membranes. However, the effect can be greater on hardwoods where the percent of non-lignified cellular elements is higher.

In order to degrade cellulose in wood, white rot fungi and fungi from other taxonomic groups often produce a range of endoglucanases (EC 3.2.14), cellobiohydrolases (EC 3.2.1.91) and β-glucosidases (EC 3.2.1.21). Endoglucanases catalyse indiscriminant cleavage of cellulose along the backbone of the glucose chains, cellobiohydrolases attack the chains from the reducing or non-reducing ends and glucosidases can utilize any cellobiose or cello-oligosaccharides released to produce glucose. In a complete cellulose decay system, the enzymes are thought to work in cooperation and thus could be expected to have some close proximity. Several white rot fungi (e.g. *Phanerochaete chrysosporium, Sporotrichium pulverutentum, Heterobasidium annosus, Schizophyllum commune, Trametes versicolor, Pcynoporus cinnabarinus*) are also known to produce cellobiose dehydrogenase (CDH) for oxidizing products of hydrolysis produced by cellulolytic and other enzymes. In comparison to the work carried out on lignin degrading enzymes produced by white rot fungi, much less has been done on the cellulase and hemicellulase systems, with the possible exception of *P. chrysosporium* (*50*). For an overview of the carbohydrate degrading enzymes produced by white rot and other fungi, readers should consult the CAZy database: www.cazy.org. The database outlines all the known families of enzymes that both cleave or build complex polysaccharides. It is of considerable interest that recent genomic analysis has shown mould and litter fungi like *Aspergillus nidulans* and *Fusarium gramineum* and most surprising *T. reesei* to encode lesser numbers of cellulase enzymes than the white rot fungus *Phanerochaete chrysosporium* (*49, 51, 52*). Presumably, for the moulds this reflects their ecological niche and primary attack of non-lignified substrates in nature in contrast to white rot fungi. Genomic work on the white rot fungus *Schizophyllum commune* has shown it rich in genes encoding enzymes that degrade hemicelluloses, pectins and cellulose (*53*). However, *S. commune* has fewer genes encoding lignin degrading enzymes (*54*). On softwoods, *S. commune* normally causes only weak attack of wood fibres producing low weight losses, presumably reflecting an inability to access the cellulose through the lignin barrier. In addition, S. commune frequently causes a special type of attack of the secondary S2 cell wall layer of softwoods forming thin concentric slits (openings) in the wall (*55, 56*), thought to reflect aspects of the native wood wall structure.

Lignin Degrading Enzymes

Unlike cellulose, degradation of lignin is restricted by its chemical, macromolecular and heterogeneous structure that can vary considerably not only between hardwoods and softwoods but also within the secondary cell wall and middle lamella regions between different cellular elements (*57*). Although the overall distribution of lignin in wood cells walls at the nanostructure level complements cellulose, the lignin polymer is not linear but rather randomly orientated with variable linkages. Lignin degradation is therefore known to be carried out by random oxidative reactions (i.e. not hydrolytic) by a limited number of extracellular oxidative enzymes primarily peroxidases and laccases together with associated enzymes and cofactors (*27, 58, 59*). But even within this group of efficient lignin degrading fungi their ability varies and can be reflected by the different morphological decay patterns produced in wood cells. To date three peroxidase enzymes have been isolated and characterized from a range of white rot fungi namely: 1) manganese peroxidase (MnP; EC 1.11.1.13) discovered by Gold and co-workers (*48, 60*); 2) Lignin peroxidase (LiP; EC1.11.1.14) found by Tien and co-workers (1983, (*47*)); and more recently 3) the versatile peroxidases (VPs, EC 1.11.1.16) discovered by Martinez and co workers in the 1990's (*61–64*). An equivalent CAZyme database known as FOLy(Fungal Oxidative Lignin Enzymes (http://foly-db.esil.univ-mrs.fr/) (*54*) to the carbohydrate degrading enzymes was also set up in 2008 for fungal oxidoreductases. More recently, the CAZy database (http://www.cazy.org/Auxiliary-Activities.html) was updated to include auxiliary activities (AA) to cover redox enzymes that act in conjunction with CAZymes since it was found that several members of the CBM33 and family GH61 were lytic polysaccharide monooxygenases. The AA class currently includes eight families of ligninolytic enzymes and two families of lytic polysaccharide mono-oxygenases. The main aim of the three databases is to provide an overview for possibilities of exploiting cellulose- and lignin degrading enzymes in biotechnological processes.

All three peroxidases are heme-containing (i.e. protoporphyrin) glycoproteins with variable molecular weight (MnP: 38-60 kDa, LiP, 35-48 kDa; VP 42-45 kDa) and require H_2O_2 as the oxidant (*65*). Under liquid culture conditions, normally several isoenzymes are secreted (e.g. (*46*)). Lignin peroxidase (LiP) referred also as ligninase or 1, 2-bis (3, 4-dimethoxyphenyl) propane 1,3-diol: hydrogen-peroxide oxidoreductase carries out one electron oxidation of non-phenolic lignin producing aryl cation radicals (*66*). The enzyme has been extensively studied since its discovery (*47*) and has been shown to catalyse a variety of reactions including Cα-Cβ cleavage and ring opening in lignin (*59, 67, 68*) using both synthetic and lignin model compounds (*59*). Of the three known peroxidases, LiP has the highest redox potential. Manganese peroxidase (MnP) or Mn (II): hydrogen peroxide oxidoreductase oxidizes Mn(II) to Mn(III) in a catalytic cycle in which Mn (III) is first stabilized by organic acids (e.g. malonate, oxalate, lactate, (*69*)) with the chelated Mn(III), then oxidizing different compounds including phenolic rings and degradation of lignin. Considerable study has been conducted on Mn(III) as a possible oxidant that can easily penetrate into wood and in the presence of unsaturated fatty acids (e.g. linoleic acid) is able to cause lipid peroxidation with the peroxyl radicals formed able to act themselves as oxidants ((*70–74*), see below).

MnP was discovered at the same time as LiP (*48*). However, the numerous enzyme screening studies carried out on the ability of a diverse range of white fungi to produce MnP and LiP has shown MnP to be more frequently detected possibly suggesting a greater importance (*46, 58, 75*). However, it should also be recognized that both LiP and MnP likely differ significantly between white rot species and strains in their properties as well the amounts produced under different environmental conditions. In contrast with LiP and MnP, the versatile peroxidases (*reactive-black-5:hydrogen-peroxide oxidoreductase*) were not reported until the late nineties in white rot fungi (*76, 77*). VP's combine the substrate characteristics of LiP and MnPs and are able to oxidize a variety of low and high redox potential dyes (i.e. azo-dye Reactive Black 5) as well as phenols and hydroquinones, To date, VP's have only been reported from the edible white rot fungi *Bjerkandera* (*adusta*) and *Pleurotus* (*eryngii*) genera and their full distribution and importance are unknown. All three peroxidases are released during white rot decay and both MnP and LiP have been localized extracellularly associated with both the fungal hyphae and decayed wood during attack (see below).

Laccases

Laccases (EC 1.10.3.2; p-diphenol:oxygen oxidoreductase) are copper containing oxidases that can carry out one-electron oxidation of phenolic rings producing phenoxy radicals (*78*). The occurrence of laccases in white rot fungi has been known for a long time with respect to wood decay (e.g. (*79–81*)), although they are much better known for their involvement in fungal physiological processes such as fungal fruit body formation and melanin pigmentation (*78*). Laccases have been shown to catalyze a diverse range of native compounds and model dimers involving Cα-Cβ cleavage, although their effect on macromolecular lignin is limited and therefore their true delignification ability on wood is still controversial (*23*). Although work in the 1990's showed that through the use of synthetic mediators such as 2, 2'-azino-bis (3-ethylbenzothiazoline-6-sulfonate) and 1-hydroxybenzotriazole, laccase oxidation of lignin non-phenolic compounds was possible (*82, 83*), to date the only natural laccase mediator described is 3-hydroxyanthranilic acid from the simultaneous white rot fungus *Pycnoporus cinnabarinus* (*84*). Recent genomic studies would also seem to strengthen a case against its importance in lignin degradation with no laccase encoding genes found in *P. chrysosporium*, large numbers found in *Coprinus cinereus* (i.e. a dung fungus) and presence in the brown rot fungus *Postia placenta* (*85*). Despite questions about its true biochemical role in lignin biodegradation, the enzyme is secreted extracellularly and has been localized at sites of wood cell wall degradation for both simultaneous and preferential white rot fungi (*30, 35*). Generally from the screening studies conducted, laccases are commonly found produced together with MnP (*27*).

Non-Enzymatic Processes

One of the most interesting and complicated aspects with white rot decay concerns the actual biochemical mechanism(s) used *in-situ* for mineralization of lignocellulose within wood cell walls. The fact that these mechanisms tend to vary with wood type, fungal species, time and even along a single hypha when a substrate is being degraded adds to difficulties and the likelihood of multiple mechanisms. As indicated above an array of hydrolytic and oxidative enzymes can be secreted by white rot fungi during wood decay. Many of these enzymes have been proven present during wood decay *in-situ* by variety of methods including detection directly on sections using TEM-immunocytochemistry (*30*) or in extracts from degraded wood (*86, 87*). While presence of enzymes can explain the direct attack of wood cell walls that can be visualized using microscopy, their location cannot explain lignocellulose decay distant from fungal hyphae (*88, 89*). Morphologically, brown- and white rot show some similarities in that in the former the cellulose and hemicelluloses are degraded and demethylated lignin left as a skeleton, while in preferential white rot the lignin and hemicelluloses are degraded leaving modified delignified cellulose fibres. Simultaneous white rot is somewhat similar to preferential white rot in that-- a thin zone of decay in which all the lignocellulose components are degraded nearest to the cell lumen is produced-- that slowly progresses across the wood cell wall during mineralization. The restricting factor with all three morphological decay patterns is the wood cell wall porosity *–in both dry and wet states-* that limits the penetration and diffusion of known oxidoreductase and hydrolytic enzymes. Thus in order to explain this phenomen of "decay at a distance" a range of oxidants have been postulated and studied (e.g. (*26*)). TEM has provided evidence for the involvement of non-enzymatic agents in both white- and brown rot fungi simply by showing mineralization (i.e. as a change in wall structure at micro- and ultrastructural levels) of the cell wall remote from the fungal hyphae. A perplexing problem with proving the involvement of oxidants in wood decay is to show their occurrence in wood under conditions allowing for the complete fungal decay system to be expressed. This for example is not always the case when fungi are grown in liquid cultures that may be optimal for enzyme production but not for oxidant evolution.

From the studies conducted, we know that a non-enzymatic wood decay system involving oxidants/reductants should have the following credentials, they: 1) should be generated remote to fungal hyphae thereby protecting it from attack; 2) should be sufficiently stable and produced inside the wood cell wall before they react, and 3) should have a self regenerating system. Electron micrographs of white rotted cell walls indicate that the radicals may have to travel distances of over several microns (e.g. pine, Figures 1, 3), therefore a system where they are produced in the wall in close proximity would be the most realistic. Over the last twenty years a number of non-enzymatic systems have been proposed but currently no one system is fully accepted. Most of the work has been done on *Phanerochaete chrysoporium* and *Pleurotus spp.* (*26, 76, 90, 91*). These extracellular fungal systems include: 1) the MnP system where the enzyme reacts with H_2O_2 and Mn(II) organic chelates in lignocelluloses to produce Mn(III) chelates that can cleave phenolic structures in lignin (*92, 93*) or MnP and the

Mn (III) chelates oxidize organic acids and unsaturated lipids to produce reactive oxygen species such as peroxyl radicals (.OOR) (*71*); 2) The LiP system with use of H_2O_2 for removing electrons from non-phenolic lignin structures in wood by producing convalent bond cation intermediates (*47, 66, 91, 94*); 3) the cellobiose dehydrogenase (CDH) system which reduces Fe(III) chelates in lignocellulose to Fe(II) oxalate chelates which then undergo Fenton reactions with H_2O_2 that may be produced by either pyranose 2-oxidase or glucose 1-oxidase (*86, 95*) or by Fe(II) autooxidation (*96*) generating hydroxyl radicals (H_2O_2 + Fe_{2+} + H^+ ---H_2O + Fe_{3+} + .OH); and 4) The veratryl alcohol (VA) (3,4-dimethoxybenzyl alcohol) system where extracellular LiP oxidizes fungal secreted VA to produce cation radicals that diffuse into the wood cell wall and cleave lignin via electron transfer (*97–99*). A recent study to visualize the spatial distribution of oxidants around hyphae during decay of spruce wood by *P. chrysosporium* using novel oxidizing beads provided evidence for oxidation gradients from the hyphae (*100*). The dominating oxidant had a half life of 0,1 s and was best fitted to the cation radical and metabolite of VA. The hypothesis was therefore consistent with known extracellular distribution of LiP during decay by the fungus (*34*).

There is little doubt determining the true biochemical mechanisms during enzymatic/non-enzymatic white rot decay is very difficult and probably can only be understood with studies where the fungus is actually grown on wood rather than in artificial liquid culture environments. Understanding these systems is of particular importance for producing new and effective targeted wood protection systems. For more details, readers should consult reviews over the years of the non-enzymatic systems and specific studies proposed for white rot fungi (*26, 101–103*).

The advent of genomic analyses in recent years has in some ways revolutionized our possibility to understand events of lignocellulose attack and the potential battery of enzymes that can be encoded and released during decay of wood by white rot fungi. Knowing the profile of enzymes that can be secreted under native conditions offers possibilities for being able to replicate the process. Some care is however needed as this will no doubt be dependent on situation and physiological state of the fungus and could presumably vary along the length of a hypha. An early example surrounds H_2O_2 production. It was originally considered that glucose-1 oxidase (G1O) an H_2O_2 producing enzyme characteristic for Ascomycetes was the source of H_2O_2 in *Phanerochaete chrysosporium* (strain ME-446) for driving lignin degrading (e.g. MnP) enzymes (*104*). Subsequent studies with the *P. chrysosporium* K-3 strain showed pyranose 2 oxidase (P2O) to be present (*95*). The fact that the two H_2O_2 producing oxidases could be present and thus be utilized in a tandem series in which the primary products of the two enzymes (i.e. 2-keto-D-glucose (*105*) and D-glucono-1,5-lactone (*106*) are subsequently oxidized to 2-keto-D-gluconate offered a possibility of producing double the H_2O_2 per mol of glucose. This possibility was investigated in optimized (nitrogen starved) liquid cultures and in extracts removed from birch wood degraded by the two Phanerochaete strains. Using, chromatographic, electrophoretic, and immunological methods only evidence for P2O was derived and G1O was not detected (*95*). Recent genomic analyses of P. chrysosporium (strain RP78) (*107*) however, provides evidence for single genes encoding both

G1O and P2O as well as glyoxal oxidase and cellobiose dehydrogenase. This would indicate that *P. chrysosporium* has several possibilities for H_2O_2 production but that its enzymatic source is likely to be highly regulated and most probably could change over time during decay of the substrate. An ability to encode the enzyme therefore may not necessarily mean it will be produced. There is also a further possibility that minor genomic differences in the different *Phanerochaete* strains could also be important in the profiles of enzymes produced.

Detection of White Rot Enzymes in Situ in Wood

Using immunocytochemical techniques and antibodies with gold labeled probes in conjunction with TEM, it has been possible to localize ligninolytic (lignin peroxidase (LiP), Mn-peroxidase (MnP), laccase (Li)), cellulolytic (*30, 108–112*) hemicellulolytic and glucose oxidizing enzymes (*35, 86, 113*) associated with the thin zones of decay produced by different white rot fungi (Figures 5, 6). Using a similar approach, the same enzymes were confirmed associated with the fungal hyphae within the periplasmic space and associated with the cell wall, extracellular slime and tripartite membranes in wood during decay (Figures 5, 6) (*34, 36, 113*). This approach has proved the presence of the enzymes associated with degrading wood cell walls and thus provided indirect evidence of a probable involvement in wood cell wall mineralization processes *in-situ*. Immunocytochemical studies applying double labeling approaches with *Phlebia radiata* and *Phanerochaete chrysosporium* have further shown evidence for the presence of different oxidoreductase enzymes in close proximity to one another (e.g. LiP and MnP and laccase) suggesting interactions and cooperative attack (Figure 5) (*30, 35*). In particular, demonstration of pyranose oxidase capable of producing H_2O_2 from sugars together with MnP which requires this oxidant testifies to an intricate and well adapted process. Of the enzymes implicated in mineralization, less work has been carried out on the localization of hydrolytic enzymes in wood decay capable of cellulose and hemicellulose degradation apart from that reported by Daniel (*30*). However, both endoglucanases and cellobiohydrolases (exoglucanases) are thought to be always involved. Few immunocytochemical studies have been carried out to date on the native and cellulase-less preferential white rot fungi. Although the work thus far shows a similar labeling pattern for enzymes associated with the fungus as that observed with wild type simultaneous white rot fungi, detection of enzymes within the broad decay zones of preferential fungi in wood cells has not been proven unequivocally (*24, 34*). Two possible explanations include loss of the enzymes or antigenicity during specimen preparation for microscopy or that the enzymes are indeed absent and that decay within the cell wall is carried out through low molecular weight agents ((*114*) i.e. less than 5,7 kD,) in non-enzymatic reactions (as described above). Absence of the enzymes in the decay zones is further consistent with the difficulties for large proteins such as cellulases and peroxidases with molecular weights of ca 40-70,000 Kda in penetrating the zones and earlier studies on changes in porosity of wood cell walls during decay (*88, 114, 115*).

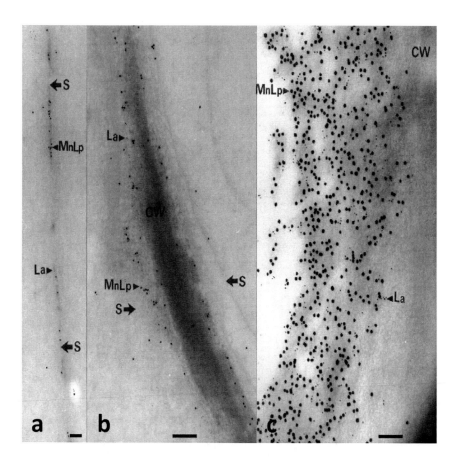

Figure 5. TEM micrographs showing aspects of simultaneous white rot of birch with localization of extracellular lignin- (Lp) and manganese peroxidases (Mn) and laccase (La) with degrading cell walls and extracellular slime using gold labeling and specific antibodies (MnLp, large black dots; laccase, small black dots). a) Lignin- and Mn peroxidases and laccases associated with extracellular slime distant to hyphae; b, c) Lignin- and Mn peroxidases and laccases associated with degrading wood secondary cell walls (b, inner S1, primary and middle lamella regions; c) secondary S2 layer) with enzymatic penetration only in opened cell wall regions. Bars: a, b, c, 1.0 μm.

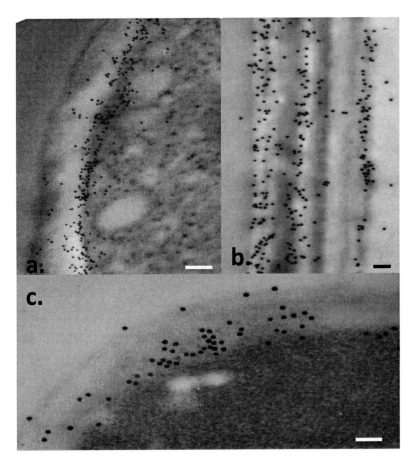

Figure 6. TEM micrographs showing periplasmic localization of cellulase (cellobiohydrolase) and Mn-peroxidase with Phanerochaete chrysosporium hyphae and extracellular cellobiohydrolase associated with the cell walls of degraded birch wood using immunogold labeling. a, b) Cellobiohydrolase (black dots) present in the peripheral cell cytoplasm (periplasmic space) and fungal cell wall and extracellular distribution on highly degraded birch wood cell wall; c) Typical distribution of Mn-peroxidase in the periplasmic space and fungal cell wall during decay of wood. Bars: a, b, c, 1.0 μm.

Importance of Slime in White Rot Decay

A major morphological feature observed with wood decay by most white rot fungi as well as other rots and bacteria (see below) is the characteristic presence of extracellular slime materials. Association of extracellular slime with fungal hyphae is a general feature for most fungi, allowing not only for adhesion to substrates (*116*) and retention of moisture, but has been variously implicated in wood decay processes (*30, 34, 35, 109, 113, 117, 118*). TEM studies using ruthenium red staining showed the presence of tripartite membranes

(*24, 119, 120*) associated with wood lumen cell walls during attack by *Trametes versicolor* and *Phanerochaete chrysosporium*, with morphological evidence for decay beneath the membranes in the wood cell wall (*30*). Other immuno- and cytochemical studies with the same fungi showed LiP (*34, 121*) and the hydrogen peroxide producing enzyme pyranose-2 oxidase (*86*) associated with slime and the tripartite membranes lining the lumen of partially degraded wood fibres (*30, 118*). Further work showed laccase to be bound with the extracellular slime produced by *Rigidoporus lignosus* and *Lentinus edodes* hyphae during wood decay (*112, 122, 123*). Oxidoreductase enzymes have also been reported associated with the extracellular slime of brown rot fungi (e.g. *Gloeophyllum trabeum* (*25*)). Using NMR and other characterization approaches, the extracellular slime of *P. chysosporium* has been characterized as a β-1,3-1,6-D-glucan (*117, 118, 124*), a high molecular weight polymer similar to that found in fungal cell walls. Polysaccharide polymers can change considerably in morphology through aggregation and skrinkage during fixation, dehydration and drying (e.g. critical point drying, freeze drying) processes and be even lost during specimen preparation for SEM and TEM. Thus more extensive studies were conducted on *Phlebia radiata* (wild and cellulase-mutant) using both high resolution Cryo-FE-SEM involving the rapid freezing of degraded samples in LN_2 slush (-210°C) followed by deep-etching, coating with Au/Pd with observations at -120°C (*35*). Since rapid freezing retains the hydrated state of samples, the slime has a more gel/sheath-type appearance and remains more in the native state. Similar samples were also processed for conventional TEM and a range of oxidoreductase enzymes detected using immunocytochemical techniques (*35*). FEM-SEM x-ray microanalysis and Cryo-FE-SEM x-ray analysis has further shown the association of calcium oxalate and manganese oxide associated with hyphae, the latter directly associated with wood cell wall decay (*30*). Both the FE-SEM and TEM studies confirmed the association of characteristic extracellular slime materials with the wood cell wall layers and remaining middle lamella corners at all stages of decay (Figures 2, 5). Immunocytochemical studies further showed a dual association of LiP, MnP and laccases with the extracellular slime and wood cell walls regions under attack (Figure 5). FE-SEM approaches showed evidence for two morphological forms of the extracellular slime produced in concentric orientation in the cell lumen and in radial orientation in the degrading cell wall; the latter confirmed with TEM. Extracellular slime can be produced by fungi as a means of removing components like sugars from the local environment in order not to repress decay. It is also well known to be produced under conditions of nitrogen starvation (*125*). Based on the decay patterns evolved with *P. radiata* decay and spatial distribution of the extracellular slime, some important roles for slime were hypothesized (e.g. (*24*)): 1) that the extracellular slime is produced progressively by lumina hyphae and conveys the enzymes (i.e. oxidoreductases in this case) involved in decay onto the fibre wall; 2) that the slime material accumulates on the lumen wall and penetrates into the fibre cell wall following the native architecture causing progressive decay across the wall.; 3) enzymes associated with the slime carryout local decay of the fibre wall components causing delamination of the wood cell wall layers; and 4) other components associated with the slime (e.g. calcium oxalate, MnII) could provide a means for initial and prolonged non-enzymatic attack and opening of the fibre wall for subsequent enzyme attack. Theoretically, extracellular slime has additional advantages including a means of progressive transfer to and retention of high molecular enzymes at sites of cell wall decay as well as providing a medium for retaining contact with the fibre in advance stages of decay. It would also explain how remote attack of the lumen wall distant to hyphae can take place at early stages of white rot as seen in species such as *Trametes versicolor* (*30*). Although further research is needed on the role of slime during decay, it does offer a number of attractive arguments to help explain how enzymes and non-enzymatic processes could operate in the wood cell wall distant to lumena hyphae.

The genomic approach developed over the last 10 years provides a whole new means of characterization of the whole spectrum of genes available and encoded proteins that can be produced (e.g. (*126*)). For example, with comparative studies on the genomic profiles of the simultaneous white rot fungus *Phanerochaete chrysosporium* and preferential white rotter *Ceriporiopsis subvermispora*, evidence was obtained for a larger inventory of oxidoreductase enzymes (i.e. MnP), as well as desaturase-encoding genes for lipid metabolism capability (*126*). A genomic approach provides overview on the profile of protein types available for encoding, thus it is interesting that the *P. chrysosporium* strain was shown not to contain genes encoding laccases, which is both consistent and contrasts with earlier reports from biochemical assays. The genomic approach gives an extra dimension to our understanding of wood decay processes so that inference can be made on the types of enzymes and multiple enzyme complexes that may be produced *in-vivo* during the two main morphological forms of white rot decay. However, this approach cannot explain decay mechanisms at the wood cell wall level and there will still be a need to understand enzyme spatial distributions in wood cells. Most importantly, it will provide knowledge of the enzymes likely involved and how their profiles could change over time. Naturally, whether or not an enzyme is encoded or not will depend on the physiological state of the fungus and its ability to degrade the wood substrate. In this respect, the genomic approach opens up new possibilities for wood protection as genomic analysis of decay fungi grown on modified or metal treated wood could provide profiles of enzymes produced and thus possible targets of decay prevention. Additionally, the protein profiles could be used for optimized and environmental friendly lignocellulose disassembly processes.

Other Types of Morphological White Rot Decay

In addition to the normal simultaneous and preferential decay patterns produced in wood, a number of white rot fungi have been reported to produce cavities in the S2 cell wall in the fibre longitudinal direction as opposed to bore holes running in transverse orientation (*24, 127–130*). Typical examples include *Odumansiella mucida* (*127*) that can produce very thin hyphae (e.g. 0.3-0,5 µm) and longitudinal orientated soft-rot like cavities within the secondary cell walls of wood and *Auricularia auriculajudea* that forms highly branched cavities (*130*).

The longitudinal cavities differ from those of traditional soft rot cavities in that they may or may not be aligned with the cellulose microfibrils and normally do not show angled cavities. However, the cavities may arise through L/T-branching reminiscent of that of soft-rot decay as seen with *O. mucida* (*127*). Thin hyphal development has also been shown for a range of litter degrading white rot fungi including *Mycena spp.* (*131*), as well as fungi attacking hard substrates such as wood impregnated with silica (Daniel unpub. obs). The development of very thin mycelia in such divergent groups in Ascomycetes and Basidiomycetes therefore suggests a common adaptation for the attack and penetration into wood cell walls. This for example is similar to appresorium development and mechanism for fungal pathogens to penetrate leaves (*132*).

Brown Rot Decay of Wood and Lignocellulose

Decay by brown rot fungi has also received considerable study over the last several decades. This group and the mechanisms of decay are reviewed later in the book so only brief aspects on the morphological aspects of decay are discussed (*24*).

From a morphological point of view it is frequently reported that the first signs of decay by brown rot fungi in wood occur at the interfaces of the S1 and S2 cell wall layers. This needs further study since it may rather represent diffusion of active agents along the longitudinal axis of fibres rather than across the transverse fibre wall, with initiation and penetration taking place at sites of bordered pits. It is generally accepted that brown rot fungi colonize and develop throughout the wood structure by growth in the cell lumena of fibres and other wood cells. However, brown rot fungi can produce hyphae that grow within the secondary cell walls of both soft- and hardwoods as indicated above for white rot fungi (e.g. *O. mucida*) that could be advantagous for access to the substrate. Typical examples include G. trabeum (Daniel unpub. obs) and *Coniophora puteana* (*133, 134*). Brown rot fungi also frequently develop very thin hyphae that develop and grow into very hard substrates like silica impregnated wood which could offer advantages. In this respect, development is not too dissimilar to the weathering of a variety of stone substrates by saprophytic fungi (*135*).

Soft Rot Decay

Soft rot is a type of wood decay that occurs primarily under conditions of excessive moisture and produced by Ascomycetes and Fungi Imperfecti. This form of attack is of considerable interest and economic importance since several species have the ability to attack wooden constructions including preservative treated wood (*20, 134, 136, 137*). It occurs under both aquatic and terrestrial situations and often together with other types of wood decay (*31*). Soft rot has been characterized as one if not the major form of attack of preservative treated (e.g. waterborne preservatives, CCA, creosote) utility poles in Sweden and several other countries in the 1970's (*20, 138–142*). The term "soft-rot" was originally coined by Savory (*143*) to describe the soft appearance and texture of wood degraded by fungi when wet, although characteristic features of soft rot were known very much earlier (*33, 144*). Wood degraded by soft rot tends to have a greyish discoloration and cracks on the surface similar to brown rot.

The most characteristic feature of soft rot is the morphology of attack of wood cell walls. Two distinct morphological types of attack known as Type I and Type II decay may be produced (*31*). Type I results in the formation of characteristic cavities by hyphae within the secondary cell walls that align themselves with the cellulose microfibrils (i.e. along the microfibril angle), while Type II attack results from a form of wood cell wall thinning quite similar to that observed for certain higher Ascomycetes (e.g. *Hypoxylon, Daldina sp.*, (*30*)) and white rot basidiomycetes. The main difference compared with white rot fungi is that soft rot fungi do not degrade the middle lamella regions even in advanced stages of decay (*31*). Soft rot erosion normally occurs under very high moisture conditions such as under aquatic situations. Type I and II attack may also occur in the same wood cells. A further morphological form known as Type III has been described where after penetration and cavity formation within wood secondary cell walls (e.g. within the S2 layer), solubilisation of polysaccharides takes place in a manner reminiscent of that of brown rot with modified lignin left over (*145*).

Possibly the most researched aspect concerning soft rot are the mechanisms behind cavity formation in wood cell walls and the ability of certain species and strains to tolerate heavy metals (*142, 146*). Numerous studies on aspects of cavity formation have shown how specialized microhyphae (ca 0.5 µm) or fine penetration hyphae penetrate from the cell lumen into the S2 cell wall layer of wood cells (*20, 134, 136, 147, 148*). Once in the S2 layer, the microhyphae can either pass straight through the cell wall expanding in size once again in the lumen of the adjacent cell rather like blue stain fungi or reorientate in the wall along the cellulose microfibrils either by L-bending in one direction or by producing a T-shaped branch to produce microhyphae that can develop in both axial directions. When growth of the microhyphae ceases, cavity formation begins along the proboscis (*147, 148*). After cavity enlargement, the hyphae produce another fine hyphae aligned with the cellulose microfibrils (i.e. along the cellulose microfibril angle, MFA) and the process repeats itself. The fact the hyphae always follow the orientation of the cellulose microfibrils is shown by hyphae following the circular orientation of microfibrils in pit chamber walls and between bordered pits as well as encircling the apex of softwood tracheids (*149, 150*). Studies have further shown that the orientation of soft rot fungi along cellulose microfibrils is so consistent that it can be used for measuring and comparing the MFA's in different wood tissues and cell types and gives results equivalent to confocal ellipsoidal microscopy approaches (*149–154*).

Frequently soft rot fungi are included or classified together with mould fungi. While they may belong to the same taxonomic group (Fungi Imperfecti) the ability of mould fungi (e.g. *Trichoderma reesei*) to degrade wood is normally very limited despite having often well-developed cellulolytic ability. Here the principle reason lies with the inability of true mould fungi to overcome the lignin barrier that protects the cellulose within cell walls from attack.

Bacterial Decay of Wood and Lignocelluloses

The effects of bacterial attack on wood have been known for a very long time (e.g. *(155)*). Primarily this has been recognized from the ability of aerobic and anaerobic bacteria to cause pronounced increase in the permeability of round wood particularly pine and refractory species (e.g. spruce) during ponding or water sprinkling of roundwood. Economically, this still represents possibly the most important effect of bacterial attack of wood although the logistics of modern forestry with quick turn-around from the cutting of trees to saw milling and kiln drying means its significance is much less today than previously. Only when major storms arise such as occurred in Europe in the late nineties and 2000's has there been a need for major water sprinkling over longer periods of time (e.g. in this case years). Over the last ca 30 years however, unequivocal evidence for bacterial degradation of lignified cell walls has been obtained primarily through the use of microscopy techniques particularly electron microscopy (TEM and SEM) for examination of wood samples degraded in the laboratory or removed from terrestrial or aquatic situations. Transmission electron microscopy in particular with its higher resolution allowed bacterial decay patterns to be distinguished from fungal decay patterns. *True wood degrading bacteria* are now known to be cosmopolitan having a tendency to be aerobic or facultative aerobic existing in a diverse range of aquatic and terrestrial environments *(31)*. True wood degrading bacteria are known to be single-celled, motile Gram negative, pleomorphic bacteria possessing considerable cell wall plasticity *(156)*. Such true wood degrading bacteria can cause significant attack of lignified fibres and tracheids in both hard- and softwoods and have an ability to degrade preservative treated wood (e.g. CCA treated, *(156)*), chemically modified timber, highly durable timbers with high extractive levels (e.g. *Eusideroxylon zwageri*, *(157)*) and/or high lignin content (e.g. *Alstonia scholaris*). In nature, bacterial decay frequently occurs with all other forms of fungal decay in both terrestrial and aquatic environments *(31)*.

Bacterial Decay Leading to Increased Permeability

The "ponding" or "water-spraying" of roundwood was previously routinely employed in many countries to protect timber from decay fungi, blue stain and insect attack by raising the moisture content above 130%. These bacteria primarily attack non-lignified tissues such as the ray parenchyma cells and pit membranes of the sapwood causing considerable improvements in permeability in all grain orientations particularly the radial direction. Similar tissues and structures in heartwood are normally less affected due to partial lignification or the encrusting of extractives. However, refractory softwoods with lignified parenchyma (e.g. Norway spruce) tend also to be less affected compared to pines (e.g. *Pinus sylvestris*) and therefore longer treatment times are required. Commercial trials have shown however that degradation generally results in an irregular and often over absorption of preservative solutions, varnishes, stains, paints etc during ponding, thereby imparting limitations on the final use of the wood material (*158, 159*) and for example such material is unacceptable for joinery (e.g. house facades). Long-term water sprinkling and ponding of poles was also attempted earlier as a possible method for reducing bleeding of creosote impregnated poles (e.g. (*160, 161*)) and increased penetration of preservatives into refractory wood species like spruce. Ray parenchyma and bordered pits are the sites of most rapid attack. Both rod-shaped (*ca* 1-2 µm long) and cocci bacteria have been reported causing decay (*162, 163*) as well as aerobic and anaerobic forms. The ponding of wood in Scandinavian countries for example was discontinued many years ago and water sprinkling is more temporary due to logistics and environmental restrictions where closed systems are only now allowed. In tropical countries where wood is still transported via water (rivers, sea) and where low lignified trees are involved, increases in permeability and even decay of wood fibres may result depending on time exposed.

True Bacterial Decay of Lignified Tissues

Two main forms of bacterial degradation namely: tunneling (TB) and erosion (EB) attack of lignified tissues are now well known having been described in a plethora of different environments over the last 30 years (*31*). These two types of attack are still based on the microscopy classification of the decay patterns produced in lignified fibres and not taxonomic features of the organisms. Despite their cosmopolitan existence, the preparation of pure cultures of active wood degrading bacteria have proved to be very difficult and thus their true taxonomic origin is still unknown. However, mixed cultures from bacterial degraded wood primarily of erosion bacteria from different habitats have been produced under laboratory conditions and decay patterns replicated from native conditions (*164*). From a concentrated effort in the early 2000's (*165–168*), mixed cultures of erosion bacteria were established and using DNA analysis, information on their possible taxonomic affiliation was obtained (see below). In the last decade, most efforts have been focused on erosion bacteria decay as interest in the restoration of archaeological artifacts and the attack of foundation pilings has intensified as this form of bacteria wood decay under conditions of low oxygen has been shown to be the most serious and consistently important. In contrast, studies on tunneling bacteria have been less apart from reports on its presence in wood exposed in terrestrial or aquatic environments. Electron microscopy observations have confirmed that both types of decay are caused by Gram-negative bacteria.

Tunneling bacteria produce a very characteristic attack of the wood secondary cell walls where single-celled pleomorphic bacteria (ca 1-2 µm long) produce very thin branching and radiating decay patterns. Initially, such thin branching patterns were thought produced by thin fungal hyphae but was unequivocally confirmed that bacteria were involved using TEM (Figures 1, 7) (*156*). Tunneling bacteria can cause direct penetration of wood cell walls either from the cell lumen or from the surface of wood and unlike erosion bacteria (see below) and soft rot fungi (see earlier) but consistent with white rot decay (see earlier) can penetrate and decay *-at least to some extent-* the highly lignified middle lamella regions joining cells (Figures 1, 7). Two further characteristics include that only a single bacteria is found within tunnels, with cell division resulting in bifurcation of tunnels and that the motile bacteria produce concentrically orientated slime secretions (presumably polysaccharide-protein conjugates) as the bacteria progressively move forward in the wood by the sloughing of the outer bacterial cell wall material (Figure 7). The extracellular slime secretions bind heavy metals (e.g. CCA) during attack of preservative treated wood and dyes during processing for TEM (e.g. ruthenium red, uranyl acetate) suggesting an anionic character (*156*). The fact that tunneling bacteria are often found degrading highly lignified wood (e.g. *Homalium foetium, Alstonia scholaris* (*169*)) and wood species with high extractive content (e.g. *Eusideroxylon, zwageri, Ocotea rodiaei,* (*157, 170*)) as well as the middle lamella regions testifies to a ligninolytic and aromatic degrading capacity. This will however only be truly confirmed when pure cultures are established. Apart from the occasional reporting for the presence of tunneling bacteria in exposed test wood samples in recent years (e.g. (*171*)), very little work if any has been done on their mechanisms of decay. Most of the work on tunneling bacteria was done in the 1980's and 1990's where the morphological characteristics of decay were demonstrated and early studies done on the mineralization of lignin model compounds (DHP's) (*172*). Most research in recent times has only been to verify decay patterns either in new environmental situations or attack of new preservative types during testing or from in-service situations. Thus, the true importance of tunneling bacteria is still unknown. However, its cosmopolitan distribution and occurrence together with other rot types during decay of both treated- and untreated wood in-service in both terrestrial and aquatic situations indicates an important contribution in biomineralization in nature (*31, 32*). Compared with fungi and erosion bacteria (see below) however, its full significance in the decay of wood is unkown. Presumably, its importance will increase when wood treatments are of such a kind (e.g. toxicity, high lignin/extractive) to produce competitive free environments where fungal decay is somewhat limited and where aerobic conditions exist. Apart from its important attack of fence poles in vine yards in NZ (*173*) its true significance is unclear unlike that of erosion bacteria (see below).

In contrast to tunneling bacteria, research on erosion bacteria decay of wood particularly its importance for waterlogged wood from buried ship wrecks or terrestrial archaeological excavations and its effect on the physical properties of foundation piles has escalated in recent years (*174–181*). Under oxygen limiting conditions such as with buried shipwrecks (*29, 182, 183*) terrestrial archaeological artifacts (*29, 179, 180, 184, 185*) and building foundations (*174–177, 179, 181, 186, 187*), fungal attack is suppressed and erosion bacteria proliferation may represent the only major type of decay and can thus cause serious problems. The problem is not only important for wood but also significant for non-wood materials particularly waterlogged archaeological bamboo as reported from Asia (*188*). Although, bacterial attack of foundation piles has serious economic implications, since millions of wooden foundations are still in-service, the piles provide excellent material for studies on the true effects on wood almost entirely degraded by erosion bacteria. This material can thus be used for providing macro- and micro studies on the effects of bacterial attack on strength (*186, 187,*

189) as well as providing materials for chemical analysis (*190, 191*). In general, studies have shown foundation piles to have high moisture content, low and variable compressive strength and density that reflect by in large the effects of bacterial erosion attack over many 100's of years. Studies have further shown new possibilities for correlating physical anatomical and chemical parameters of bacterial decay using a multidisciplinary approach with grading systems (*174*) and databases based on observations of thousands of poles (*175, 176*).

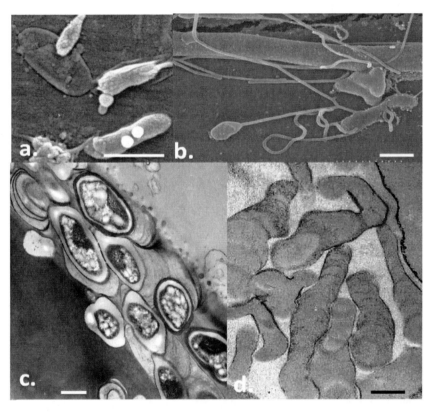

Figure 7. Aspects of tunneling bacterial attack of wood as visualized using SEM and TEM. a) Typical slime secretion that secures the entrance of the bacterial tunnels on the wood cell lumen; b) The bacteria are frequently enclosed in extracellular slime material (tubes) and attached to the lumen wood cell; c) Section of spruce tracheids with the bacteria attacking the secondary cell wall and middle lamella joining cells; d) Characteristic nature of tunneling bacteria decay of all types of wood cell walls with concentric slime layers left behind in the tunnels (also seen in c) during attack. These slime secretions typically bind metals in preservative treated woods. Bars: a, b, 1.0 µm; c, d, 0,5 µm. (Images a, c, d reproduced from reference (24). Copyright 2003 American Chemical Society).

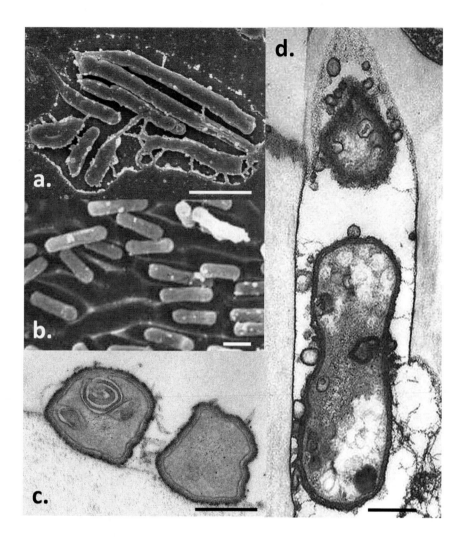

Figure 8. SEM and TEM micrographs showing erosion bacteria attack of wood. a) Early stage of decay with rod-shaped erosion bacteria encapsulated in slime attached to the lumen wood cell wall; b) Attack of the secondary cell wall with erosion bacteria forming characteristic erosion channels within the S2 layer; c) TEM micrograph showing a cross section of the bacteria adpressed to the S2 cell wall as would be observed from b); d) Longitudinal section of erosion bacteria from within an erosion channel showing the characteristic secretion of small extracellular vesicles that are assumed to contain wood degrading enzymes. Bars: a, 0.4µm; b, c, 1.0 µm; d, 0,2 µm. (Figure reproduced from reference (24). Copyright 2003 American Chemical Society).

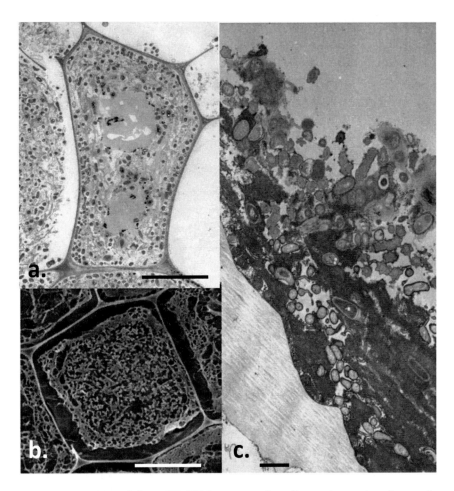

Figure 9. TEM and Cryo-FE-SEM micrographs of bacteria erosion decay of wood cells. a) TEM micrograph showing almost total removal of the secondary cell walls of birch with only the middle lamella regions (i.e. not degraded) remaining in advanced stages of attack. The cell lumen appears full of slime as does regions of the former secondary cell walls; b) Cryo-FE-SEM showing similar features as a) with only the middle lamellae remaining and the lumen and former cell walls filed with bacteria, slime and breakdown products; c) TEM longitudinal sections showing bacteria adpressed to the wood cell wall and typical erosion attack. Bars: a, b, 5.0 µm; c, 1.0 µm. (Images b, c reproduced from reference (24). Copyright 2003 American Chemical Society).

Erosion bacteria produce a morphological decay pattern that has both similarities to white rot (simultaneous) and soft rot (i.e. Type 2) in that the bacteria tend to initiate decay from the cell lumen through the S3 layer (Figures 1, 8, 9). In addition, they frequently cause decay in wood secondary cell walls from sites of bordered pits particularly in softwoods producing characteristic angular cavities perpendicular to the fibre axis. Unlike white rot fungi, but similar to soft rot fungi, erosion bacteria tend not to actively degrade the high lignin containing middle lamella regions suggesting more limited ligninolytic ability than tunneling bacteria although bleaching of this region has been reported (*24, 192*). Wood cell wall degradation results from the combined activity of numerous bacteria progressively eroding the underlying cell wall, producing discrete, but characteristic channels along which the bacteria move and divide as the colony progressively expands from its ends (Figure 8). This often gives rise to a "striped" decay pattern which is readily seen in longitudinal sections with polarized light where a loss in birefringence indicates attack of underlying cellulose and hemicelluloses (Figure 9). Bacteria erosion decay is similar to soft rot attack in that the bacteria normally show alignment like soft rot cavity hyphae with the underlying cell wall cellulose microfibrils (Figure 8) and which may represent a method to overcome the lignin barrier protecting the cellulose. Frequently, attack spreads from the rays and thus in cross-sections of wood samples, fibres are seen at various different stages of attack. In TEM sections, erosion bacteria often appear closely addressed to the underlying cell wall encapsulated in slime. Studies so far have shown EB to have a rod-shaped appearance (1-2 µm long) often with pointed ends (Figures 8, 9).

A characteristic feature of both erosion and tunneling decay is the apparent encapsulation of the bacteria in extracellular slime materials (Figures 7-9) (*24, 156, 192, 193*) that presumably provide the organism with protection (e.g. from wood preservatives, Figure 7), desiccation, etc. as well as providing a media for motility and passage of enzymes (Figures 7, 8) to the wood substrate. However, very little is known on the mechanisms of decay, again hindered because of the lack of pure cultures. TEM observations indicate that both erosion and tunneling bacterial produce extracellular secretions in the form of vesicles (ca 0.003-0.06 µm) (*24*) or have surface complexes that appear morphologically similar with that reported for some cellulolytic bacteria such as *Bacteroides succinogenes* (*194*) and for the multicellulase "cellulosomes" found on the outer cell wall of anaerobic bacteria like *Clostridium thermocellulum*. These structures have been noted on both the outer cell wall of erosion bacteria and associated with the decay channels (Figure 8). Gross chemical analyses (*190, 191*) of foundation piles and archaeological wood degraded by erosion bacteria has shown an increase in the relative lignin (as klason lignin) concentration consistent with loss of cellulose and hemicelluloses, reduction in phenolic compounds and increase in inorganic materials such as phosphorus. Recent observations with a more rapid approach and using sections with FT-IR have further confirmed the changes in lignin which are consistent with microscope observations (*190*).

Despite the progress in understanding bacteria decay, establishment of pure cultures remain an enigma both for understanding the true mechanisms of decay and for taxonomic affiliation. Bacterial degraded wood from archaeological and foundation piles have been used as sources in attempts to obtain pure cultures of erosion bacteria (*166, 167*). While isolated bacterial consortia under conditions of reduced oxygen were capable of replicating decay patterns from native situation, pure cultures were not obtained emphasizing the difficulties involved. The isolation of wood degrading bacteria appears problematic and use of traditional techniques of isolation (e.g. streaking, dilution) have all failed (*31, 166*). Using molecular analyses of the most purified cultures (*167*) as well as DNA isolated from degraded archaeological waterlogged wood (*168*), sequences suggested the erosion bacteria may belong to the cytophaga-flavobacterium-bacteroides (CFB) complex. Based on cellular and decay characteristics in common with gliding bacteria, the attack of cellulose fibres by wood degrading erosion bacteria were also previously thought to belong to the Myxobacterales or Cytophagales (*31, 178, 195*).

Despite numerous reports over the years, no new morphological forms of bacterial decay of wood have been described. Thus we can assume that the two types described which can be differentiated through levels of oxygen availability, represent the major forms that exist and that any slight differences that appear in decay pattern may reflect minor variations in enzymatic ability and strain differences. For a more detailed account on bacterial decay the reader should consult one of several reviews available (*24, 31, 178*).

Other Bacteria Colonizing Wood

A variety of other bacteria forms have also been shown present in woody tissues particularly when samples have been taken from native terrestrial and aquatic (marine and lakes) situations and examined using microscopy (*183, 196–198*). The role of these bacteria in decay is unknown and probably they exist as secondary feeders and cause only slight cell wall modification (*197*). A diverse range of bacteria forms are also normally found associated with wood materials in the guts of marine borers (e.g. *Limnoria spp.* (*199*), (*200*)*, Chelura terebrans, (201)*) and in the frass (*199*).

Earlier reports of actinomycete (i.e. filamentous bacteria) soft rot decay of lime wood (*202*) have not been verified and apart from ultrastructural evidence for the etching of wood cell walls by *Streptomyces* spp (*30, 203*) and the decay of birch fibres by a so far unidentified filamentous bacteria (*30*), the role of these filamentous bacteria in wood and lignocellulose biomineralization remains obscure. Their known ability to produce a wide variety of polysaccharide degrading enzymes (*204*), decay lignin model compounds (*205–207*) and none-less by their frequent isolation and cosmopolitan distribution in soils would suggest a role (*208, 209*) at least for non-treated wood.

References

1. Eriksson, K.-E.; Kirk, T. K.Biopulping, biobleaching and treatment of kraft bleaching effluents with white-rot fungi. In *Comprehensive Biotechnology: The Principles, Applications and Regulations of Biotechnology in Industry, Agriculture and Medicine*; Young, M. M., Cooney, C. I., Humphrey, A. E., Eds.; Pergamon Press: New York, 1985.
2. Kirk, T. K.; Burgess, R. R.; Koning, J. W., Jr. Use of fungi in pulping wood: an overview of biopulping research. In *Frontiers in Industrial Mycology*, Springer: New York, 1992; pp 99−111.

3. Messner, K.; Srebotnik, E. *FEMS Microbiol. Revi.* **1994**, *13* (2), 351–364.
4. Akhtar, M.; Blanchette, R.; Myers, G.; Kirk, T. In *Enzyme Applications of Fiber Processing*; Eriksson, K.-E. L., Cavaco-Paulo, A., Eds.; ACS Symposium Series 687; American Chemical Society: Washington, DC, 1998.
5. Mai, C.; Kues, U.; Militz, H. *Appl. Microbiol. Biotechnol.* **2004**, *61*, 477–494.
6. Majcherczyk, A.; Hüttermann, A.; Bruce, A.; Palfreyman, J. *For. Prod. Biotechnol.* **1997**, 129–140.
7. Shuen, S. K.; Buswell, J. *Lett. Appl. Microbiol.* **1992**, *15* (1), 12–14.
8. Cerniglia, C. E. *J. Ind. Microbiol. Biotechnol.* **1997**, *19* (5), 324–333.
9. Otte, M.-P.; Comeau, Y.; Samson, R.; Greer, C. W. *Biorem. J.* **1999**, *3* (1), 35–45.
10. Schubert, M.; Engel, J.; Thöny-Meyer, L.; Schwarze, F.; Ihssen, J. *Appl. Environ. Microbiol.* **2012**, *78* (20), 7267–7275.
11. Lehringer, C.; Richter, K.; Schwarze, F. W.; Militz, H. *Wood Fiber Sci.* **2009**, *41* (4), 373–385.
12. Schwarze, F. W.; Schubert, M. *Appl. Microbiol. Biotechnol.* **2011**, *92* (3), 431–440.
13. Schwarze, F. W.; Landmesser, H. *Holzforschung* **2000**, *54* (5), 461–462.
14. Eaton, R. A.; Hale, M. D. *Wood: decay, pests and protection*; Chapman and Hall Ltd: New York, 1993.
15. Eaton, R. *Preserv. Timber Trop.* **1985**, *17*, 157.
16. Daniel, G.; Bergman, Ö. *Holz Roh- Werkst.* **1997**, *55* (2-4), 197–201.
17. Van Acker, J.; Stevens, M.; Rijchaert, V. *International Research Group on Wood Preservation*, IRG/WP/95-10125, Ghent; 1995.
18. Schmidt, O. *Mater Org.* **1996**, *30*.
19. Schmidt, O.; Schmitt, U.; Moreth, U.; Potsch, T. *Holzforschung Int. J. Biol., Chem., Phys., Technol. Wood* **1997**, *51* (3), 193–200.
20. Leightley, L. E. Ph.D. Thesis, Department of Biological Sciences, University of Portsmouth, Portsmouth, England, 1977.
21. Leightley, L. *Bot. Mar.* **1980**, *23* (6), 387–395.
22. Mouzouras, R. *Biol. Mar. Fungi* **1986**, *4*, 341.
23. Eriksson, K.-E. L.; Blanchette, R. A.; Ander, P. *Microbial and Enzymatic Degradation of Wood and Wood Components*; Springer-Verlag: Berlin, 1990.
24. Daniel, G. Microview of wood under degradation by bacteria and fungi. In *Wood Deterioration and Preservation*; ACS Symposium Series 845; ACS Publications: Washington DC, 2003; pp 34−72.
25. Daniel, G.; Volc, J.; Filonova, L.; Plíhal, O.; Kubátová, E.; Halada, P. *Appl. Environ. Microbiol.* **2007**, *73* (19), 6241–6253.
26. Hammel, K. E.; Kapich, A. N.; Jensen, K. A.; Ryan, Z. C. *Enzyme Microb. Technol.* **2002**, *30* (4), 445–453.
27. Hatakka, A.; Hammel, K. E. Fungal biodegradation of lignocelluloses. In *Industrial Applications*; Springer: New York, 2011; pp 319−340.
28. Goodell, B.; Jellison, J.; Liu, J.; Daniel, G.; Paszczynski, A.; Fekete, F.; Krishnamurthy, S.; Jun, L.; Xu, G. *J. Biotechnol.* **1997**, *53* (2), 133–162.

29. Blanchette, R. A.; Nilsson, T.; Daniel, G.; Abad, A.; Rowell, R.; Barbour, R. Biological degradation of wood. In *Archaeological Wood: Properties, Chemistry, and Preservation*; Advances in Chemistry 225; American Chemical Society: Washington DC, 1990; pp 141–174.
30. Daniel, G. *FEMS Microbiol. Rev.* **1994**, *13* (2), 199–233.
31. Daniel, G.; Nilsson, T. *For. Prod. Biotechnol.* **1998**, 37–62.
32. Nilsson, T. D.; Daniel, G. *International Research Group on Wood Preservation*, IRG/WP 1443, 1990.
33. Schacht, H. *Jahrb. Wiss. Bot.* **1863**, *3*, 442–483.
34. Daniel, G.; Nilsson, T.; Pettersson, B. *Appl. Environ. Microbiol.* **1989**, *55* (4), 871–881.
35. Daniel, G.; Volc, J.; Niku-Paavola, M.-L. *C. R. Biol.* **2004**, *327* (9), 861–871.
36. Daniel, G.; Pettersson, B.; Nllsson, T.; Volc, J. *Can. J. Bot.* **1990**, *68* (4), 920–933.
37. Nyhlen, L.; Nilsson, T. *Colloq. INRA* 1987.
38. Srebotnik, E.; Messner, K. *Appl. Environ. Microbiol.* **1994**, *60* (4), 1383–1386.
39. Blanchette, R.; Reid, I. *Appl. Environ. Microbiol.* **1986**, *52* (2), 239–245.
40. Blanchette, R.; Otjen, L.; Carlson, M. *Phytopathology* **1987**, *77* (5), 684–690.
41. Otjen, L.; Blanchette, R. A. *Can. J. Bot.* **1986**, *64* (5), 905–911.
42. Blanchette, R. A. *Ann. Rev. Phytopathol.* **1991**, *29* (1), 381–403.
43. Kirk, T. K.; Chang, H.-m.; Lorenz, L. *Wood Sci. Technol.* **1975**, *9* (2), 81–86.
44. Ander, P.; Eriksson, K.-E.; Yu, H.-S. *J. Gen. Microbiol.* **1984**, *130* (1), 63–68.
45. Nilsson, T.; Daniel, G. *Holzforschung Int. J. Biol., Chem., Phys., Technol. Wood* **1989**, *43* (1), 11–18.
46. Hatakka, A. *FEMS Microbiol. Rev.* **1994**, *13* (2), 125–135.
47. Tien, M.; Kirk, T. K. *Proc. Natl. Acad. Sci. U.S.A.* **1984**, *81* (8), 2280–2284.
48. Glenn, J. K.; Gold, M. H. *Arch. Biochem. Biophys.* **1985**, *242* (2), 329–341.
49. Martinez, D.; Berka, R. M.; Henrissat, B.; Saloheimo, M.; Arvas, M.; Baker, S. E.; Chapman, J.; Chertkov, O.; Coutinho, P. M.; Cullen, D. *Nat. Biotechnol.* **2008**, *26* (5), 553–560.
50. Baldrian, P.; Valášková, V. *FEMS Microbiol. Rev.* **2008**, *32* (3), 501–521.
51. Wymelenberg, A. V.; Gaskell, J.; Mozuch, M.; Kersten, P.; Sabat, G.; Martinez, D.; Cullen, D. *Appl. Environ. Microbiol.* **2009**, *75* (12), 4058–4068.
52. Wymelenberg, A. V.; Gaskell, J.; Mozuch, M.; Sabat, G.; Ralph, J.; Skyba, O.; Mansfield, S. D.; Blanchette, R. A.; Martinez, D.; Grigoriev, I. *Appl. Environ. Microbiol.* **2010**, *76* (11), 3599–3610.
53. Ohm, R. A.; de Jong, J. F.; Lugones, L. G.; Aerts, A.; Kothe, E.; Stajich, J. E.; de Vries, R. P.; Record, E.; Levasseur, A.; Baker, S. E. *Nat. Biotechnol.* **2010**, *28* (9), 957–963.
54. Levasseur, A.; Piumi, F.; Coutinho, P. M.; Rancurel, C.; Asther, M.; Delattre, M.; Henrissat, B.; Pontarotti, P.; Asther, M.; Record, E. *Fungal Genet. Biol.* **2008**, *45* (5), 638–645.
55. Nilsson, T. D.; Daniel, G. *International Research Group on Wood Preservation*, IRG/WP 1184; 1983.

56. Daniel, G.; Nilsson, T. *Studies on the S2 layer of Pinus sylvestris*; Department of Forest Products, report. Uppsala 154: 34, 1984.
57. Hon, D. N.-S.; Shiraishi, N. *Wood and Cellulosic Chemistry, Revised, and Expanded*; CRC Press: Boca Raton, FL, 2000.
58. Hofrichter, M. *Enzyme Microb. Technol.* **2002**, *30* (4), 454–466.
59. Hammel, K. E.; Cullen, D. *Curr. Opin. Plant Biol.* **2008**, *11* (3), 349–355.
60. Glenn, J. K.; Akileswaran, L.; Gold, M. H. *Arch. Biochem. Biophys.* **1986**, *251* (2), 688–696.
61. Pérez-Boada, M.; Ruiz-Dueñas, F. J.; Pogni, R.; Basosi, R.; Choinowski, T.; Martínez, M. J.; Piontek, K.; Martínez, A. T. *J. Mol. Biol.* **2005**, *354* (2), 385–402.
62. Ruiz-Duenas, F.; Camarero, S.; Perez-Boada, M.; Martinez, M.; Martinez, A. *Biochem. Soc. Trans.* **2001**, *29* (Pt 2), 116–122.
63. Ruiz-Dueñas, F. J.; Martínez, M. J.; Martínez, A. T. *Mol. Microbiol.* **1999**, *31* (1), 223–235.
64. Ruiz-Dueñas, F. J.; Morales, M.; García, E.; Miki, Y.; Martínez, M. J.; Martínez, A. T. *J. Exp. Bot.* **2009**, *60* (2), 441–452.
65. Martínez, A. T. *Enzyme Microb. Technol.* **2002**, *30* (4), 425–444.
66. Kersten, P. J.; Tien, M.; Kalyanaraman, B.; Kirk, T. K. *J. Biol. Chem.* **1985**, *260* (5), 2609–2612.
67. Kirk, T. K.; Farrell, R. L. *Annu. Rev. Microbiol.* **1987**, *41* (1), 465–501.
68. Schoemaker, H. E.; Piontek, K. *Pure Appl. Chem.* **1996**, *68*, 2089.
69. *Environmentally friendly technologies for the pulp and paper industry*. Kirk, T. K., Cullen, D., Young R, Akhtar M, Eds.; John Wiley and Sons, Inc: New York, 1998.
70. Moen, M. A.; Hammel, K. E. *Appl. Environ. Microbiol.* **1994**, *60* (6), 1956–1961.
71. Kapich, A. N.; Jensen, K. A.; Hammel, K. E. *FEBS Lett.* **1999**, *461* (1), 115–119.
72. Kapich, A. N.; Steffen, K. T.; Hofrichter, M.; Hatakka, A. *Biochem. Biophys. Res. Commun.* **2005**, *330* (2), 371–377.
73. Watanabe, T.; Katayama, S.; Enoki, M.; Honda, Y.; Kuwahara, M. *Eur. J. Biochem.* **2000**, *267* (13), 4222–4231.
74. Enoki, M.; Watanabe, T.; Nakagame, S.; Koller, K.; Messner, K.; Honda, Y.; Kuwahara, M. *FEMS Microbiol. Lett.* **1999**, *180* (2), 205–211.
75. Orth, A. B.; Royse, D.; Tien, M. *Appl. Environ. Microbiol.* **1993**, *59* (12), 4017–4023.
76. Martínez, M. J.; Ruiz-Dueñas, F. J.; Guillén, F.; Martínez, Á. T. *Eur. J. Biochem.* **1996**, *237* (2), 424–432.
77. Heinfling, A.; Ruiz-Dueñas, F. J.; Martínez, M. a. J.; Bergbauer, M.; Szewzyk, U.; Martínez, A. T. *FEBS Lett.* **1998**, *428* (3), 141–146.
78. Thurston, C. F. *Microbiology* **1994**, *140* (1), 19–26.
79. Käärik, A. *Stud. For. Suec.* **1965**, *31*, 1–80.
80. Bollag, J.-M.; Leonowicz, A. *Appl. Environ. Microbiol.* **1984**, *48* (4), 849–854.
81. Baldrian, P. *FEMS Microbiol. Rev.* **2006**, *30* (2), 215–242.
82. Bourbonnais, R.; Paice, M. G. *FEBS Lett.* **1990**, *267* (1), 99–102.

83. Call, H.; Mücke, I. *J. Biotechnol.* **1997**, *53* (2), 163–202.
84. Eggert, C.; Temp, U.; Eriksson, K.-E. *Appl. Environ. Microbiol.* **1996**, *62* (4), 1151–1158.
85. Martinez, D.; Challacombe, J.; Morgenstern, I.; Hibbett, D.; Schmoll, M.; Kubicek, C. P.; Ferreira, P.; Ruiz-Duenas, F. J.; Martinez, A. T.; Kersten, P. *Proc. Natl. Acad. Sci. U.S.A.* **2009**, *106* (6), 1954–1959.
86. Daniel, G.; Volc, J.; Kubatova, E. *Appl. Environ. Microbiol.* **1994**, *60* (7), 2524–2532.
87. Mishra, C.; Leatham, G. F. *J. Ferment. Bioeng.* **1990**, *69* (6), 374–379.
88. Srebotnik, E.; Messner, K.; Foisner, R. *Appl. Environ. Microbiol.* **1988**, *54* (11), 2608–2614.
89. Srebotnik, E.; Messner, K. *Holzforschung Int. J. Biol., Chem., Phys., Technol. Wood* **1991**, *45* (2), 95–101.
90. Martínez, Á. T.; Ruiz-Dueñas, F. J.; Martínez, M. J.; del Río, J. C.; Gutiérrez, A. *Curr. Opin. Biotechnol.* **2009**, *20* (3), 348–357.
91. Hammel, K. E.; Kalyanaraman, B.; Kirk, T. K. *Proc. Natl. Acad. Sci. U.S.A.* **1986**, *83* (11), 3708–3712.
92. Wariishi, H.; Akileswaran, L.; Gold, M. H. *Biochemistry* **1988**, *27* (14), 5365–5370.
93. Wariishi, H.; Valli, K.; Gold, M. H. *Biochem. Biophys. Res. Commun.* **1991**, *176* (1), 269–275.
94. Harvey, P.; Schoemaker, H.; Bowen, R.; Palmer, J. *FEBS Lett.* **1985**, *183* (1), 13–16.
95. Volc, J.; Kubátová, E.; Daniel, G.; Přikrylová, V. *Arch. Microbiol.* **1996**, *165* (6), 421–424.
96. Kersten, P.; Cullen, D. *Fungal Genet. Biol.* **2007**, *44* (2), 77–87.
97. Harvey, P.; Schoemaker, H.; Palmer, J. *FEBS Lett.* **1986**, *195* (1), 242–246.
98. Candeias, L. P.; Harvey, P. J. *J. Biol. Chem.* **1995**, *270* (28), 16745–16748.
99. Bietti, M.; Baciocchi, E.; Steenken, S. *J. Phys. Chem. A* **1998**, *102* (38), 7337–7342.
100. Hunt, C. G.; Houtman, C. J.; Jones, D. C.; Kitin, P.; Korripally, P.; Hammel, K. E. *Environ. Microbiol.* **2013**.
101. Wood, P. M. *FEMS Microbiol. Rev.* **1994**, *13* (2), 313–320.
102. Evans, C. S.; Dutton, M. V.; Guillén, F.; Veness, R. G. *FEMS Microbiol. Rev.* **1994**, *13* (2), 235–239.
103. Joseleau, J.-P.; Gharibian, S.; Comtat, J.; Lefebvre, A.; Ruel, K. *FEMS Microbiol. Rev.* **1994**, *13* (2), 255–263.
104. Kelley, R.; Reddy, C. A. *Arch. Microbiol.* **1986**, *144* (3), 248–253.
105. Janssen, F. W.; Ruelius, H. W. *Biochim. Biophys. Acta, Enzymol.* **1968**, *167* (3), 501–510.
106. Neidleman, S. L.; Amon Jr, W. F.; Geigert, J. *Production of 2-keto-D-gluconic acid and hydrogen peroxide*; Patent US4351902 A, 1982.
107. Martinez, D.; Larrondo, L. F.; Putnam, N.; Gelpke, M. D. S.; Huang, K.; Chapman, J.; Helfenbein, K. G.; Ramaiya, P.; Detter, J. C.; Larimer, F. *Nat. Biotechnol.* **2004**, *22* (6), 695–700.
108. Blanchette, R.; Abad, A.; Farrell, R.; Leathers, T. *Appl. Environ. Microbiol.* **1989**, *55* (6), 1457–1465.

109. Ruel, K.; Odier, E.; Joseleau, J. *Biotechnol. Pulp Pap. Manuf.* **1990**, 83–98.
110. Daniel, G.; Goodell, B.; Jellison, J.; Paszyzyński, A.; Crawford, R. *Appl. Microbiol. Biotechnol.* **1991**, *35* (5), 674–680.
111. Daniel, G.; Pettersson, B.; Volc, J.; Nilsson, T. *Proceedings of Biotechnology in Pulp and Paper Manufacture*; Kirk, T. K., Chang, H. M., Eds.; Butterworth-Heinemann: Boston, MA, 1990; 99−111.
112. Evans, C. S.; Gallagher, I. M.; Atkey, P. T.; Wood, D. A. *Biodegradation* **1991**, *2* (2), 93–106.
113. Daniel, G.; Volc, J.; Kubatova, E.; Nilsson, T. *Appl. Environ. Microbiol.* **1992**, *58* (11), 3667–3676.
114. Messner, K.; Fackler, K.; Lamaipis, P.; Gindl, W.; Srebotnik, E.; Watanabe, T.; Goodell, B.; Nicholas, D.; Schultz, T. In *Overview of white-rot research: where we are today*, Current knowledge of wood deterioration mechanisms and its impact on biotechnology and wood preservation; Symposium at the 221st National Meeting of the American Chemical Society, San Diego, CA, April 1-5, 2001; American Chemical Society: Washington, DC, 2003; pp 73−96.
115. Flournoy, D. S. K.; T., K.; Highley, T. *Holzforschung* **1991**, *45*, 383–388.
116. Jones, E. *Mycol. Res.* **1994**, *98* (9), 961–981.
117. Ruel, K.; Joseleau, J.-P. *Appl. Environ. Microbiol.* **1991**, *57* (2), 374–384.
118. Ruel, K.; Joseleau, J.-P. *Food Hydrocolloids* **1991**, *5* (1), 179–181.
119. Foisner, R.; Messner, K.; Stachelberger, H.; Roehr, M. *J. Ultrastruct. Res.* **1985**, *92* (1), 36–46.
120. Foisner, R.; Messner, K.; Stachelberger, H.; Röhr, M. *Trans. Br. Mycol. Soc.* **1985**, *85* (2), 257–266.
121. Srebotnik, E.; Messner, K.; Foisner, R.; Pettersson, B. *Curr. Microbiol.* **1988**, *16* (4), 221–227.
122. Goodell, B.; Yamamoto, K.; Jellison, J.; Nakamura, M.; Fujii, T.; Takabe, K.; Hayashi, N. *Holzforschung Int. J. Biol., Chem., Phys., Technol. Wood* **1998**, *52* (4), 345–350.
123. Nicole, M.; Chamberland, H.; Rioux, D.; Lecours, N.; Rio, B.; Geiger, J.-P.; Ouellette, G. *Appl. Environ. Microbiol.* **1993**, *59* (8), 2578–2588.
124. Buchala, A. J.; Leisola, M. *Carbohydr. Res.* **1987**, *165* (1), 146–149.
125. Bes, B.; Pettersson, B.; Lennholm, H.; Iversen, T.; Eriksson, K.-E. *Biotechnol. Appl. Biochem.* **1987**, *9* (4), 310–318.
126. Fernandez-Fueyo, E.; Ruiz-Dueñas, F. J.; Ferreira, P.; Floudas, D.; Hibbett, D. S.; Canessa, P.; Larrondo, L. F.; James, T. Y.; Seelenfreund, D.; Lobos, S. *Proc. Natl. Acad. Sci. U.S.A.* **2012**, *109* (14), 5458–5463.
127. Daniel, G.; Volc, J.; Nilsson, T. *Mycol. Res.* **1992**, *96* (1), 49–54.
128. Schwarze, F.; Lonsdale, D.; Fink, S. *Mycol. Res.* **1995**, *99* (7), 813–820.
129. Schwarze, F. W.; Engels, J.; Mattheck, C. *Fungal strategies of wood decay in trees*; Springer: New York, 2000.
130. Worrall, J. J.; Anagnost, S. E.; Zabel, R. A. *Mycologia* **1997**, *89* (2), 199–219.
131. Nilsson, T.; Daniel, G. *International Research Group on Wood Preservation*, IRG/WP1358; 1988.
132. Shaw, B.; Kuo, K.; Hoch, H. *Mycologia* **1998**, *90*, 258–268.
133. Kleist, G.; Schmitt, U. *Holzforschung* **2001**, *55* (6), 573–578.

134. Daniel, G.; Nilsson, T. *J. Inst. Wood Science* **1989**, *11*, 162–171.
135. Hoffland, E.; Kuyper, T. W.; Wallander, H.; Plassard, C.; Gorbushina, A. A.; Haselwandter, K.; Holmström, S.; Landeweert, R.; Lundström, U. S.; Rosling, A. *Front. Ecol Environ.* **2004**, *2* (5), 258–264.
136. Crossley, A. Ph.D. Thesis, University of London, London, England, 1979.
137. Hale, M. D. C. Ph.D. Thesis, Department of Biological Sciences, University of Portsmouth, Portsmouth, England, 1983.
138. Gersonde, M.; Kerner-Gang, W. *Int. Biodeterior. Bull* **1976**, *12* (1), 5–13.
139. Zabel, R. A.; Wang, C. J.; Anagnost, S. E. *Wood Fiber Sci.* **1991**, *23* (2), 220–237.
140. Nilsson, T.; Henningsson, B. *Mater. Org.* **1978**, *13* (4), 297–313.
141. Leightley, L.; Eaton, R. In *Mechanisms of decay of timber by aquatic micro-organisms*, Record of the 1977 annual convention of the british wood preserving association, cambridge, june 28th july 1st, 1977, British wood preserving association: 1978; pp 221-246.
142. Henningsson, B.; Nilsson, T. *Mater. Org.* **1976**.
143. Savory, J. *Ann. Appl. Biol.* **1954**, *41* (2), 336–347.
144. Bailey, I. W.; Vestal, M. R. *J. Arnold Arbor., Harv. Univ.* **1937**, *18*, 196–205.
145. Anagnost, S.; Worrall, J.; Wang, C. *Wood Sci. Technol.* **1994**, *28* (3), 199–208.
146. Daniel, G.; Nilsson, T. *Int. Biodeterior.* **1988**, *24* (4), 327–335.
147. Hale, M. D.; Eaton, R. A. *Mycologia* **1985**, 447–463.
148. Hale, M. D.; Eaton, R. A. *Mycologia* **1985**, 594–605.
149. Khalili, S.; Nilsson, T.; Daniel, G. *Holz Roh- Werkst.* **2001**, *58* (6), 439–447.
150. Anagnost, S. E.; Mark, R. E.; Hanna, R. B. *Wood Fiber Sci.* **2000**, *32* (1), 81–87.
151. Anagnost, S.; Mark, R. E.; Hanna, R. B. *IAWA J.* **2005**, *26* (3), 325.
152. Anagnost, S. E.; Mark, R. E.; Hanna, R. B. *Wood Fiber Sci.* **2002**, *34* (2), 337–349.
153. Brandstrom, J.; Bardage, S. L.; Daniel, G.; Nilsson, T. *IAWA J.* **2003**, *24* (1), 27–40.
154. Bergander, A.; Brändström, J.; Daniel, G.; Sahnen, L. *J. Wood Sci.* **2002**, *48* (4), 255–263.
155. Liese, J. *Mahlke-Troschel, Handbuch der Holzkonservierung* **1950**, *3*, 44.
156. Daniel, G. F., Nilsson, T. *International Research Group Wood Preservation*, IRG/WP/1260, 1985.
157. Nilsson, T.; Singh, A.; Daniel, G. *Holzforschung Int. J. Biol., Chem., Phys., Technol. Wood* **1992**, *46* (5), 361–368.
158. Boutelje, J.; Jonsson, U. *STFI - Meddelande Serie A* 1976, *nr 376 (TT52)*.
159. Boutelje, J.; Henningsson, B.; Lundström, H. *STFI Meddelande Serie A,* 1977, *Nr 435 (TTA:70)*.
160. Bergman, Ö.; Henningsson, B.; Persson, E. *Svenska Träskyddsinstitutet Meddelande, Report No. 132*, 1975.
161. Bergman, Ö.; Martinsson, S. *Svenska Träskyddsinstitutet Meddelande, 119*, 1979.
162. Dunleavy, J.; McQuire, A. *J. Inst. Wood Sci.* **1970**, *5* (2), 20–8No. 26.

163. Daniel, G. E., T.; Nilsson, T.; Singh, A.; Liukko, K. *International Research Group on Wood Preservation*, IRG/WP93-10033, 1993
164. Nilsson, T.; Daniel, G. *International Research Group on Wood Preservation*, 1992.
165. Nilsson, T.; Björdal, C. *Int. Biodeterior. Biodegrad.* **2008**, *61* (1), 11–16.
166. Nilsson, T.; Björdal, C. *Int. Biodeterior. Biodegrad.* **2008**, *61* (1), 3–10.
167. Nilsson, T.; Björdal, C.; Fällman, E. *Int. Biodeterior. Biodegrad.* **2008**, *61* (1), 17–23.
168. Landy, E. T.; Mitchell, J. I.; Hotchkiss, S.; Eaton, R. A. *Int. Biodeterior. Biodegrad.* **2008**, *61* (1), 106–116.
169. Singh, A. P.; Nilsson, T.; Daniel, G. F. *International Research Group Wood Preservation*, IRG/WP/1462, 1990.
170. Pitman, A.; Sawyer, G.; Daniel, G. *International Research Group on Wood Preservation* 1995.
171. Rehbein, M.; Koch, G.; Schmitt, U.; Huckfeldt, T. *Micron* **2013**, *44*, 150–158.
172. Nilsson, T.; Daniel, G. Proceedings of 3rd Biotechnology in Pulp and Paper Industry; Eriksson, K.E., Ander, P., Eds.; Swedish Forest Products Research Lab: Stockholm, 1986; pp 54−56.
173. Nilsson, T. *International Research Group on Wood Preservation*, IRG/WP 1234, 1984.
174. Huisman, D.; Manders, M.; Kretschmar, E.; Klaassen, R.; Lamersdorf, N. *Int. Biodeterior. Biodegrad.* **2008**, *61* (1), 33–44.
175. Klaassen, R. K. *Int. Biodeterior. Biodegrad.* **2008**, *61* (1), 45–60.
176. Klaassen, R. K. *Int. Biodeterior. Biodegrad.* **2013**.
177. Klaassen, R. K.; Creemers, J. G. *J. Cult. Heritage* **2012**, *13* (3), S123–S128.
178. Singh, A. P. *J. Cult. Heritage* **2012**, *13* (3), S16–S20.
179. Macchioni, N.; Capretti, C.; Sozzi, L.; Pizzo, B. *Int. Biodeterior. Biodegrad.* **2013**, *84*, 54–64.
180. Björdal, C.; Daniel, G.; Nilsson, T. *Int. Biodeterior. Biodegrad.* **2000**, *45* (1), 15–26.
181. Macchioni, N.; Pizzo, B.; Capretti, C.; Giachi, G. *J. Archaeol. Sci.* **2012**, *39* (10), 3255–3263.
182. Kim, Y. S.; Singh, A. P. *IAWA J.* **2000**, *21* (2), 135–155.
183. Riess, W.; Daniel, G. *Int. J. Naut. Archaeol.* **1997**, *26* (4), 330–338.
184. Kim, Y. S. *Holzforschung* **1990**, *44* (3), 169–172.
185. Björdal, C.; Nilsson, T.; Daniel, G. *Int. Biodeterior. Biodegrad.* **1999**, *43* (1), 63–73.
186. Boutelje, J.; Bravery, A. F. *J. Inst. Wood Science* **1968**, *4*, 47–57.
187. Paajanen, L., Viitanen, H. *International Research Group on Wood Preservation*, IRG/WP 1370, 1988.
188. Cha, M. Y.; Lee, K. H.; Kim, Y. S. *Int. Biodeterior. Biodegrad.* **2014**, *86*, 115–121.
189. Kretschmar, E.; Gelbrich, J.; Militz, H.; Lamersdorf, N. *Int. Biodeterior. Biodegrad.* **2008**, *61* (1), 69–84.
190. Gelbrich, J.; Mai, C.; Militz, H. *Int. Biodeterior. Biodegrad.* **2008**, *61* (1), 24–32.

191. Gelbrich, J.; Mai, C.; Militz, H. *J. Cult. Heritage* **2012**.
192. Daniel, G. F.; Nilsson, T. *International Research Group Wood Preservation*, IRG/WP/1283, 1986.
193. Daniel, G. F.; Nilsson, T.; Singh, A. *Can. J. Microbiol.* **1987**, *33* (10), 943–948.
194. Forsberg, C. W.; Beveridge, T. J.; Hellstrom, A. *Appl. Environ. Microbiol.* **1981**, *42* (5), 886–896.
195. Daniel, G. F.; Nilsson, T *International Research Group Wood Preservation*, IRG/WP/1283, 1987.
196. Holt, D. M. *J. Inst. Wood Science* **1983**, *9*, 212–223.
197. Powell, K.; Pedley, S.; Daniel, G.; Corfield, M. *Int. Biodeterior. Biodegrad.* **2001**, *47* (3), 165–173.
198. Holt, D. M. *J. Inst. Wood Science* **1983**, *9*, 212–223.
199. Daniel, G.; Nilsson, T.; Cragg, S. *Holz Roh- Werkst.* **1991**, *49* (12), 488–490.
200. Daniel, G.; Cragg, S., Nilsson, T. *International Research Group on Wood Preservation*, IRG/WP/4169, 1991.
201. Cragg, S.; Daniel, G. *International Research Group on Wood Preservation*, IRG/WP/4180-92, 1992.
202. Baecker, A.; King, B. In *Decay of Wood by Actinomycetales*; Oxley, T. A., et. al., Eds.; Proc. 4th Int. Biodetn. Symp. Berlin; 1978.
203. Nilsson, T.; Bardage, S.; Daniel, G. F. *International Research Group on Wood Preservation*, IRG/WP93-1444, 1990.
204. McCarthy, A. J.; Williams, S. T. *Gene* **1992**, *115* (1), 189–192.
205. Crawford, D. L.; Barder, M. J.; Pometto, A. L., III; Crawford, R. L. *Arch. Microbiol.* **1982**, *131* (2), 140–145.
206. Ramachandra, M.; Crawford, D. L.; Hertel, G. *Appl. Environ. Microbiol.* **1988**, *54* (12), 3057–3063.
207. Phelan, M. B.; Crawford, D. L.; Pometto, A. L., III *Can. J. Microbiol.* **1979**, *25* (11), 1270–1276.
208. Cavalcante, M. S. *The role of actinomycetes in timber decay*. Ph.D. Thesis, University of Portsmouth, Portsmouth, England, 1981.
209. Safo-Sampah, S. Ph.D. Thesis, University of California, Berkeley, California, 1985.

Chapter 3

Omics and the Future of Sustainable Biomaterials

Juliet D. Tang[*,1,2] and Susan V. Diehl[2]

[1]USDA ARS Corn Host Plant Resistance Research, P.O. Box 5367, 810 Highway 12E, Mississippi State, Mississippi 39762
[2]Forest Products, Mississippi State University, 201 Locksley Way, Starkville, Mississippi 39759
*E-mail: juliet.tang@ars.usda.gov.

With global focus on the conversion of biomass into products, fuels, and energy, there is a strong need for information that will lead to new sustainable products, applications, and biotechnological advances. The omics approach to biology is a discovery-driven method that may deliver solutions to these overarching problems. It gives scientists the ability to obtain a systems-level understanding of life that begins with identifying the genome or genes in an organism. The pivotal technology that enabled this revolutionary approach is next generation DNA sequencing. New fields bring new jargon and new analytical methods making it difficult to appreciate the technology or the significance of the science. Our goal in this review is to present an introduction to the most important omics approaches that have been used to gain insight into how wood decay fungi convert lignocellulose into energy and why certain species are metal-tolerant. The expectation is that once relevant genes are identified, biotechnological methods will give rise to novel solutions that will advance wood protection and utilization.

© 2014 American Chemical Society

The Genomics Era

The ability to use high throughput technology to sequence a genome, which is the complete set of genetic material or DNA (deoxyribonucleic acid) in an organism, has revolutionized the rate of gene discovery. There are four nucleotide bases that make up the linear sequence of DNA. They are G for guanine, A for adenine, T for thymine, and C for cytosine. Despite this apparent simplicity, it is the sequence of these four nucleotides that governs all cellular activity. Thus, by sequencing genomes, the process of decoding life begins.

The race to sequence genomes did not gain momentum until the Human Genome Project was completed in 2003 (*1*). The project involved hundreds of scientists worldwide, took 13 years to complete, and cost $3 billion. At project completion, (i) millions of DNA fragments or reads were sequenced, (ii) computational and visualization tools were developed to map and assemble the reads from unique overlapping regions into a consensus sequence, (iii) a draft of the human genome, which was 3 billion nucleotides long, was generated, (iv) 20,000 to 25,000 genes were identified and annotated both in terms of their structure and function, and (v) a precedence of open access to data and software tools was established through public repositories like the National Center for Biotechnology Information (NCBI), the European Bioinformatics Institute (EBI), and the DNA Databank of Japan (DDBJ).

Sequencing Technology History

The Sanger method of sequencing and separation by capillary electrophoresis was developed in 1977 and became the major sequencing technology for over 25 years (*2*). The sequenced nucleotides called reads are generated from fragments of genomic DNA that have been cloned into appropriate bacterial vectors. The Sanger method produces highly accurate reads (500 to 900 bases long), but the maximum instrument output is limited to 96 reads per run or 86 Kbp (kilobase pair) of DNA sequence. In 2005 and 2006, the first next generation sequencing (NGS) prototypes were released. These platforms incorporated three major technological improvements: (i) PCR (polymerase chain reaction) replaced bacterial cloning; (ii) DNA was sequenced during the nucleotide addition step rather than by chain termination; and (iii) by spatially separating the microscopic clusters DNA fragments on a solid surface, millions of genomic DNA fragments could be sequenced in parallel during one run (*3*). As a result, DNA sequencing was transformed into a high throughput technology. This drastically accelerated the speed at which genomes could be sequenced and reduced the costs.

Early concerns arose because the first NGS systems produced much shorter reads with higher error rates. These drawbacks were partially compensated for by the large increase in depth of coverage. Depth of coverage refers to the number of times a nucleotide in a particular position of the genome is represented in the reads. Manufacturers have continually pushed the technology to increase read length, maximize output, and reduce base-calling errors. At the same time, new computational algorithms addressed the problem of read error and solved the daunting problem of assembling the millions of short reads into a continuous linear sequence. These algorithms like Velvet (*4*), Euler (*5*), and SOAPdenovo (*6*) are based on the de Bruijn graph, a mathematical concept of graph theory. The momentum of using NGS to sequence microbial genomes really gained speed, however, when convincing evidence showed that NGS alone or in combination with the Sanger method could produce *de novo* (no known reference sequence) assemblies of two fungal genomes, *Grosmannia clavigera* (*7*) and *Sordaria macrospora* (*8*). Because of these technological advances, genomics rapidly rose to the forefront of all aspects of biological science.

Today, there are three major NGS platforms in popular use: Roche, Illumina, and Life Technologies. Each relies on different detection methods to determine the identity of the added nucleotide during the sequencing reaction, and all three have multiplexing capability to increase the number of libraries or samples that can be simultaneously sequenced. The Roche GS-FLX+ offers the longest read length (700 bp), which has the advantage of reducing the number of gaps in the assembly and reducing the depth of coverage needed to get an assembly. The disadvantages are lower output (400 Mbp) and higher error rates for runs of the same nucleotide that are more than 7 bases long. The Illumina HiSeq2000 outstrips the others in terms of output (600 Gbp) but requires higher coverage when assembling a genome without a reference sequence. It is also by far the cheapest for reagents. The major drawback is the short read length (100 bp), which produces shorter contigs and a draft assembly with many gaps. The Life Technologies Ion Torrent is intermediate in terms of output (10 Gbp) and read length (200 bp), but has the fastest run times. The Ion Torrent like the Roche system has problems reading tracts of homopolymers. All three NGS platforms sell affordable personal or compact sequencers. If current cost trends hold, it will not be long before DNA sequencing and *de novo* assembly of microbial genomes will become routinely affordable (*9*).

Genomes of Fungi that Decay Wood

Basidiomycota

Fungi are a premier resource for bioprospecting because they have a tremendous metabolic diversity that includes many novel gene products and enzymes. To facilitate discovery of gene products with practical applications for energy and the environment, the Joint Genome Institute (JGI) of the Department of Energy has sequenced many basidiomycete genomes (*10–14*). Their sequencing pipeline uses the Sanger method or a hybrid approach that mixes Sanger with the Roche and Illumina platforms. Understanding the role genes play in lignocellulose degradation is relevant for the wood protection industry because it identifies potential targets that can be inhibited to prevent decay. Prior to 2010, there were only two sequenced wood decay genomes, *Phanerochaete chrysosporium* (*13*), a white rot fungus, and *Postia placenta* (*14*), a brown rot fungus. Now, three years later, there are sixteen. Seven are brown rot fungi and ten are white rot fungi (Table 1). Their average genome size is 44.59 Mbp (SD = 11.77) and they contain an average of 13,070 genes (SD = 2926). Others, like the genome of *Ophiostoma piceae*, a causative agent of bluestain, have been assigned Bioproject numbers in NCBI and should be released in the near future.

Table 1. Sequenced Genomes of Brown and White Rot Fungi *

Fungus	Decay Type	Genome (Mbp)	Predicted Genes	Year	Cited
Coniophora puteana	BR	43.0	13,761	2012	*(11)*
Fibroporia radiculosa	BR	33.6	9,262	2012	*(15)*
Fomitopsis pinicola	BR	46.3	14,724	2012	*(11)*
Gloeophyllum trabeum	BR	37.2	11,846	2012	*(11)*
Postia placenta MAD698-R v1	BR	42.5	12,541	2009	*(14)*
Serpula lacrymans S7.9 v2	BR	42.8	12,917	2011	*(10)*
Serpula lacrymans S7.3 v2	BR	47.0	14,495	2011	*(10)*
Wolfiporia cocos	BR	50.5	12,746	2012	*(11)*
Auricularia delicata	WR	75.1	23,577	2012	*(11)*
Ceriporiopsis subvermispora	WR	39.0	12,125	2012	*(18)*
Dichomitus squalens	WR	42.8	12,290	2012	*(11)*
Fomitoporia mediterranea	WR	74.9	11,333	2012	*(11)*
Ganoderma lucidum	WR	39.9	12,080	2012	*(16)*
Heterobasidion irregulare v2	WR, P	33.6	11,464	2012	*(11)*
Phanerochaete chrysosporium v2	WR	35.1	10,048	2006	*(13)*
Punctularia strigosozonata	WR	34.2	11,538	2012	*(11)*
Stereum hirsutum	WR	46.5	14,072	2012	*(11)*
Schizophyllum commune	WR	38.5	13,210	2010	*(12)*
Trametes versicolor	WR	44.8	14,296	2012	*(11)*

* Abbreviations: BR, brown rot; WR, white rot; P, pathogen; Mbp, megabase pair.

The genomes of *S. lacrymans* S7.3 (*10*), *F. radiculosa* (*15*), and *G. lucidum* (*16*) were among the first fungal genomes to be sequenced entirely by NGS. The species, platforms, and strategies were: *S. lacrymans,* Roche/454, single end (400 bp reads from a 3000 bp library); *F. radiculosa,* Illumina, paired end (76 bp reads from a 300 bp library); and *G. lucidum,* Illumina, paired end (100 bp reads from 200 bp and 6000 bp libraries). Single and paired end strategies indicate whether the library fragments were sequenced from one or both ends, respectively. With very short reads, paired end sequencing effectively increases the short read lengths to the fragment size of the library, even if the intervening sequence is unknown. Furthermore, by varying the insert size of the library, larger gaps between known sequence fragments can be bridged. Therefore, when using NGS to sequence a genome *de novo*, longer assemblies are possible by (i) increasing read length, (ii) using a paired end strategy, (iii) generating sequence from libraries of different size, and (iv) combining sequencing data from more than one NGS platform. The draft assembly contains contigs or regions of contiguous sequence that are connected across gaps of known size into scaffolds. Successful assemblies have been obtained with the following technologies and depth of coverage: 8x for Sanger (*S. lacrymans* 7.9, (*10*)), 40x for a hybrid approach that uses Sanger, Roche, and Illumina (*Fomitopsis pinicola* and *F. mediterranea* (*11*)), 150x for Illumina (*F. radiculosa* (*15*)) and 60x for Roche (*O. novo-ulmi* (*17*)).

Genome Annotation

Structural Annotation

The genetic information embedded in the sequence of the four nucleotides (G, A, T, and C) can be decoded because enough biological knowledge exists to predict protein structure and function from the genes in the genomic sequence. All eukaryotic genes are divided into alternating exons and introns, which are the coding and non-coding regions, respectively. When a gene is expressed, it is transcribed into a precursor messenger RNA (mRNA), which is a complementary copy of the gene without the introns. The mature mRNA is transported out of the nucleus into the cytoplasm where it is read by the ribosomes and translated into protein. Proteins, in turn, are responsible for cell structure and function. Except for a few minor variations, the genetic code, or the translation of codon by a 3 letter nucleotide sequence to amino acid, is universal for ALL organisms. The genetic code also includes codons for start and stop translation sites.

When gene prediction only requires the genomic sequence as input, the method is called *ab initio*. Examples of popular *ab initio* gene prediction tools are: GENSCAN (*19*), AUGUSTUS (*20*), FGENESH (*21*), and GeneMark (*22*). GENSCAN, AUGUSTUS, and FGENESH require a training set of about 1000 validated genes to estimate the model parameters. GeneMark, on the other hand, can learn directly from the input sequence. GeneMark-ES, which was used for gene prediction in *F. radiculosa* (*15*), can improve gene prediction accuracy because it adds a parameter based on the intron branch point sequence (*22*), which is conserved in fungi (*23*). An alternative approach to gene prediction uses extrinsic information from products of gene expression like ESTs (expressed sequence tags), cDNA (complementary DNA sequence copied from mRNA), and protein homologues from other species to predict the location of the intron/exon splice sites. Two widely used homology-based gene prediction tools are FGENESH+ (*21*) and Genewise (*24*). While *ab initio* gene predictions have higher sensitivity, homology-based gene predictions have higher specificity. Therefore, many pipelines will produce an initial set of gene predictions *ab initio*, then improve the predictions using homology-based tools. Combiner tools such as Combiner (*25*) and GLEAN (*26*) then choose the best gene structure from the collection of predictions. This kind of structural annotation pipeline was used by JGI (*10–14*) and for *G. lucidum* (*16*).

Table 2 summarizes the gene structure for the genomes listed in Table 1. As a group, the averages for gene and transcript lengths and number of exons per gene are moderately conserved. Exon and intron lengths are the most and least highly conserved, respectively. Since exons encode proteins, constraints related to function are under the most selection pressure to keep exon length relatively unchanged. The accumulation of nucleotide substitutions, insertions, and deletions in introns, on the other hand, generally have less impact on function since introns are non-coding, and eventually lead to greater variation in intron length. This holds true in a comparison with the plants, *Arabidopsis thaliana*, rice, and maize. The average exon lengths are similar: 237 bp for the wood decay fungi (Table 2) and 217, 254, and 259 bp, respectively, for the plants (*27*). Intron lengths, however, are more variable: 82 bp for the wood decay fungi (Table 2) and 167, 413, and 607 bp, respectively, for the plants (*27*).

Table 2. Summary of Gene Structure from the Genomes Listed in Table 1

Gene Feature	Mean	SD	% CV
Gene length bp	1804	159	9
Transcript length bp	1419	96	7
Exon length bp*	237	12	5
Intron length bp	82	10	12
Exons per gene	5.95	0.46	8

* Mean exon length excluded the value for *H. irregulare* (561 bp), which appeared to be an outlier.

Functional Annotation

Assigning names to genes and determining what they do is the process of genome functional annotation. It begins by using the exon sequences in the genes to predict the amino acid sequence of the polypeptide chain(s) that form the proteins. Chemical properties of the amino acids dictate the biochemical activity of a protein. For example, non-covalent forces generated by the amino acid sequence cause regions of the polypeptides to fold into three dimensional structures that look like coils and sheets. Thus, protein structure is essentially modular where different regions within the protein form specific functional and structural domains. Domains may also contain short amino acid sequences that have binding or catalytic activity. Activity of these site-specific features hinges upon its location within a domain. Thus, how a protein interacts with other molecules is determined not only by its amino acid sequence, but also by the spatial geometry of localized sites and domains within the protein.

Decades of protein analysis have shown that domains and site-specific features are conserved within families of proteins that share the same ancestral gene. Therefore, if a newly predicted protein shows high sequence similarity to another experimentally validated protein, they are assumed to be homologs. The workhorse for performing pairwise sequence similarity searches is BLAST or basic local alignment search tool (28). One version of BLAST is blastp. It compares a protein query sequence against every sequence in a public or custom protein database, returning the best hits in the alignment with their alignment scores and E-values (the probability of getting the same hit by chance when searching a database of similar size). The algorithm starts by finding seeds or small "word-size" matches then uses dynamic programming to extend the match to find the local alignments. Statistics are then calculated and the high scoring segment pairs are used to rank the best hits. As of April 9, 2013 for fungi alone, there were 734,575 protein records covering 643 taxa in the NCBI protein database that can be used for alignment comparisons (29).

Although evolutionary relationships can be inferred from protein sequence homology, it may or may not imply functional similarity. Small changes in a binding or catalytic site may go undetected in a pair-wise sequence alignment comparison, but could significantly alter protein function. Therefore, additional evidence for protein function is drawn from comparisons of protein domains and site-specific features with the signature or predictive models of all described protein families. Generally, combined evidence from sequence homology and family membership is sufficient to assign a putative function to a predicted gene product.

The major tool for determining the similarity of protein signatures is InterProScan (30). InterProScan attempts to classify the function of a query sequence by identifying its protein signature and then comparing it against the signatures of known protein families. InterProScan depends upon InterPro (31), which is a single searchable resource that integrates protein information from eleven different databases around the world. Each member database has its own process for identifying and using protein domains and/or site-specific features to classify proteins into protein families. For instance, PROSITE models are based on a collection of biologically significant sites (32), while those in Pfam are based on domains (33). Position-specific scores can be based on a multiple sequence alignment of functionally validated proteins from different species or on hidden Markov Models (HMM) that calculates a transitional probability for one amino acid given the presence of one or more previous amino acids. At present, there are 1668 documented site-specific features in PROSITE (34) and 14,831 functionally annotated protein families in Pfam (35).

Organizing and adding context to the functional annotations of proteins is the mission of the Gene Ontology (GO) Consortium (36). They developed structured vocabularies or ontologies to describe proteins in terms of three domains or categories: cellular component, biological process, and molecular function. Each term within an ontology has a defined relationship with one or more other terms in the same domain. The relationships (is a type of, is a part of, or is regulated by) are described by directed acyclic graphs, meaning that there are child and parent relationships among terms, but the relationships are not necessarily hierarchichal, as one child term can have multiple parents from different levels.

GO annotations complement sequence homology and InterPro annotations plus have the advantage of being easily queried and analyzed by geneset enrichment analysis. The latter determines which GO terms are more highly represented when comparing genesets from different treatments or conditions. For example, during active lignocellulose degradation, one would expect the number of differentially expressed genes with carbohydrate-active enzyme function to be significantly greater compared to glucose-based growth media or non-decay treatments when a preservative was still protecting the wood. Currently, many databases like InterPro (*31*), KEGG pathways (*37*), MetaCyc (*38*), and the Enzyme Commission (EC (*39*)) have already been mapped to GO and the GO terms are retrieved when a protein search of the individual database is performed. In addition, bioinformatics tools like Blast2GO will automate much of the process (*40*). It performs the blastp alignments to the NCBI nr (non-redundant) protein database, retrieves the GO terms, scans InterPro, gets the EC numbers, and will perform geneset enrichment analysis in a single, easy to learn interface.

Predictions for the cellular location of a gene product can also come directly from the protein sequence. A tool that predicts the mitochondrial or extracellular localization of a gene product is TargetP (*41*). The companion tool, SignalP detects the presence of a signal peptide or signal anchor (*41*). The former suggests that a protein may be secreted and the latter, that a protein may be anchored in a membrane. Since fungi secrete many compounds for defense and for digestion of their food, which occurs extracellularly, most fungal genomes include annotations produced by TargetP and SignalP.

Annotations of wood decay fungal genomes show that the majority of the predicted proteins are homologous with proteins in the NCBI nr database, e.g. 83% for *G. lucidum* (*16*), 89% *F. radiculosa* (*15*), and 68% for *S. lacrymans* S7.3 (*10*). The percentage of genes that map to GO terms for these three species ranged from 40 to 58%. The percentage that actually receive a putative functional assignment is lower and only occurs when there is strong similarity to a documented gene product and its corresponding protein family signature. More often than not, a predicted protein will not match an entire signature, but matches only one domain such as kinase. In this case, the annotation is informative but not specific since kinases are one of the largest superfamilies in eukaryotes. A protein with a hypothetical assignment means it had no database match or matched another hypothetical protein. Extrinsic evidence of gene transcription (e.g. from mRNA and/or protein detection) gives further indication that the gene is doing something in the cell and has a biological role. If there is no evidence of gene expression, then it may be a matter of incorrect timing with respect to developmental stage or environmental conditions, or the gene may be a pseudogene.

There is no doubt that automated genome annotation is a powerful predictive technology that can identify the entire functional potential of an organism. Predictions, though, still need expert review or curation and experimental validation. In additions, portions of genes could be missed or misassembled and functional annotations could be incomplete or incorrect. Nevertheless, genomics lays a comprehensive foundation for subsequent systems-level and gene-by-gene investigations.

Other Omes

NGS was selected as Method of the Year in 2007 (*3*) because it has transformed scientific progress in the area of functional genomics, that is, determining how genomes control cell function at the molecular level. Besides genomics, some of the major applications that arose directly or indirectly from NGS are transcriptomics, proteomics, metabolomics, and metagenomics (Table 3). These emerging omics approaches have presented scientists with the challenge of exploring life as a system of interconnected networks of biochemical reactions within one organism or community of organisms. Since the networks are dynamically changing, it is critically important to link the molecular omics data with indicators like physical, morphological, or chemical changes to the wood. These measurements are more accurate indicators of decay stage than time of exposure to the fungus and provide a meaningful biological context for interpreting and comparing different omics datasets.

Table 3. Fields of Study That Have Advanced Because of NGS and Genomics*

Technology	*Discipline*	*Application*
RNA-Seq	Transcriptomics	Identification and quantification of gene expression
2D-LC MS/MS	Proteomics	Identification and quantification of proteins or gene products
LC-MS/MS, GC-EI-MS	Metabolomics	Identification and quantification of metabolites that are products of enzyme activity
NGS	Metagenomics	Identification and quantification of microbial genomes present in a biological sample without culturing

* Abbreviations: RNA-Seq, RNA-sequencing; 2D, 2 dimensional; MS/MS, tandem mass spectrometry; LC, liquid chromatography; GC-EI-MS, gas chromatography-electron ionization-mass spectrometry.

Transcriptomics

Because DNA and mRNA are complementary, the transcriptome, which is the collection of genes transcribed into mRNA at a particular time in a particular cell or tissue type, can be sequenced and quantified by NGS in a technique called

RNA-Seq. Quantitative analysis of differential gene expression is based on counts of the number of reads that align to the predicted CDS (coding DNA sequence) of a reference genome. *De novo* transcriptome assembly should also be possible when a reference genome is not available and has been demonstrated in yeast (*42*).

Exploring dynamic changes in gene expression is a critical first step to address the questions of what, when, where, why, and how genes are turned on. Because mRNA has a half-life of minutes to hours (*43*), the transcriptome is a direct product of gene regulation. By varying the experimental conditions, genes that show differential regulation and correlated expression can be identified. By studying these relationships, scientists gain insights into the signaling pathways that turn genes on or off, the transcription factors that regulate gene activities, and the biological roles played by networks of biochemical pathways. The sequence of the transcriptome also serves to update the genome structural annotations, which tend to evolve as more information is gathered.

Commercial solutions for the analysis of RNA-Seq data are becoming more widespread (e.g. DNASTAR, Madison, WI and Golden Helix, Bozeman, MT). Alternatively, open source solutions can be downloaded separately to perform alignment, differential gene expression analysis, GO enrichment, etc. in individual steps. Despite differences in alignment algorithms and statistical models, a comparison of several tools (three for alignment and five for differential gene expression) showed good agreement among the results (*42*).

Proteomics

The predicted protein sequences generated by genomics has fueled the growth of proteomics, another investigative tool of functional genomics. Shotgun proteomics is a popular method of identifying the proteins in complex mixtures and uses multidimensional protein identification technology (MudPIT) (*44*). After the proteins are extracted, reduced, and trypsin-digested, the peptide fragments are separated and analyzed by 2D-LC MS/MS (2 dimensional-liquid chromatography tandem mass spectrometry). During MS1, peptide masses are identified from the charged molecular ions. In MS2, ions are fragmented to create a fragment ion spectrum. Fragmentation tends to occur at preferred sites along the peptide backbone, which generates a characteristic pattern for the peptide. However, since fragmentation is not consistent, the amino acid sequence of the peptide may only partially be identified. Scoring algorithms like Mascot (*45*) and Sequest (*46*) use a combination of the peptide mass, amino acid sequence, and the fragment ion spectrum to rank the best matches against searches of protein databases. Imcreasing both mass accuracy and coverage will improve the confidence of a protein identification from a tryptic digest.

In quantitative proteomics, global differences in protein abundance are measured. Label and label-free approaches have been developed, but labeling with isobaric tags is the most reproducible and accurate (*47*). Isobaric tags have three regions, a reporter, a balance, and a reactive site. Lighter reporters are paired with heavier balances so that their combined masses are equal or isobaric. After digestion, the peptides from the different samples are covalently attached to a different tag at the reactive site. The samples are combined in equal amounts and analyzed by 2D-LC MS/MS. During MS1 analysis, the labeled peptides co-elute and have the same mass, which removes bias due to the separation method. Label detection occurs during fragmentation in the MS2 stage, when the reporter is cleaved off the peptide. The relative peak intensities of the reporter ions approximate the relative amount of protein in the original samples. Two types of tags are available: tandem mass tags (*48*), which can be used in a multiplex of up to 10 samples, and iTRAQ labels (*49*), which have an 8-plex option.

Because proteins are more stable than mRNA, cells have other mechanisms besides synthesis for controlling their activity. These include post-translational modifications, such as phosphorylation, glycosylation, ubiquitination, methylation, and acetylation, that regulate when and where proteins are active. Alternative splicing of the exons from a single gene can also produce proteins with different properties. Thus, when interpreting proteomics data, quantitative differences suggest but do not always equate with differences in activity.

Metabolomics

The only ome that gives a true snapshot of the physiological state of a cell or organism is the metabolome, the collection of small molecule metabolites that are the intermediates and end products of metabolism. Metabolism is the sum of the chemical reactions that sustain life and is a functional manifestation of the genome, transcriptome, and proteome combined. Untargeted metabolic profiling is still an emerging technology that relies mostly on MS-based techniques. As many metabolites as possible are identified and quantified down to the pico- and femtomole levels.

Each MS technique has its strengths and weaknesses (*50, 51*). GC-EI-MS (gas chromatography-electron ionization-mass spectrometry) is quantitative and provides retention and peak-rich fragmentation spectra, but mass accuracy is generally too low to be useful. High resolution LC-MS/MS instruments like the Orbitrap and quadrupole time of flight mass spectrometers give accurate mass values, but relatively few peaks are found in the MS/MS spectra and the method is only semi-quantitative. Metabolite identifications rely on matching masses, retention indices, and MS/MS spectra to reference compounds. METLIN (*52*), a public database from the Scripps Institute, currently holds 56,612 tandem MS spectra for 11,208 ions (*53*). They also offer Xcms Online, which is an easy to use interface for uploading MS data, searching METLIN, and viewing results (*54*). One feature of the software is the cloud plot (*55*), which gives a global representation of the data in one diagram. For each sample analyzed, the plot simultaneously shows the metabolites, the direction and magnitude of the fold change, its p-value, whether matches were found in the METLIN database, m/z values, elution time, solvent profile, and the total ion chromatogram. The NIST Mass Spectral Library (NIST 11) is more comprehensive than METLIN, but access requires a paid subscription. It has 212,961 EI spectra, 121,586 tandem MS spectra for 15,180 ions, and 346,757 retention indices for 70,835 compounds (*56*).

Integrating Omics Data

Mapping omics datasets to biochemicals pathways to determine the link between metabolic networks and biological function can be performed with resources like KEGG (*37*) and MetaCyc (*38*). For example, BioCyc, which is based on the MetaCyc collection of metabolic pathways, uses Pathway Tools (*57*) to create an organism-specific pathway genome database. The metabolic pathways and regulatory networks are predicted from the genomic annotations by automated and/or manual curation. The Omics Viewer in Pathway Tools simplifies the interpretation of large-scale data sets by overlaying transcriptomics, proteomics, and metabolomics data onto the pathway networks so that quantitative differences of the transcripts, proteins, and metabolites can be viewed in their entire metabolic context. Pathway tools also incorporates a genome browser that provides graphical visualization of genomic features like the genes, their orientation, positional coordinates, and transcriptional evidence.

For practical applications like biocide or drug targeting, Pathway Tools is designed to find chokepoint reactions. A chokepoint reaction is the only producer or consumer of a metabolite in a metabolic network. Search modifiers can restrict the output to targets that are specific, selective, safe for humans, and pronounced in their effects. Thus, these kinds of functional genomics studies have the potential to greatly accelerate the discovery rate of targets for small molecule inhibition. This has clear benefits for the wood protection industry. By using a rational approach to the development of wood protection chemicals, molecular targets can be pinpointed and inhibitor effects on both target and non-target species can be predicted.

Metagenomics

A combined approach using Sanger and Roche NGS technology (or one that produces reads at least 400 bases long) has begun to transform studies on microbial ecology. This application, known as metagenomics, eliminates the bias introduced by culturing, since DNA is extracted directly from the environmental sample. It can be used semi-quantitatively to compare the functional potential of different microbial communities and to broadly identify the responsible taxonomic groups. At this point, the technology has not been used to investigate fungal decay communities, but it has been successful for describing key degraders in an anaerobic community on poplar chips (*58*). Deducing structural and functional annotations from genomes of non-eukaryotes is more straightforward because bacteria and *Archaea* lack introns, their protein sequences are more highly conserved, even among distantly related species, and more reference genomes have been sequenced. Programs like Phylogenetic Marker COGs (IMG/M-ER from the JGI website) automate the process of identifying COGs or clusters of orthologous groups from genomic sequence, and, in certain cases, can link the contigs based on their COGs to a phylogenetic group. Phylogenetically identified contigs can then be used to train metagenomic gene classifier algorithms (*59*) to bin the remaining unclassified contigs. Databases that focus on single groups of molecules like CAZy (Carbohydrate-Active enZYmes (*60*)) can also be searched to assign genes more detailed functional annotations regarding polysaccharide degradation. Thus, inferences about the functional role of each taxonomic cluster in a decay community can be made. Given the current read lengths, average exon and intron sizes of fungal genomes, growth of KOGs (Eukaryotic Orthologous Groups of Proteins) and number of fungal genomes sequenced, metagenomics of fungal decay communities is technologically feasible.

Many of the detected genomes will not have a reference genome in a public database, but a more specific inventory of the taxonomic diversity in an environmental sample can be deduced by combining metagenomics with another NGS application called targeted amplicon sequencing. For fungi, the most widely used targeted amplicon is the variable ITS (internal transcribed spacer) region of the ribosomal DNA repeat unit (*61*), which has been extensively sequenced. This approach has been used to profile 1914 OTUs (operational taxonomic units) present on 343 samples taken from decaying Norway spruce logs in two different environments in Sweden (*62*). OTU composition between logs was very different and only a few OTUs were present in most logs; basidiomycetes were more abundant than ascomycetes, but ascomycetes were more widespread; and species diversity increased with stage of decay. Sporocarp production was also determined to be a poor indicator of OTU composition and diversity. Moreover, new primer sequences have been published that are better suited for NGS technology (*63*). They target the ITS2 region of the ribosomal DNA repeat, introduce less bias against species with longer amplicons, and capture more diversity than the traditionally used ITS1F primer (*64*).

Accelerating Discovery

Today's scientific and technological environment is highly competitive and rapidly changing both for university researchers and industries. With a global focus on the conversion of biomass into products, fuels, and energy, there is a strong need for information and understanding that will lead to new sustainable products and applications. Wood decay fungi are unique. They are one of the few groups of organisms that can fully degrade wood. These fungi have evolved unique non-specific yet complex mechanisms to break down the lignocellulosic matrix of wood. The white rot decay fungi preferentially remove lignin or cause simultaneous degradation removing lignin, cellulose and hemicellulose together. Furthermore, white rot fungi only remove wood that is in direct contact with their hyphae, relying on the localized action of their secreted enzymes to digest the wood polymers down to their individual components. Brown rot decay fungi, on the other hand, selectively remove cellulose and hemicelluloses, leaving behind a modified lignin that constitutes humus. They initiate decay more pervasively using hydroxyl free radicals produced by the Fenton reaction. Because of the disruption caused by the free radicals, the secreted carbohydrate-active enzymes like glycoside hydrolases (GH) can reach their substrates even in regions not in direct contact with hyphae and catalyze the breakdown of cellulose and hemicelluloses down to their composite sugars. Comparisons of these fungal genomes through genomics, transcriptomics, proteomics, and metabolomics will open the door to many new discoveries and a better understanding of how these complex decay mechanisms differ among gene family members, species, wood substrates, and environmental factors.

Comparative Genomics

Results from phylogenomics indicate that wood decay fungi have literally changed the landscape of our planet (*11*). During the late Carboniferous Period into the early Permian, large deposits of black coal were formed (*65*). During the Permian Period, coal deposition dramatically decreased, setting limits to the world's coal supply. Using a molecular clock approach to phylogenetic analysis, Floudas et al. (*11*) deduced that the first ligninolytic peroxidase gene arose about 295 million years ago, overlapping with the Permian period. These data led the authors to postulate a causal relationship between the evolution of lignin-degradation in white rot fungi and the decline in black coal formation (*66*).

A comparative study of 27 gene families revealed major differences among the 31 wood decay genomes that could explain the functional differences in decay between white and brown rot fungi (*11*). The genomes of the white rot fungi contained more carbohydrate-active genes (61-148 genes) than the brown rot fungi (32-68 genes). In particular, genes encoding the group of enzymes that act on crystalline cellulose were present in all white rot genomes, but were absent in all brown rot genomes. For lignin degradation the peroxidases, such as lignin peroxidase (LiP), manganese peroxidase (MnP), and versatile peroxidase (VP), were found in the white rot genome but were absent in the brown rot genomes. An evolutionary analysis suggested that the most recent ligninolytic ancestor was a white rot fungus containing a MnP gene, and that LiP arose from a single origin while VP arose twice. While there was expansion of the peroxidase genes in the white rot lines, there appeared to be contraction and eventual loss of the peroxidase genes in the different brown rot lineages.

Comparative genomics has also been used to gain insight into the two modes of white rot decay. Fernandez-Fueyo et al. (*18*) compared the genomes of the white rot fungi, *Ceriporiopsis subvermispora* and *P. chrysosporium*. The species that exhibits selective delignification, *C. subvermispora*, contained over twice the number of MnP genes, ten-fold fewer LiP genes, and seven laccase genes versus no laccase in the simultaneous decay species *P. chrysosporium*. This implies that MnP and laccase may govern selective delignification, while lignin removal during simultaneous decay depends more on LiP.

Suzuki et al. (*67*) used comparative genomics to understand how *P. carnosa* can survive exclusively on softwoods, while other *Phanerochaete* species like *P. chrysosporium* prefer hardwoods. Some differences were found in the carbohydrate-active gene families like the GH5 mannanases. Differences in abundances of lignin-degrading genes were also detected. *P. chrysosporium* had more LiP and fewer MnP genes, which was consistent with its ability to grow better on pulp with a high lignin content. The most striking discovery was the large number of cytochrome P450 monooxygenase genes in *P. carnosa* (266) versus *P. chrysosporium* (149). The authors suggest that *P. carnosa* is able to use the high number of P450 genes to detoxify wood extractives, and in conjunction with its lignolytic genes, degrade the heartwood of softwood.

Functional Genomics of Wood Decay

The discovery that brown rot fungi have multiple functional genes for laccase, but lack the typical cellulases and peroxidases found in white rot fungi arose from research by Martinez et al. (*14*). They combined genomics with gene expression microarrays and proteomics to elucidate the lignocellulose degradation mechanisms of *P. placenta*. With respect to carbohydrate degradation, they found genes for numerous hemicellulases and one endoglucanase, but none had the predicted functional domains and sites to bind and degrade crystalline cellulose directly. Expression of these gene products were confirmed by the microarray study and proteomics analysis of the secretome. They also uncovered numerous expressed genes that had putative functions related to Fenton chemistry like iron reductases, quinone reductase, and oxidases, specifically copper radical oxidases and GMC oxidoreductases.

MacDonald et al. (*68*) used RNA-Seq and qRT-PCR (*69*) to compare the transcriptome of the white rot fungus *P. carnosa* during growth on different wood species and nutrient media. Three MnPs genes were highly expressed on wood, but no unique differences were found when comparing growth on softwood or hardwood. Some of the peroxidase genes, though, were expressed at different levels on the different wood species. The qRT-PCR study revealed a higher abundance of MnPs and LiPs at early stages of cultivation and higher levels of carbohydrate-active enzymes at later stages, suggesting a sequential mode of degradation where lignin is degraded to some extent prior to the carbohydrates.

Using genome screening tools, directed mutagenesis and heterologous expression, Fernandez-Fueyo et al. (*70*) identified 16 peroxidase genes in the selective ligninolytic white rot fungus *C. subvermispora*. The list included 13 putative MnP genes, one generic peroxidase and two unusual LiP/VP genes. The latter were phylogenetically and catalytically intermediate between the standard LiPs and VPs, suggesting the existence of a VP-LiP transitional stage. Other species of white rot fungi also exhibit large MnP gene families. Salame et al. (*71*) used homologous recombination (*72*) to selectively inactivate different *mnp* genes. They found that inactivation (i) did not affect expression of the non-targeted MnPs, (ii) did not reduce decolorization of the substrate orange II, (iii) but did significantly reduce total MnP activity in enzyme assays. They concluded that there was functional redundancy among these genes and that a reduction of one gene was compensated by other gene family members. Less functional redundancy was observed for the laccase gene family in *P. ostreatus*. Pezzella et al. (*73*) used qRT-PCR to quantify transcription of the laccase genes under different conditions and developmental stages. The results depicted a picture of very complex regulation. Some laccase genes were strongly expressed under certain culture conditions, others were expressed only during sporocarp formation, but the majority were poorly expressed under all tested conditions.

Another noteworthy study employed metabolomics to determine which metabolite was the iron reductant that drives Fenton chemistry of the brown rot fungus *S. lacrymans* (Boletales) (*74*). It had been proposed that the catechol variegatic acid was the main reductant in this fungus (*10*), although brown rot fungi from other lineages (Gloeophyllales and Polyporales) use 2,5-dimethoxyhydroquinone (DMHQ) (*75*). Korripally et al. (*74*) found that decaying wood contained no variegatic acid but had significant levels of DMHQ, supporting the involvement of this metabolite in Fenton chemistry in all three divergent brown rot lineages.

Functional Genomics of Wood Preservative Tolerance

Certain brown rot fungi like *F. radiculosa* have the ability to overcome copper-based wood preservatives and decay pressure-treated wood (*76*). However, genetic control of these mechanisms is poorly understood. Using RNA-Seq on the transcriptome of *F. radiculosa*, Tang et al. (*77*) identified 917 transcripts that were differentially regulated during decay of wood treated with a copper-based preservative. Over 100 of these genes had functions related to wood decay. Fungal metabolism appeared to change dramatically depending upon whether the RNA was taken when the preservative was protecting the wood or when the preservative lost its efficacy and the wood exhibited high strength loss. Using the results of the transcriptomics analysis as a guide, Tang et al. (*77*) then used qRT-PCR (quantitative reverse transcription-PCR) to perform a time-course study of ten of the differentially regulated genes. They found that the copper-based preservative induced higher than normal expression for several genes in the shortcut pathway of oxalate biosynthesis, oxalate breakdown, and for laccase (*77*). Laccase is a multi-copper oxidase that is induced by copper (*78*). These results reinforce the theory that copper oxalate crystal formation is the mechanism of copper tolerance in brown rot fungi (*79, 80*) and supports a role for laccase in initiating the Fenton reaction of high-oxalate producing brown rot fungi (*81*). Expression of a GH5 and GH10 gene with putative roles in the degradation of cellulose and hemicelluloses, respectively, were repressed by the preservative until the preservative lost its efficacy and wood showed high strength loss (*77*). The significance of these results is that it provides several targets that could be used to prevent decay initiation in copper-tolerant species of brown rot fungi.

Functional Genomics of Wood Attacking Insects

Proteomic analysis of *Fusarium solani*, isolated from the gut of the longhorned beetle, detected the presence of cellulases, glycosyl hydrolases, xylanases, laccases, and Mn-independent peroxidases (*82*). Tartar et al. (*83*) compared the digestive contributions of the gut of the termite, *Reticulitermes flavipes*, versus its symbionts using a parallel metatranscriptome analysis. They attributed phenoloxidase activity to host-derived laccase genes and found that termite laccases were phylogenetically similar to fungal laccases. Cellulase genes with complementary activity for cellulose degradation were contributed cooperatively by host and symbionts, while genes for hemicellulose degradation were all symbiont-derived. A later investigation used Roche NGS and proteomics to evaluate the metagene expression of the termite gut and its symbionts when fed diets of different lignin complexity (*84*). The findings revealed two detoxification enzyme families not previously associated with lignin digestion in termites: aldo-keto reductases and catalases. Recombinant versions of these host enzymes showed they significantly enhanced lignocellulose breakdown.

Future Perspectives

It is truly an exciting time to be studying wood biodegradation. Because of NGS, gene discovery is no longer a rate limiting step. Once gene targets and their inhibitors are identified, strategies to prevent decay might involve including the target-specific inhibitors in existing preservative formulations, covalently attaching them to wood by chemical modification, or even genetically engineering the trees to express and concentrate the inhibitors in the wood cell wall layers. All these options stem from a rational approach that uses our understanding of systems biology to develop environmentally sustainable approaches to enhance wood protection and utilization.

References

1. About the Human Genome Project. http://www.ornl.gov/hgmis/home.shtml (accessed June 22, 2013).
2. Sanger, F.; Nicklen, S.; Coulson, A. R. *Proc. Natl. Acad. Sci. U.S.A.* **1977**, *74*, 5463–7.
3. Kiermer, V. Method of the Year. *Nat. Methods* **2008**, *5* DOI: 10.1038/nmeth1153.
4. Zerbino, D. R.; Birney, E. *Genome Res.* **2008**, *18*, 821–29.
5. Chaisson, M. J.; Brinza, D.; Pevzner, P. A. *Genome Res.* **2009**, *19*, 336–46.
6. Luo, R.; Liu, B.; Xie, Y.; Li, Z.; Huang, W.; Yuan, J.; He, G.; Chen, Y.; Pan, Q.; Liu, Y.; et al. *Gigascience* **2012**, *1*, 18.
7. Diguistini, S.; Liao, N. Y.; Platt, D.; Robertson, G.; Seidel, M.; Chan, S. K.; Docking, T. R.; Birol, I.; Holt, R. A.; Hirst, M.; et al. *Genome Biol.* **2009**, *10*, R94.
8. Nowrousian, M.; Stajich, J. E.; Chu, M.; Engh, I.; Espagne, E.; Halliday, K.; Kamerewerd, J.; Kempken, F.; Knab, B.; Kuo, H. C.; et al. *PLoS Genet.* **2010**, *6*, e1000891.
9. Chin, C. S.; Alexander, D. H.; Marks, P.; Klammer, A. A.; Drake, J.; Heiner, C.; Clum, A.; Copeland, A.; Huddleston, J.; Eichler, E. E.; et al. *Nat. Methods* **2013**, *10*, 563–9.
10. Eastwood, D. C.; Floudas, D.; Binder, M.; Majcherczyk, A.; Schneider, P.; Aerts, A.; Asiegbu, F. O.; Baker, S. E.; Barry, K.; Bendiksby, M.; et al. *Science* **2011**, *333*, 762–5.
11. Floudas, D.; Binder, M.; Riley, R.; Barry, K.; Blanchette, R. A.; Henrissat, B.; Martinez, A. T.; Otillar, R.; Spatafora, J. W.; Yadav, J. S.; et al. *Science* **2012**, *336*, 1715–9.

12. Ohm, R. A.; de Jong, J. F.; Lugones, L. G.; Aerts, A.; Kothe, E.; Stajich, J. E.; de Vries, R. P.; Record, E.; Levasseur, A.; Baker, S. E.; et al. *Nat. Biotechnol.* **2010**, *28*, 957–63.
13. Vanden Wymelenberg, A.; Minges, P.; Sabat, G.; Martinez, D.; Aerts, A.; Salamov, A.; Grigoriev, I.; Shapiro, H.; Putnam, N.; Belinky, P.; et al. *Fungal Genet. Biol.* **2006**, *43*, 343–56.
14. Martinez, D.; Challacombe, J.; Morgenstern, I.; Hibbett, D.; Schmoll, M.; Kubicek, C. P.; Ferreira, P.; Ruiz-Duenas, F. J.; Martinez, A. T.; Kersten, P.; et al. *Proc. Natl. Acad. Sci. U.S.A.* **2009**, *106*, 1954–59.
15. Tang, J. D.; Perkins, A. D.; Sonstegard, T. S.; Schroeder, S. G.; Burgess, S. C.; Diehl, S. V. *Appl. Environ. Microbiol.* **2012**, *78*, 2272–81.
16. Liu, D.; Gong, J.; Dai, W.; Kang, X.; Huang, Z.; Zhang, H. M.; Liu, W.; Liu, L.; Ma, J.; Xia, Z.; et al. *PLoS One* **2012**, *7*, e36146.
17. Forgetta, V.; Leveque, G.; Dias, J.; Grove, D.; Lyons, R., Jr.; Genik, S.; Wright, C.; Singh, S.; Peterson, N.; Zianni, M.; et al. *J. Biomol. Technol.* **2013**, *24*, 39–49.
18. Fernandez-Fueyo, E.; Ruiz-Duenas, F. J.; Ferreira, P.; Floudas, D.; Hibbett, D. S.; Canessa, P.; Larrondo, L. F.; James, T. Y.; Seelenfreund, D.; Lobos, S.; et al. *Proc. Natl. Acad. Sci. U.S.A.* **2012**, *109*, 5458–63.
19. Burge, C.; Karlin, S. *J. Mol. Biol.* **1997**, *268*, 78–94.
20. Stanke, M.; Waack, S. *Bioinformatics* **2003**, *19* (Suppl 2), ii215–25.
21. Salamov, A. A.; Solovyev, V. V. *Genome Res.* **2000**, *10*, 516–22.
22. Ter-Hovhannisyan, V.; Lomsadze, A.; Chernoff, Y. O.; Borodovsky, M. *Genome Res.* **2008**, *18*, 1979–90.
23. Kupfer, D. M.; Drabenstot, S. D.; Buchanan, K. L.; Lai, H.; Zhu, H.; Dyer, D. W.; Roe, B. A.; Murphy, J. W. *Eukaryot. Cell* **2004**, *3*, 1088–100.
24. Birney, E.; Clamp, M.; Durbin, R. *Genome Res.* **2004**, *14*, 988–95.
25. Allen, J. E.; Pertea, M.; Salzberg, S. L. *Genome Res.* **2004**, *14*, 142–8.
26. Elsik, C. G.; Mackey, A. J.; Reese, J. T.; Milshina, N. V.; Roos, D. S.; Weinstock, G. M. *Genome Biol.* **2007**, *8*, R13.
27. Haberer, G.; Young, S.; Bharti, A. K.; Gundlach, H.; Raymond, C.; Fuks, G.; Butler, E.; Wing, R. A.; Rounsley, S.; Birren, B.; et al. *Plant Physiol.* **2005**, *139*, 1612–24.
28. Altschul, S. F.; Madden, T. L.; Schaffer, A. A.; Zhang, J.; Zhang, Z.; Miller, W.; Lipman, D. J. *Nucleic Acids Res.* **1997**, *25*, 3389–402.
29. NCBI RefSeq Release 59 Statistics. ftp://ftp.ncbi.nih.gov/refseq/release/release-statistics/ (accessed June 22, 2013).
30. Quevillon, E.; Silventoinen, V.; Pillai, S.; Harte, N.; Mulder, N.; Apweiler, R.; Lopez, R. *Nucleic Acids Res.* **2005**, *33*, W116–20.
31. Hunter, S.; Jones, P.; Mitchell, A.; Apweiler, R.; Attwood, T. K.; Bateman, A.; Bernard, T.; Binns, D.; Bork, P.; Burge, S.; et al. *Nucleic Acids Res.* **2012**, *40*, D306–12.
32. Sigrist, C. J.; de Castro, E.; Cerutti, L.; Cuche, B. A.; Hulo, N.; Bridge, A.; Bougueleret, L.; Xenarios, I. *Nucleic Acids Res.* **2013**, *41*, D344–7.
33. Punta, M.; Coggill, P. C.; Eberhardt, R. Y.; Mistry, J.; Tate, J.; Boursnell, C.; Pang, N.; Forslund, K.; Ceric, G.; Clements, J.; et al. *Nucleic Acids Res.* **2012**, *40*, D290–301.

34. PROSITE. http://prosite.expasy.org (accessed June 22, 2013).
35. Pfam. http://pfam.sanger.ac.uk/ (accessed June 22, 2013).
36. Ashburner, M.; Ball, C. A.; Blake, J. A.; Botstein, D.; Butler, H.; Cherry, J. M.; Davis, A. P.; Dolinski, K.; Dwight, S. S.; Eppig, J. T.; et al. *Nat. Genet.* **2000**, *25*, 25–9.
37. Kanehisa, M.; Goto, S.; Sato, Y.; Furumichi, M.; Tanabe, M. *Nucleic Acids Res.* **2012**, *40*, D109–14.
38. Caspi, R.; Altman, T.; Dreher, K.; Fulcher, C. A.; Subhraveti, P.; Keseler, I. M.; Kothari, A.; Krummenacker, M.; Latendresse, M.; Mueller, L. A.; et al. *Nucleic Acids Res.* **2012**, *40*, D742–53.
39. Enzyme Commission. http://www.chem.qmul.ac.uk/iubmb/enzyme (accessed June 22, 2013).
40. Götz, S.; García-Gómez, J. M.; Terol, J.; Williams, T. D.; Nagaraj, S. H.; Nueda, M. J.; Robles, M.; Talón, M.; Dopazo, J.; Conesa, A. *Nucleic Acids Res.* **2008**, *36*, 3420–35.
41. Emanuelsson, O.; Brunak, S.; von Heijne, G.; Nielsen, H. *Nat. Protoc.* **2007**, *2*, 953–71.
42. Nookaew, I.; Papini, M.; Pornputtapong, N.; Scalcinati, G.; Fagerberg, L.; Uhlen, M.; Nielsen, J. *Nucleic Acids Res.* **2012**, *40*, 10084–97.
43. Kebaara, B. W.; Nielsen, L. E.; Nickerson, K. W.; Atkin, A. L. *Genome* **2006**, *49*, 894–9.
44. Washburn, M. P.; Wolters, D.; Yates, J. R., 3rd *Nat. Biotechnol.* **2001**, *19*, 242–7.
45. Perkins, D. N.; Pappin, D. J.; Creasy, D. M. *Electrophoresis* **1999**, *20*, 3551–67.
46. Eng, J. K.; McCormick, A. L.; Yates, J. R., 3rd *J. Am. Soc. Mass Spectrom.* **1994**, *5*, 976–89.
47. Li, Z.; Adams, R. M.; Chourey, K.; Hurst, G. B.; Hettich, R. L.; Pan, C. *J. Proteome Res.* **2012**, *11*, 1582–90.
48. Thompson, A.; Schafer, J.; Kuhn, K.; Kienle, S.; Schwarz, J.; Schmidt, G.; Neumann, T.; Johnstone, R.; Mohammed, A. K.; Hamon, C. *Anal. Chem.* **2003**, *75*, 1895–904.
49. Ross, P. L.; Huang, Y. N.; Marchese, J. N.; Williamson, B.; Parker, K.; Hattan, S.; Khainovski, N.; Pillai, S.; Dey, S.; Daniels, S.; et al. *Mol. Cell Proteomics* **2004**, *3*, 1154–69.
50. Scheubert, K.; Hufsky, F.; Bocker, S. *J. Cheminf.* **2013**, *5*, 12.
51. Lei, Z.; Huhman, D. V.; Sumner, L. W. *J. Biol. Chem.* **2011**, *286*, 25435–42.
52. Tautenhahn, R.; Cho, K.; Uritboonthai, W.; Zhu, Z.; Patti, G. J.; Siuzdak, G. *Nat. Biotechnol.* **2012**, *30*, 826–8.
53. METLIN. http://metlin.scripps.edu/ (accessed June 22, 2013).
54. Tautenhahn, R.; Patti, G. J.; Rinehart, D.; Siuzdak, G. *Anal. Chem.* **2012**, *84*, 5035–9.
55. Patti, G. J.; Tautenhahn, R.; Rinehart, D.; Cho, K.; Shriver, L. P.; Manchester, M.; Nikolskiy, I.; Johnson, C. H.; Mahieu, N. G.; Siuzdak, G. *Anal. Chem.* **2013**, *85*, 798–804.
56. NIST Standard Reference Database 1A. http://www.nist.gov/srd/nist1a.cfm (accessed June 22, 2013).

57. Karp, P. D.; Paley, S. M.; Krummenacker, M.; Latendresse, M.; Dale, J. M.; Lee, T. J.; Kaipa, P.; Gilham, F.; Spaulding, A.; Popescu, L.; et al. *Briefings Bioinf.* **2010**, *11*, 40–79.
58. van der Lelie, D.; Taghavi, S.; McCorkle, S. M.; Li, L. L.; Malfatti, S. A.; Monteleone, D.; Donohoe, B. S.; Ding, S. Y.; Adney, W. S.; Himmel, M. E.; et al. *PLoS One* **2012**, *7*, e36740.
59. Mande, S. S.; Mohammed, M. H.; Ghosh, T. S. *Briefings Bioinf.* **2012**, *13*, 669–81.
60. Cantarel, B. L.; Coutinho, P. M.; Rancurel, C.; Bernard, T.; Lombard, V.; Henrissat, B. *Nucleic Acids Res.* **2009**, *37*, D233–8.
61. Lindahl, B. D.; Nilsson, R. H.; Tedersoo, L.; Abarenkov, K.; Carlsen, T.; Kjoller, R.; Koljalg, U.; Pennanen, T.; Rosendahl, S.; Stenlid, J.; et al. *New Phytol.* **2013**, *199*, 288–99.
62. Kubartova, A.; Ottosson, E.; Dahlberg, A.; Stenlid, J. *Mol. Ecol.* **2012**, *21*, 4514–32.
63. Ihrmark, K.; Bodeker, I. T.; Cruz-Martinez, K.; Friberg, H.; Kubartova, A.; Schenck, J.; Strid, Y.; Stenlid, J.; Brandstrom-Durling, M.; Clemmensen, K. E.; et al. *FEMS Microbiol. Ecol.* **2012**, *82*, 666–77.
64. Gardes, M.; Bruns, T. D. *Mol. Ecol.* **1993**, *2*, 113–8.
65. Thomas, L. *Coal Geology*; John Wiley & Sons, Ltd: West Sussex, England, 2002; p 384.
66. Robinson, J. M. *Geology* **1990**, *18*, 607–10.
67. Suzuki, H.; MacDonald, J.; Syed, K.; Salamov, A.; Hori, C.; Aerts, A.; Henrissat, B.; Wiebenga, A.; VanKuyk, P. A.; Barry, K.; et al. *BMC Genomics* **2012**, *13*, 444.
68. MacDonald, J.; Doering, M.; Canam, T.; Gong, Y.; Guttman, D. S.; Campbell, M. M.; Master, E. R. *Appl. Environ. Microbiol.* **2011**, *77*, 3211–8.
69. Macdonald, J.; Master, E. R. *Appl. Environ. Microbiol.* **2012**, *78*, 1596–600.
70. Fernandez-Fueyo, E.; Ruiz-Duenas, F. J.; Miki, Y.; Martinez, M. J.; Hammel, K. E.; Martinez, A. T. *J. Biol. Chem.* **2012**, *287*, 16903–16.
71. Salame, T. M.; Knop, D.; Levinson, D.; Yarden, O.; Hadar, Y. *Appl. Environ. Microbiol.* **2013**, *79*, 2405–15.
72. Salame, T. M.; Knop, D.; Tal, D.; Levinson, D.; Yarden, O.; Hadar, Y. *Appl. Environ. Microbiol.* **2012**, *78*, 5341–52.
73. Pezzella, C.; Lettera, V.; Piscitelli, A.; Giardina, P.; Sannia, G. *Appl. Microbiol. Biotechnol.* **2013**, *97*, 705–17.
74. Korripally, P.; Timokhin, V. I.; Houtman, C. J.; Mozuch, M. D.; Hammel, K. E. *Appl. Environ. Microbiol.* **2013**, *79*, 2377–83.
75. Suzuki, M. R.; Hunt, C. G.; Houtman, C. J.; Dalebroux, Z. D.; Hammel, K. E. *Environ. Microbiol.* **2006**, *8*, 2214–23.
76. Clausen, C. A.; Jenkins, K. M. *Chronicles of Fibroporia radiculosa (=Antrodia radiculosa) TFFH 294*; 240; U.S. Department of Agriculture, Forest Service, Forest Products Laboratory: Madison, WI, 2011; pp 1–5.
77. Tang, J. D.; Parker, L. A.; Perkins, A. D.; Sonstegard, T. S.; Schroeder, S. G.; Nicholas, D. D.; Diehl, S. V. *Appl. Environ. Microbiol.* **2013**, *79*, 1523–33.

78. Piscitelli, A.; Giardina, P.; Lettera, V.; Pezzella, C.; Sannia, G.; Faraco, V. *Curr. Genomics* **2011**, *12*, 104–12.
79. Green, F.; Clausen, C. A. *Int. Biodeterior. Biodegrad.* **2003**, *51*, 145, 9.
80. Clausen, C. A.; Green, F. *Int. Biodeterior. Biodegrad.* **2003**, *51*, 139–44.
81. Wei, D.; Houtman, C. J.; Kapich, A. N.; Hunt, C. G.; Cullen, D.; Hammel, K. E. *Appl. Environ. Microbiol.* **2010**, *76*, 2091–7.
82. Scully, E. D.; Hoover, K.; Carlson, J.; Tien, M.; Geib, S. M. *PLoS One* **2012**, *7*, e32990.
83. Tartar, A.; Wheeler, M. M.; Zhou, X.; Coy, M. R.; Boucias, D. G.; Scharf, M. E. *Biotechnol. Biofuels* **2009**, *2*, 25.
84. Sethi, A.; Slack, J. M.; Kovaleva, E. S.; Buchman, G. W.; Scharf, M. E. *Insect Biochem. Mol. Biol.* **2013**, *43*, 91–101.

Chapter 4

Genetic Identification of Fungi Involved in Wood Decay

Grant T. Kirker*

USDA-FS, Forest Products Laboratory, Madison, Wisconsin 53726
*E-mail: gkirker@fs.fed.us.

Wood decay is a complex process that involves contributions from molds, bacteria, decay fungi, and often insects. The first step in the accurate diagnosis of decay is identification of the causal agents, but wood decay in the strictest sense (white and brown rot) is caused by cryptic fungal species that are very difficult to identify using traditional methods. Genetic methods offer fast, reliable, and accurate means to identify microbes from infected woody material. The purpose of this chapter is to summarize the available first generation DNA based techniques for identification of microorganisms, primarily fungi, involved in the decay process and to discuss their strengths and limitations.

Introduction

In the 2003 ACS text (*Wood Deterioration and Preservation*), Jellison, Jasalavich, and Ostrofsky gave an overview of the past and current DNA-based technologies that have been used to study fungi involved in the decay process. The intent of this chapter is to build on their overview by providing additional background about several of the first generation molecular techniques that have been used over the past two decades, discuss advantages and limitations specific to the given methodologies, and attempt to simplify the often confusing terminology associated with molecular analysis of micro-organisms.

© 2014 American Chemical Society

There have been significant advances made in sequencing technology, including accuracy and efficiency. Throughput and maximum read lengths are being pushed to new limits. The use of next-generation sequencing platforms, such as pyrosequencing, sequencing by synthesis, and semiconductor sequencing, will be covered in another chapter by Tang and Diehl.

Traditional Methods for Identification of Wood Decay Micro-Organisms

Morphological

Fungal morphology has been the classical means to identify decay fungi obtained from wood. There are several available keys for identification (*1–3*), which contain most of the brown and white rot fungi. However, the morphological keys require some familiarization with specialized fungal structures and hyphal morphology that are not always readily visible when decay fungi are propagated on artificial media. In addition, isolation of decay fungi from woody material is challenging due to the additional non-decaying stains, molds, and yeasts also present on the surface and interior of the wood.

Cultural

Nobles (*4*) developed a key for identification of wood decay fungi using cultural characteristics. The key was based on enzyme chemistry, oxidation, culture growth rates on select media, and some morphological considerations. A second edition was published in 1968 that expanded the key to 252 species from the original 149 (*5*). The key used a diagnostic species code where numerical values were assigned at each diagnostic step and the ID was determined by the unique species code obtained at the end. This method was also used (*6*) in a punch card analysis of the Aphyllophorales. The Nobles key remains an important contribution to our understanding of decay fungi and how they react to different cultural conditions. In its prime, the Nobles Laboratory in Ottawa was identifying 3000 isolates per year using this method and was regarded as a worldwide authority in this field. The descriptions are detailed and reliable, but this method requires the time involved with using specialized media as well as specialized expertise in the diagnosis of the many different tests involved.

Phospholipid Analysis

Analysis of fatty acids from the phospholipid bilayers of microbial cells can be a useful tool for soil microbial communities and identification of individual organisms. Fatty acids are extracted, saponified, separated, and often derivatized before analysis using GC-MS or HPLC. The main premise behind phospholipid lipid analysis is that an individual bacterial or fungal species has a unique make-up of fatty acids in their phospholipid bilayer of the cell membrane. There are two types of phospholipid analyses typically used for soil microbial characterization,

[1] Phospholipid Fatty Acid Analysis (PLFA) or [2] Total Soil Fatty Acid Methyl Esters (TSFAMES). TSFAMES have been shown to provide better yield than PLFAs, but are more complex to analyze (*7*). The PLFA technique has been used to characterize and quantify bacterial communities in soils (*8*), as well as fungi in decaying wood (*9*). One potential limitation of this approach is that the identification requires standardization of unknown cultures on artificial media prior to analysis, which is not suitable for organisms that do not grow well on traditional media. This method also involves a considerable amount of preparatory steps and reagents.

Immunological Assays

There are several available methods that use antibodies derived from wood decay fungi to identify early stages of decay, and include techniques such as particle agglutination, immunofluorescence assays, dot blots, ELISA tests, "dipstick assays" and chromatographic assays (*10*). Immunological detection relies on the presence of an antigen which previously required specialized cultivation techniques to produce sufficient antigens and antibodies for the tests. Biological supply companies now produce a wide range of antibodies commercially, but cost can be a limiting factor. Immunological based tests can also be confounded by the presence of inhibitory compounds that occur in later stages of decay (*11*).

Genetic Methods

General Considerations

The Central Dogma

The basis for all genetic analyses revolves around the central dogma of molecular biology (*12*). It simplistically shows the proper flow of genetic information in all biological systems beginning with deoxyribonucleic acid (DNA). DNA is transcribed into ribonucleic acid (RNA) which is the active messenger that in turn translates into proteins. This is important to keep in mind when interpreting molecular data. DNA based analyses should always be considered for the potential microbial inhabitants of a given system since DNA is a very stable inactive state of nucleic acids and can persist in the environment in resting structures, whereas RNA is the active messenger involved in metabolic processes and gives a better representation of what is actively causing degradation. The organisms detected using a DNA based assay aren't necessarily all metabolically active and contributing to active deteriorative processes; some may simply exist as fragments of mycelia, spores, latent resting structures, and other inactive forms. There are additional methodologies developed that attempt to address these but require additional techniques (*13*).

Polymerase Chain Reaction

A major development in molecular biology that serves as the basis for most of the analyses discussed in this chapter is the polymerase chain reaction (PCR). PCR enabled scientists to multiply DNA on an exponential scale (*14*) so that a specific gene or region of DNA can be isolated, differentially amplified and studied. The premise of PCR involves three basic steps: denaturing the template DNA, annealing of the primers, and extension of the DNA. Denaturing is the melting or loosening of the DNA helix that allows the primers access to a strand of DNA. Annealing is the process of attaching nucleotide specific primers that flank the areas of the DNA to be copied. Extension involves the actual reading of the original strand and attaching the matching bases on a new strand of DNA based on the genetic code. Taq DNA polymerase is a thermally stable enzyme that is used to copy and build the strands of the DNA region of interest. PCR is the underlying principle that drives all of these emergent technologies for molecular ID and which prompted the development of numerous technologies that can be used to identify and characterize organisms based on their genetic information.

PCR Primer Selection

An important consideration when applying molecular approaches to characterizing fungi is selection of amplification targets, which will determine what regions of DNA will be copied in the subsequent PCR. The aforementioned DNA based procedures can also be used as useful tools in characterization of fungal population genetics (mating systems, mutant detection, and countless other possibilities), but in this review they are only discussed in the context of fungal identification and how they have previously been used for wood associated fungi. There are several commonly used priming targets used in fungal genetics for identification, with the most common of these being the internal transcribed spacer (ITS) region. ITS is a conserved region of ribosomal DNA that can be used to differentiate between species of fungi. There are two commonly used primers for amplifying ITS, the general fungal primer ITS1-ITS 4 primer pair (*15*) and the ITS1-F and ITS4-B primer sets (*16*). These are commonly used for sequencing, community analysis, and are a reliable target for routine amplification of fungal DNA from wood. The general primer amplifies for all fungi (includes mold, stains, yeast, etc.) while the basidiomycete specific primer amplifies only fungi that belong in the basidiomycota. Basidiomycete fungi also include those fungi that are key components of the wood decay cycle. There are also Ascomycete specific ITS primer sets that only amplify DNA from members of the phylum Ascomycota (*17*). The use of selective primers is one way to exclude some of the generalist micro-organisms that commonly predominate environmental samples. Additionally, large subunit (LSU) ribosomal DNA, small subunit (SSU) ribosomal, intergenic spacer regions (IGS) and beta-tubulin have all been used to amplify and differentiate species. There have been efforts to standardize and develop universal loci for DNA barcoding of fungi and perspective targets include cytochrome oxidase, translation elongation factors, and ribosomal polymerase B2 (*18*). At this time the ITS region the most widely used DNA region for routine molecular analysis.

Methods Based on DNA Sequence Information

Cloning And Sequencing

The most straight forward and common method of molecular identification of decay fungi is through direct sequencing of conserved regions (*15*) and subsequent Basic Local Alignment Search Tool (BLAST) search through the NCBI database which compares the sequences to known sequences in the database. There is an exponentially growing amount of genbank entries that can be matched with sequenced data. Sequencing of PCR products has been used by many research labs to study wood decay fungi (*19–29*).

Sequencing may be difficult from severely decayed samples due to the presence of inhibitory compounds (i.e. humic acids polyphenols). The main downside to direct sequencing is that it requires a pure culture. To address this, many researchers use cloning to propagate their PCR products. In cloning, the PCR products are inserted into vectors, such as plasmid or other circular DNA form, and that way can be stabilized for future study and manipulation. Cloning requires some specialized equipment and incubator space, but does make retaining reference material and downstream applications much simpler.

Species-Specific Probes

A more targeted approach for detection is through the use of species-specific oligonucleotide probes (SSOP). These can be incorporated onto an array set-up where multiple species may be screened for presence/absence. Moreth and Schmidt (*30*) developed species specific probes for *Serpula lacrymans* and were able to detect the fungus in wood samples. Oh et al. (*31*) developed species specific probes for 11 wood decay fungi based on sequence specific probes from sequence data from the ITS 1 and ITS2 region of several wood rotting basidiomycetes and developed a highly sensitive "reverse southern blot" procedure that could successfully identify the target fungi in both laboratory samples and naturally decayed wood. SSOPS are a highly targeted approach to detection of wood decay fungi, but probes have to be first made and this requires sequence data and successful incorporation onto the SSOP filters.

Quantitative PCR (Q-PCR)

Q-PCR was made possible by developments in real-time PCR technology, and is simply an adaptation of conventional PCR that incorporates fluorescent dyes or probes that gives earlier, more sensitive quantification of target copies produced in a sample. Q-PCR has numerous applications and several formats are currently available. Horisawa et al (*32*) developed a species specific real-time PCR assay for detection and identification of five different wood decay fungi and was able to quantify these fungi in mixed samples from as little as 0.01ng of genomic DNA. Eikenes et al. (*33*) used a qPCR assay to monitor colonization of birch blocks through the course of an EN 113 decay test and compared them to other routinely used methods for quantifying fungal biomass (ergosterol and chitin assays) and found excellent correlations in early stages of decay but concluded that the qPCR assay was not suited for late stage decay, possibly due to inhibitory compounds, extraction efficiency, and high background from highly decayed samples.

Multiplex PCR Methods

Multiplex PCR allows for simultaneous amplification of multiple targets within a single PCR reaction. Guglielmo et al. (*34*) developed a multiplex PCR detection method for 11 specific taxa using ribosomal DNA, which included two variable domains (D1, D2), the conserved ITS1, 5.8S, and ITSII regions, and mitochondrial DNA, for fungi that that occur on hardwoods. This method could detect fungi with as low as 1 picogram of fungal material and had an 82% success rate for identification. An extensive validation of the method was first performed using spiked wood samples and finally on increment cores. Interestingly, the method also identified additional fungi not visually confirmed in 35% of the samples. One potential drawback to multiplex-PCR is that each component of the PCR reaction has to be optimized and often different PCR products will be incompatible based on their PCR settings (PCR settings are highly dependent on nucleotide composition of the template and primer composition).

Restriction Fragment Length Polymorphisms (RFLP)

RFLPs were the earliest technology developed for DNA fragment analysis, and PCR-RFLP is derived from this earlier procedure and is now more widely used. The basis of most of these fingerprinting methods rely on polymorphisms to yield information on the genetic make-up. Polymorphisms are different forms of the genotype of an organism that exist in a natural population. Polymorphisms are used by evolutionary biologists to observe speciation and natural selection, but can also yield information about the relatedness of individuals within a population (closely related individuals will share more polymorphisms than un-related ones). These cuts create fragments of smaller DNA pieces which create a characteristic DNA fingerprint based on the differences in nucleotide composition of the fungi in the sample. These are PCR products that are digested using multiple restriction enzymes, and banding patterns are visualized on high resolution agarose gels. The patterns are usually characteristic to a certain species or strain and are used for comparison and characterization. Jellison and Jasalvich (*35*) have used RFLP for identification and characterization of wood decay fungi in spruce and were able to detect and identify both white and brown rots in both early and late stages of decay. Some difficulties were noted with later stage decayed samples due to the presence of inhibitory compounds produced in the decay process. Prewitt et al. (*36*) used the fragment pattern from digests of the ITS region to construct phylogenetic trees based on multiple enzyme digests. They concluded that phylogenies based on multiple digests of the ITS region was not sufficient to properly resolve species identifications as the RFLP phylogenetic trees did not agree with trees generated using sequence data. They concluded that while RFLP was useful for distinguishing species, it is not well suited for phylogenetic analysis. Adair et al. (*37*) used PCR-RFLP using an ITS1-F-2NL primer combination to detect fungi in chip piles of hemlock and lodgepole pine, and reported being able to detect and identify fungi in chip piles 4 days after inoculation. Their method was able to differentiate ascomycetes from basidiomycetes at early stages of decay. A potential drawback to using RFLP is that pure cultures are required and it cannot be used for characterizing mixed cultures or environmental samples without additional preparatory steps (i.e. cloning). RFLP can also have difficult in resolving closely related species and may require additional restriction digests to differentiate them.

- **Amplified Ribosomal DNA Restriction Analysis (ARDRA)** –ARDRA is an extension of the RFLP procedure molecular technique that was specifically developed for polymorphsims encoded in the small (16S) ribosomal subunit of bacteria for distinguishing species of bacteria (*38*). This method has also been used to study changes in microbial communities in contaminated soils (*39*) and efforts have been made to bridge prokaryote and eukaryote domains using a Universal Amplified Ribosomal Region (UARR) (*40*). Schmidt and Moreth (*41*) used ARDRA for detection of *Serpula lacrymans* in indoor environments in order to differentiate it from *Serpula himantoides* and found that ARDRA was successful using specific enzyme combinations even though the ITS fragments were of identical size. Potential drawbacks to using ITS-ARDRA are that there are few restriction sites contained in the ITS1 and ITS 2 regions, and several fungi produce closely sized fragments that can't be resolved on the output gels. Also, certain fungi have been found to undergo DNA methylation as they age and this methylation can cause difficulties with certain restriction enzymes (*31*).

Amplified Fragment Length Polymorphisms (AFLP)

AFLP was developed in the 1990's by Keygene (*42*) and is a PCR based analysis that relies on selective PCR amplification of restriction fragments from a total digest of genomic DNA. AFLP uses primers that correspond to the restriction digest recognition sites so that the fragments are selectively amplified. The banding pattern indicates presence or absence of restriction sites, and individual species or mutants present unique banding patterns. These patterns can be used to compare closely related genera to discriminate and to observe genetic changes over different gradients. The number of fragments can be increased or decreased based on the selectivity of the primers. AFLP has been used to study species compatibility among *Armillaria* (*43*), long distance dispersal of *Serpula himatoides* (*44*), and hybridization in Coniophora (*45*).

Terminal Restriction Fragment Length Polymorphisms (T-RFLP)

T-RFLP combines the RFLP methodology with fluorescent tags on the ends of the PCR products. The tagged PCR products are digested with restriction enzymes and the terminal fragments are detected using capillary electrophoresis. The resultant fragments can be used to identify species (*46, 47*) or characterize changes in microbial community structure (*48*). An important advantage of T-RFLP is that it can analyze environmental samples with multiple species and can also be multiplexed to include multiple PCR targets (*49*). Data can be exported as either binary data representing presence/absence, or relative intensity can be used to determine relative species abundance in a sample. Analysis and interpretation of T-RFLP data requires careful interpretation, but yields informative results. There are several software packages that can be used to process T-RFLP data that can also be exported for additional community or statistical analysis. Potential drawbacks of T-RFLP is the possibility of overlapping fragments leading to underestimation of total diversity or potentially missing data when looking at peak-profile data (*50, 51*). In order for T-RFLP to be used as an identification tool, prior fragments must be generated and stored in a database for reference matching. Also, careful attention must be made during interpretation to avoid polymerase errors, intraspecific ITS variation, and extra peaks due to restriction enzyme ineffecicency (*52*).

Gradient Gel Electrophoresis

Although these are not DNA based methods for identification and characterization of fungi, Gradient gel electrophoresis is included in this review because it allows for better separation between closely spaced PCR products that would normally overlap in traditional electrophoresis (*53*). These are in fact imaging techniques that are used to size and confirm DNA fragments resulting from PCR amplification, but still provide a useful tool for separation of highly diverse or mixed sample matrices. The two most common means of gradient gel electrophoresis techniques are:

- **Denaturing Gradient Gel Electrophoresis-**uses a chemical denaturant incorporated into the gel that breaks apart the DNA as it migrates through the gel and increases the separation of visualized DNA fragments on the gel. DGGE has been used to study wood decay fungi on Norway spruce stumps (*54*). Five primer pairs were investigated and final tests were performed on spruce stumps showing varying levels of decay. Highly dissimilar populations were noted when comparing the samples obtained through direct extraction of DNA compared to those obtained by culturing followed by DNA analysis, presumably due to the selective nature of artificial media. DGGE can be used to characterize complex sample matrices, but differential migration and overlapping fragments can be difficult to resolve. It is also possible to excise fragments from DGGE gels for further sequencing to obtain species information, but the techniques are challenging. PCR-DGGE, which is the direct amplification of DNA from decayed wood samples, was also used to identify *Phlebiopsis gigantea* and several other wood decay fungi in decayed conifer stumps that were pre-treated with *P. gigantea*. This method also effectively detected six species from reference samples (*55*). Subsequent sequencing from excised bands did not yield additional identifications.

- **Thermal Gradient Gel Electrophoresis** or TGGE relies on a similar process as DGGE, but uses a thermal gradient to change the structure of the samples to improve separation. Kulkabnova (*56*) used TGGE to analyze communities of decomposer fungi from different forest stands and found that tree species composition did not have an effect on species richness, but it did have a strong effect on species composition of both fungi and bacteria. TGGE has many of the same limitations as DGGE (detection limits of rare species, co-migration of similar fragments, and inability to image overlapping species). It has been suggested that the two techniques can be combined to improve the resolution of the technique as well as the incorporation of fluorescently labeled probes (*54*), but newer metagenomic methods may provide more informative results with less time and effort.

Conclusions

Genetic identification and characterization of wood decay fungi has undergone drastic changes in the last two decades and will likely undergo even more changes as new sequencing technologies become more cost effective. As a result of the reduced cost structure associated with these technologies, they are now more readily available to smaller laboratories, independent investigators, and researchers in developing countries. These molecular methods provide excellent tools for sensitive characterization of complex environments and have greatly expanded our knowledge of the fungal communities that contribute to the decay process.

References

1. Gilbertson, R. L. *Mycologia* **1980**, *72* (1), 1–49.
2. Ryvarden, L. *Genera of polypores: nomenclature and taxonomy*; Fungiflora: Oslo, Norway, 1991; Vols. 1–2.
3. Gilbertson, R. L.; Ryvarden, L. *North American Polypores*; Fungiflora: Oslo, Norway, 1986; Vol. 1 (433 pp), Vol. 2 (445 pp).
4. Nobles, M. K. *Can. J. Res.* **1948**, *26* (3), 281–431.
5. Nobles, M. K. *Can. J. Bot.* **1958**, *36* (1), 91–99.
6. Stalpers, J. A. *Identification of wood-inhabiting Aphyllophorales in pure culture*; Centraalbureau voor schimmelcultures: Baarn, The Netherlands, 1978.
7. Drenovsky, R. E.; Elliott, G. N.; Graham, K. J.; Scow, K. M. *Soil Biol. Biochem.* **2004**, *36* (11), 1793–1800.

8. Zelles, L. *Biol. Fertil. Soils* **1999**, *29* (2), 111–129.
9. Diehl, S. V.; Prewitt, M. L.; Shmulsky, F. M. In *Wood Deterioration and Preservation: Advances in Our Changing World*; Goodell, B., Nicholas, D. D., Schultz, T. P., Eds.; ACS Symposium Series 845; American Chemical Society: Washington, DC, 2003; pp 313–325.
10. Clausen, C. A. *Int. Biodeterior. Biodegrad.* **1997**, *39* (2), 133–143.
11. Goodell, B.; Jellison, J.; Liu, J.; Daniel, G.; Paszczynski, A.; Fekete, F.; et al. *J. Biotechnol.* **1997**, *53* (2), 133–162.
12. Crick, F. *Nature* **1970**, *227* (5258), 561–563.
13. Wintzingerode, F.; Göbel, U. B.; Stackebrandt, E. *FEMS Microbiol. Rev.* **1997**, *21* (3), 213–229.
14. Saiki, R. K.; Gelfand, D. H.; Stoffel, S.; Scharf, S. J.; Higuchi, R.; Horn, G. T.; Mullis, K. B.; Erlich, H. A. *Science* **1988**, *239* (4839), 487–491.
15. White, T. J.; Bruns, T.; Lee, S.; Taylor, J. In *PCR protocols: a guide to methods and applications*; Innis, M. A., Gelfand, D. H., Sninsky, J. J., White, T. J., Eds.; Academic Press: San Diego, CA, 1990; Vol. 18, pp 315–322.
16. Gardes, M.; Bruns, T. D. *Mol. Ecol.* **1993**, *2* (2), 113–118.
17. Glass, N. L.; Donaldson, G. C. *Appl. Environ. Microbiol.* **1995**, *61* (4), 1323–1330.
18. Lewis, C. A.; Bilkhu, S.; Robert, V.; Eberhardt, U.; Szoke, S.; Seifert, K. A.; Lévesque, C. A. *Open Appl. Inf. J.* **2011**, *5*, 30–44.
19. Brazee, N. J.; Lindner, D. L.; Fraver, S.; D'Amato, A. W.; Milo, A. M. *Fungal Ecol.* **2012**, *5* (5), 600–609.
20. Bruns, T. D.; Szaro, T. M.; Gardes, M.; Cullings, K. W.; Pan, J. J.; Taylor, D. L.; Li, Y. *Mol. Ecol.* **1998**, *7* (3), 257–272.
21. Glaeser, J. A.; Lindner, D. L. *For. Pathol.* **2011**, *41* (5), 341–348.
22. Johannesson, H.; Stenlid, J. *For. Ecol Manag.* **1999**, *115* (2), 203–211.
23. Johannesson, H.; Stenlid, J. *Mol. Phylogenet. Evol.* **2003**, *29* (1), 94–101.
24. Jönsson, M. T.; Edman, M.; Jonsson, B. G. *J. Ecol.* **2008**, *96* (5), 1065–1075.
25. Kauserud, H.; Schumacher, T. *Mycol. Res.* **2001**, *105* (6), 676–683.
26. Lindner, D. L.; Banik, M. T. *Mycologia* **2008**, *100* (3), 417–430.
27. Moreth, U.; Schmidt, O. *Holzforschung* **2005**, *59* (1), 90–93.
28. Moncalvo, J. M.; Wang, H. H.; Hseu, R. S. *Mycologia* **1995**, *87* (2), 223–238.
29. Schmidt, O.; Moreth, U. *Wood Sci. Technol.* **2002**, *36* (5), 429–433.
30. Moreth, U.; Schmidt, O. *Holzforschung* **2000**, *54* (1), 1–8.
31. Oh, S.; Kamdem, D. P.; Keathley, D. E.; Han, K. H. *Holzforschung* **2003**, *57* (4), 346–352.
32. Horisawa, S.; Sakuma, Y.; Doi, S. *J. Wood Sci.* **2009**, *55* (2), 133–138.
33. Eikenes, M.; Hietala, A. M.; Alfredsen, G.; Gunnar Fossdal, C.; Solheim, H. *Holzforschung* **2005**, *59* (5), 568–573.
34. Guglielmo, F.; Bergemann, S. E.; Gonthier, P.; Nicolotti, G.; Garbelotto, M. *J. Appl. Microbiol.* **2007**, *103* (5), 1490–1507.
35. Jasalavich, C. A.; Ostrofsky, A.; Jellison, J. *Appl. Environ. Microbiol.* **2000**, *66* (11), 4725–4734.
36. Prewitt, M. L.; Diehl, S. V.; McElroy, T. C.; Diehl, W. J. *For. Prod. J.* **2008**, *58* (4), 66.

37. Adair, S.; Kim, S. H.; Breuil, C. *FEMS Microbiol. Lett.* **2002**, *211* (1), 117–122.
38. Vaneechoutte, M.; Rossau, R.; De Vos, P.; Gillis, M.; Janssens, D.; Paepe, N.; Kersters, K. *FEMS Microbiol. Lett.* **1992**, *93* (3), 227–233.
39. Smit, E.; Leeflang, P.; Wernars, K. *FEMS Microbiol. Ecol.* **1997**, *23* (3), 249–261.
40. Rivas, R.; Velázquez, E.; Zurdo-Piñeiro, J. L.; Mateos, P. F.; Martínez Molina, E. *J. Microbiol. Methods* **2004**, *56* (3), 413–426.
41. Schmidt, O.; Moreth, U. *Holzforschung* **1999**, *53* (2), 123–128.
42. Vos, P.; Hogers, R.; Bleeker, M.; Reijans, M.; van De Lee, T.; Hornes, M.; Friters, A.; Pot, J.; Paleman, J.; Kuiper, M.; Zabeau, M. *Nucleic Acids Res.* **1995**, *23* (21), 4407–4414.
43. Kim, M. S.; Klopfenstein, N. B.; Hanna, J. W.; McDonald, G. I. *For. Pathol.* **2006**, *36* (3), 145–164.
44. Kauserud, H.; Stensrud, Ø.; Decock, C.; Shalchian-Tabrizi, K. N.; Schumacher, T. *Mol. Ecol.* **2006**, *15* (2), 421–431.
45. Skrede, I.; Carlsen, T.; Stensrud, Ø.; Kauserud, H. *Fungal Biol.* **2012**, *116* (7), 778–784.
46. Allmér, J.; Vasiliauskas, R.; Ihrmark, K.; Stenlid, J.; Dahlberg, A. *FEMS Microbiol. Ecol.* **2006**, *55* (1), 57–67.
47. Råberg, U.; Högberg, N. O.; Land, C. J. *Holzforschung* **2005**, *59* (6), 696–702.
48. Kirker, G. T.; Prewitt, M. L.; Schultz, T. P.; Diehl, S. V. *Holzforschung* **2012**, *66* (4), 521–527.
49. Singh, B. K.; Nazaries, L.; Munro, S.; Anderson, I. C.; Campbell, C. D. *Appl. Environ. Microbiol.* **2006**, *72* (11), 7278–7285.
50. Egert, M.; Friedrich, M. W. *Appl. Environ. Microbiol.* **2003**, *69* (5), 2555–2562.
51. Osborn, A. M.; Moore, E. R.; Timmis, K. N. *Environ. Microbiol.* **2000**, *2* (1), 39–50.
52. Avis, P. G.; Dickie, I. A.; Mueller, G. M. *Mol. Ecol.* **2006**, *15* (3), 873–882.
53. Muyzer, G. *Curr. Opin Microbiol.* **1999**, *2* (3), 317–322.
54. Vainio, E. J.; Hantula, J. *Mycol. Res.* **2000**, *104* (8), 927–936.
55. Vainio, E. J.; Hallaksela, A. M.; Lipponen, K.; Hantula, J. *Mycol. Res.* **2005**, *109* (1), 103–114.
56. Kulhánková, A.; Béguiristain, T.; Moukoumi, J.; Berthelin, J.; Ranger, J. *Ann. For. Sci.* **2006**, *63* (5), 547–556.

Chapter 5

Evolution of Fungal Wood Decay

Daniel C. Eastwood*

Department of Biosciences, Swansea University, Singleton Park, Swansea SA2 8PP, United Kingdom
*E-mail: d.c.eastwood@swansea.ac.uk.

The evolution of fungal wood decay was an event of critical importance in maintaining the planet's carbon cycle. The development of a lignocellulose-thickened secondary cell wall by plants and the subsequent swamp forests of the Carboniferous period locked up atmospheric CO_2 and created the coal seams we see today. Very few species can decompose wood due to the recalcitrance of lignin and the dense lignocellulose composite structure. An adaptation in a class II peroxidase in a clade of the Basidiomycota approximately 295 million years ago gave rise to the most efficient group of wood decay specialist, the Agaricomycetes. This chapter will consider the evolution of wood decay in the Agaricomycetes and the diversification of mechanisms and nutritional modes from a basic white rot ancestry. The convergent evolution of brown rot decay will be described where a refined suite of decay genes emerged from the loss of energetically expensive white rot mechanisms of ligninolysis. Further analysis will consider the link between saprotrophy and ectomycorrhiza formation, co-evolution with plants and avenues for future research for genomic-based studies.

© 2014 American Chemical Society

Introduction

The formation of lignocellulose was a major event in plant evolution, providing increased strength to cell walls and ultimately enabled plants to grow tall and form the forests we all recognise today. Lignocellulose is a dense composite of cellulose microfibrils tightly bound with hemicelluloses and lignin forming the woody material whose decomposition will be considered in this chapter.

Cellulose encased in a complex lignin heteropolymer is extremely difficult to break down with a limited number of organisms able to achieve it. While some bacteria, often associated with animal digestive tracts, and some Ascomycetous fungi are able to attack wood, it is a particular group in the higher fungi that have mastered the process. The primary drivers of wood decomposition in nature are saprotrophic fungi of the Basidiomycota, and specifically the subphylum Agaricomycotina.

Advancements in sequencing technologies and bioinformatic analyses have facilitated the sequencing and analysis of whole genomes of diverse wood decay species. This wealth of information continues to grow and is a valuable resource in researching fungal biology. Investigating the ancestry and evolution of wood decay in the Agaricomycotina provides insights into the different mechanisms of lignocellulose decomposition and how these processes have shaped the world around us. We can also think about how we might exploit these processes in the development of environmentally appropriate wood preservatives and to utilize these mechanisms as components of industrial biorefineries.

This chapter provides a broad overview of how genome sequencing has been used to study the evolution of wood decay and mechanisms of decomposition. Firstly the rationale and process of genome sequencing will be introduced, followed by consideration of the early events in wood decay evolution approximately 300 million years ago. Examples of divergent and convergent evolution of different decay mechanisms will be used to explore wood decomposition in the modern day and its ecological significance in a healthy forest habitat. Future developments and areas of research will also be considered.

Whole Genome Sequencing of Wood Decay Fungi

The genome is the entire DNA sequence of an organism represented by the simple four nucleotide code of adenine (A), guanine (G), cytosine (C) and thymine (T). The genome contains all the genes and regulatory elements that make a functioning cell, and it is what makes each species unique. The more closely related species are, the more similar their genomes will be. Simpler organisms tend to have smaller genomes and fewer genes (Table 1), but this is a crude guide and gene number does not always indicate the complexity of an organism. The first eukaryotic organism to be genome sequenced was the fungus *Saccharomyces cerevisiae* in 1996, yet the first Basidiomycete, the wood decay fungus *Phanerochaete chrysosporium*, was only sequenced in 2004 (*1*) and after the human genome project was published in 2001.

Table 1. The Genome Size and Gene Complement of Selected Sequenced Organisms

Organism	Approximate Genome Size (Mbp/Gbp)	Approximate Number of Genes
Escherichia coli	4.6 Mbp	4,400
Saccharomyces cerevisiae	12.1 Mbp	6,275
Neurospora crassa	42 Mbp	10,000
Phanerochaete chrysosporium	35 Mbp	10,000
Oryza sativa	500 Mbp	49,000
Drosophila melanogaster	140 Mbp	15,000
Homo sapiens	3.3 Gbp	20,000

Mbp = megabase pairs (1 million base pairs); Gbp = Gigabase pairs (1 billion base pairs).

Initial sequencing technologies relied on the labour intensive, expensive and time consuming Sanger-based sequencing methods, meaning that whole genomes would take years to complete. New technologies began to be introduced after 2005 including platforms such as Roche 454 pyrosequencing, Illumina HiSeq and MiSeq, ABI SOLiD, Helicos and Ion Torrent PGM reducing cost and time to the extent that the time to sequence a genome can be thought of as weeks and in some instances days. The accuracy of the final genome sequence is often enhanced because each section of genomic DNA is sequenced multiple times. These improved technologies are often referred to as next generation sequencing, high throughput sequencing, or massively parallel sequencing technologies, they employ novel chemistries and sequencing recording systems that is beyond the scope of this chapter (see review (2)).

Even after sequencing, there are significant bioinformatic challenges to overcome to generate a useful interface to allow researchers to better analyze the genome. Genomes are generally sequenced in small multiple overlapping chunks which must be combined together to provide an uninterrupted consensus sequence. Once constructed, computer-based algorithms are used to predict putative genes based either on the comparison with other genomes or by examining genes that have already been sequenced for that organism. Gene prediction and annotation becomes increasingly accurate as more annotated genome sequences are available to compare with one another. However, manual annotation by the human eye is still required to provide confirmation of computer-based gene predictions.

At the point of writing the genomes of 51 saprotrophic Agaricomycota covering a range of nutritional modes and diverse families were publically available, a further 16 Agaricomycete ectomycorrhizal species have also been sequenced for comparison (Table 2). The US Department of Energy's Joint Genome Institute (JGI), Walnut Creek, California, has championed the sequencing of Agaricomycetous fungi and provides an extremely useful interface through their Fungal Genomes Program Portal (http://genome.jgi.doe.gov/programs/fungi/index.jsf). The JGI has an ambitious 1000 fungal genomes sequencing target which will provide an unparalleled genetic resource for future fungal research. The rationale for sequencing Agaricomycete fungi is both ecological and biotechnological in nature. These fungi are crucial to forest health, recycling old plant matter, such as complex carbon polymers into sugars and carbon dioxide, releasing nutrients, conditioning soil, and sequestering carbon. The annual release of CO_2 from the decomposition of plant matter by fungi in the temperate and boreal forests is estimated to be equivalent to that produced by the activities of man. The fungi interact directly with plants in both mycorrhizal associations and as pathogens. It is also not understood how climate change may affect the carbon cycling in these important habitats. These fungi also produce a vast array of enzymes and secondary metabolic products that have industrial and pharmaceutical potential. In particular wood decay fungi could hold the key to biofuel production and biorefining using lignocellulosic feedstocks that are economically and environmentally sustainable. While many current industrial methods of disrupting the lignocellulose structure are energy intensive, fungi have evolved to release sugars and disrupt lignin efficiently.

Table 2. Publically Available Agaricomycota Saprotrophic and Ectomycorrhizal Species Genome Sequences

Fungus: order & species	*Family*	*Nutrition form*
Agaricales		
Agaricus bisporus var bisporus & var brunettii	Agaricaceae	Leaf litter
Amanita muscaria	Amanitaceae	Broad host ectomycorrhiza
Amanita thiersii	Amanitaceae	Leaf litter decomposer
Armillaria mellea	Physalacriaceae	Hardwood white rot and pathogen
Coprinopsis cinerea	Psathyrellaceae	Coprophilous / leaf litter
Cortinarius glaucopus	Cortinariaceae	Ectomycorrhiza
Cylindrobasidium torrendii	Physalacriaceae	White rot
Galerina marginata	Hymenogastraceae	White rot
Gymnopus luxurians	Marasmiaceae	Unknown decay of wood chips and grassland plants
Fistulina hepatica	Fistulinaceae	Brown rot of hard wood and weak parasite
Hebeloma cylindrosporum	Hymenogastraceae	White rot on pine – pioneer specie
Hypholoma sublateritium	Strophariaceae	White rot of hard wood
Laccaria amethystina	Hydnangiaceae	Poplar Ectomycorrhiza
Laccaria bicolour	Hydnangiaceae	Broad range ectomycorrhiza

Continued on next page.

Table 2. (Continued). Publically Available Agaricomycota Saprotrophic and Ectomycorrhizal Species Genome Sequences

Fungus: order & species	Family	Nutrition form
Macrolepiota fuliginosa	Agaricaceae	Leaf litter decomposer
Omphalotus olearius	Marasmiaceae	White rot of hard wood
Pleurotus ostreatus	Pleurotaceae	White rot
Schizophyllum commune	Schizophyllaceae	Transition between white rot and brown rot
Tricholoma matsutake	Tricholomataceae	Ectomycorrhiza of pine
Volvariella volvacea	Pluteaceae	Leaf litter/straw decomposer
Amylocortinales		
Plicaturopsis crispa	Amylocorticaceae	White rot early coloniser
Atheliales		
Piloderma croceum	Atheliaceae	Broad range ectomycorrhiza
Auriculariales		
Auricularia delicate	Auriculariaceae	White rot all wood
Exidia glandulosa	Auriculariaceae	White rot of hard wood
Boletales		
Coniophora puteana	Boletaceae	Brown rot mostly soft wood
Hydnomerulius pinastri	Paxillaceae	Brown rot of pine
Paxillus involutus	Paxillaceae	Ectomycorrhizal
Paxillus rybicundulus	Paxillaceae	Ectomycorrhizal pine & eucalypts
Pisolithus Microcarpus	Sclerodermataceae	Ectomycorrhizal pine & eucalypts
Pisolithus tinctorius	Sclerodermataceae	Ectomycorrhizal pine & eucalypts
Scleroderma citrinum	Sclerodermataceae	Ectomycorrhizal
Serpula lacrymans	Seprulaceae	Brown rot, mostly built environment
Serpula himantioides	Seprulaceae	Brown rot, mostly softwood in natural environment

Continued on next page.

Table 2. (Continued). Publically Available Agaricomycota Saprotrophic and Ectomycorrhizal Species Genome Sequences

Suillus brevipes	Boletaceae	Ectomycorrhiza of pine
Suillus luteus	Boletaceae	Ectomycorrhiza of pine
Cantharellales		
Botryobasidium botryosum	Botryobasidaceae	Unclear, possibly brown rot
Tulasnella calospora	Tulasnellaceae	Orchid micorrhiza
Corticales		
Punctularia strigosozonata	Punctulariaceae	White rot of hard wood
Dacrymycetales		
Calocera cornea	Dacrymycetaceae	White rot mostly of conifers
Dacryopinax sp. DJM731 SSP1	Dacrymycetaceae	Brown rot
Geastrales		
Sphaerobolus stellatus	Geastraceae	White rot of wood chip & bark
Gloeophyllales		
Gloeophyllum trabeum	Gloeophyllaceae	Brown rot
Neolentinus lepideus	Gloeophyllaceae	Brown rot of conifers
Hymenochaetales		
Formitiporia mediterranea	Hymenochaetaceae	Vine pathogen/white rot
Rickenella mellea	Repetobasidiaceae	Associated with roses
Jaapiales		
Jaapia argillaceae	Jaapiaceae	Early divergent brown rot on pine
Polyporales		
Antroda sinuosa	Formitopsidaceae	Brown rot of pine
Cerrena unicolor	Polyporaceae	White rot of hard wood
Bjerkandera adusta	Meruliaceae	White rot of hard wood
Ceriporiopsis subvermispora	Phanerochaetaceae	Sequential white rot
Daedalea quercina	Formitopsidaceae	Brown rot of hard wood
Dichomitus squalens	Polyporaceae	White pocket rot of pine
Formitopsis pinicola	Formitopsidaceae	Brown rot, mostly softwood
Ganoderma sp.10597 SS1	Ganodermataceae	White rot of bark and root
Laetiporus sulphureus	Polyporaceae	Brown rot of hard wood

Continued on next page.

Table 2. (Continued). Publically Available Agaricomycota Saprotrophic and Ectomycorrhizal Species Genome Sequences

Phanerochaete carnosa	Phanerochaetaceae	White rot of soft wood
Phanerochaete chrysosporium	Phanerochaetaceae	White rot (thermophile)
Phlebia brevispora	Meruliaceae	White rot of hard wood and conifer
Phlebiopsis gigantea	Phanerochaetaceae	White rot on conifer sap wood (pioneer)
Polyporus arcularius	Polyporaceae	White rot of hard wood
Postia placenta	Formitopsidaceae	Brown rot
Trametes versicolor	Polyporaceae	White rot
Trichaptum abietinum	Polyporaceae	White rot of conifers
Wolfiporia cocos	Formitopsidaceae	Brown rot / root parasite
Russulales		
Heterobasidion irregulare	Bondarzewiaceae	Plant pathogen / white rot
Stereum hirsutum	Stereaceae	White rot pioneer species
Sebacinales		
Piriformospora indica	Sebacinaceae	Biotrophic symbiont
Sebacina vermifera	Sebacinaceae	Orchid mycorrhiza
Trechisporales		
Sistotremastrum suecicum	Hydnodontaceae	Brown rot
Tremellales		
Tremella mesentrica	Tremellaceae	White rot

Comparative Genomics

The genes present in a particular genome, i.e. the gene complement, can be used to indicate the lifestyle of an organism, for example a wood decay fungus will have genes that mediate wood decomposition. Complex processes such as wood decay are regulated by many genes, and these may vary between species. In addition, genes can sometimes be grouped into functional classes which may be increased in number if they carry out an important role, for example some fungi contain many copies of wood decay enzymes cellobiohydrolases or peroxidases. On occasion genes or gene families may be lost in a species resulting in a loss of that gene's function. If this occurs, it could result in a change in lifestyle of the organism, for example a pathogen losing the ability to infect a host, which therefore would transform into a saprobe. The gene complement of closely related species will be more similar than that of species with a more distant ancestry. This is because with each generation small changes to the genetic code accumulate through mutation or recombination, therefore, in general, the more distant a species is, the more changes will have occurred. This forms the basis for comparative genomic analysis.

By comparing similarities and differences between genomes and considering the lifestyle, morphology or biochemistry of an organism it is possible suggest how particular traits or adaptations might have evolved, for example the ability to attack lignin, or the change from saprotrophy to mycorrhizal lifestyle. Phylogenetic analyses may also be conducted to determine the degree of relatedness between species and predict when different groups separated from one another by examining the differences in nucleotide base or amino acid composition of individual genes.

Functional Genomics

While the genome provides information on the genes present in an organism, functional genomics aims to describe what those genes actually do and how they are regulated. For a gene to be functional it must be transcribed into RNA and then translated into a protein. Transcriptomics is the study of all the genes that are transcribed by a cell or group of cells at a particular time and under specific environmental conditions, which can be detected using high throughput sequencing technologies outlined above, or through microarrays. Proteomics describes all the proteins present and are generally detected by mass spectrometry approaches. Robust experimentation would ideally use both a transcriptomic and proteomic approach because an increase in transcription does not always lead to an increase in protein levels, RNA and protein turnover are important considerations when interpreting functional genomic data. Having a sequenced reference genome is extremely useful when determining transcriptomic- and proteomic-derived data and allows gene promoters to be identified so that potential mechanisms of transcription regulation can be more fully examined.

Functional studies have been used to observe which genes are used by fungi to exploit their environments and respond to change. They can be used to reveal adaptations within species and identify mechanisms that are active during, for example, wood decay.

Early Evolution of Fungi and Plants

In order to consider evolution within the Agaricomycetes we should step back and take a wider, albeit very general view of plant and fungal evolution (for more in depth analysis see reports (*3–7*)). Early divergence events in the eukaryote lineages are not well reconciled and estimations of the time of divergence are presented with a large error ranging in the tens of millions of years at best.

In general, estimated times of divergence between species are predicted using differences in gene sequence data calibrated with appropriate fossil evidence. Different methods can be used in these analyses including penalized likelihood, maximum likelihood and Bayesian methods, e.g. relaxed clock, which will not be discussed here. Over the years researchers have increased the number of genes used to determine phylogeny from a single analysis of small subunit (SSU) rDNA sequences, to a 6 gene phylogeny, and in more recent studies 26 or 71 single-copy genes taking into account regions of fast-evolving sites. Increasing the number of genes examined increases the confidence in the phylogeny created.

It has been proposed that the fungi diverged from the animal lineage approximately 1 billion years ago. The earliest fungi were likely to have evolved in an aquatic environment with similarities to members of the present day *Rozella* and microsporidia, while the main fungal phyla and subphyla (Chytridiomycota, Blastocladiomycota, Mucoromycotina, Entomophthoromycotina, Zoopagomycotina, Kickxellomycotina, Glomeromycota, Ascomycota and Basidiomycota) evolved from a filamentous ancestry that formed approximately 800 million years ago. Filamentous growth would have been advantageous in colonising and attaching to substrates, and in the evolution of nutrition based on the secretion of extracellular digestive enzymes and absorption of simple molecules. It is predicted that diversification in the fungal kingdom occurred in terrestrial habitats and before plants colonised the land.

The progenitor of what would form the basis of the plant kingdom developed in a primarily aquatic environment from an algal-like ancestor approximately 500 million years ago. Land colonisation would follow relatively rapidly with the fossil record providing evidence of terrestrial plants 450 million years ago in the Ordorvician period. It is suggested that a close association with fungi facilitated the colonisation of land by early plants, especially members of the Glomeromycota which are hypothesised to have formed symbiotic partnerships first with cyanobacteria or algae in semi-aquatic environments and later with the evolving land plants. The Glomeromycota diverged from the lineage leading to the Dikarya approximately 550 million years ago and present day species are obligate symbionts forming arbuscular mycorrhizas with plants. Glomeromycota fossil evidence was identified from 460 million years ago and later, 400 million years ago, a member of the Ascomycota, *Paleopryrenomycites devonicus*, was found inside the stem and rhizomes of the fossil plant, *Asteroxylon mackie*.

Lignin evolution in selected plant groups is thought to have resulted from the cellular responses to stress caused by moving onto land. Increased osmotic stress and exposure to ultraviolet radiation through the loss of the protection offered by water has been suggested to have selected for the evolution of secondary metabolic pathways. The deamination of phenylalanine and subsequent aromatic ring hydroxylation allowed the accumulation of phenylpropanoids with the ability to absorb UV-B light and protect the emerging plant species from increased UV irradiation. However, plants remained small and limited to damp environments, similar to the liverworts of today, lacking the mechanical rigidity to elevate much above the ground. By the time vascular plants (tracheophytes) evolved, approximately 420 million years ago in the late Silurian period, the mechanism to deposit lignin into the cell wall had emerged. A progressive thickening of the secondary cell wall linked to water transport is recorded in the fossil record throughout the late Silurian and early Devonian periods.

Expansion of the early plants onto land aided the fixation of atmospheric carbon and the rise in oxygen levels observed during this period. The rise of the tracheophytes is linked to an expansion in class III peroxidase enzymes, whose function in polymerizing phenylpropanoid radicals into polyphenols during lignin biosynthesis is hypothesized to have emerged from an initial role of phenylpropanoid radicals in preventing oxidative stress resulting from the increased atmospheric oxygen at the time. The increase in atmospheric oxygen also encouraged leaf formation and selected for plants that could grow taller and maximise the harvest of light. Lignin-rich, wood-like biomass emerged during the Devonian and Carboniferous periods, with the first tree-like ancestors, such as Archaeopteris, identified by around 350 million years ago.

Early Evolution of Wood Decay

A little time before the plants started their move onto land, the fungi formed the Dikarya subkingdom consisting of the Ascomycota and Basidiomycota and housing approximately 98% of today's fungal species. The term Dikarya relates to the formation of dikaryotic hyphae following gamete fusion, i.e. hyphae contain two nuclei, one from each parent, for a significant part of the life cycle of the fungus. A dikaryotic system allows a rich diversity to develop resulting from many thousands, and sometimes trillions, of progeny released from a single mating. Approximately 500 million years ago the Ascomycota and Basidiomycota diverged and adapted to grow in almost every habitat known on earth, forming well over one million species we estimate currently.

The nutritional mode of these early species is subject to debate, it is likely that the ancestral nutritional mode of the first Dikarya species and Ascomycota was saprotrophic. The early Basidiomycota ancestor has been suggested to have exhibited a parasitic lifestyle, with members of the largely saprotrophic and mutualistic Agaricomycotina subphylum emerging from an ancestry shared with the Pucciniomycotinia and Ustilaginomycotina which contain the rust and smut phytopathogens. However, there is strong evidence to support the transition to and from a pathogenic and saprotrophic lifestyle has occurred more than once in the Basidiomycota.

While species within the Ascomycota evolved to efficiently decompose the cellulose, hemicellulose and pectin components of plant biomass, and soft rot wood decay is widely reported, their ability to attack lignin is limited and not well researched. In modern day forests, Ascomycete species efficiently decompose leaf litter and attack the secondary cell walls particularly of hardwood, either decomposing it completely or forming cavities. However, the middle lamella is generally not attacked. The mechanism of lignin decomposition by soft rot fungi is not well studied, although species containing extracellular peroxidases and polyphenol oxidases (laccases) are reported their gene copy number and efficiency for decay appears to be much lower than that of white rot fungi.

The main agents of wood decomposition are Agaricomycotina fungi, particularly members of the Agaricomycetes. Whole genome sequencing programmes, supported largely by the Joint Genome Institute, aimed to investigate the different mechanisms of wood decay and how they evolved. Large collaborative research teams (*8*, *9*) have conducted comparative genome approaches to ask when and how did wood decay evolve in the Agaricomycetes, what did that early ancestor look like, and how have changes in gene complement influenced nutritional lifestyles and wood decay strategies? Baysian relaxed molecular clock analyses using 31 fungal genome sequences with fossil calibrations estimated that the Agaricomycotina emerged approximately 450 million years ago and the Agaricomycetes evolved approximately 290 million years ago at the Carboniferous/Permian boundary.

The Carboniferous period is associated with the massive deposition of coal reserves formed from the vast forests at that time. It was also a period where atmospheric oxygen levels continued to rise and carbon dioxide levels fell as photosysnthesis converted CO_2 into plant matter and released oxygen. The period is often defined by extensive coastal swamp rainforests with giant tree ferns supported by shallow seas. Tectonic movement during this period has been cited as leading to glacial events and creating the supercontinent Pangea. The merging of continents caused a rise of land above sea level which formed mountains and resulted in drier conditions. The Permo-Carboniferous ice age is associated with a mass extinction event, the Carboniferous Rainforest Collapse (305 million years ago), where the vast coal forming forests were lost. The glacial event continued into the Permian period and coincides with the proposed evolution of wood decay in the Agaricomycetes.

The physical nature of the coastal swamp forests of the Carboniferous period meant that dead plant matter would be buried in largely anoxic conditions which were not conducive to lignin decomposition or to the growth of many fungi. As the mountains rose and the environment became drier, it is proposed that the ancestor of the Agaricomycetes evolved to exploit the almost unlimited woody nutrition centred on the depolymerisation of lignin through an oxygen-requiring peroxidation mechanism. Without this critical evolutionary event, even with the loss of the coastal swamps, wood decomposition would have been limited by the recalcitrance of lignin and would probably still have led to the formation of coal and continued sequestration of CO_2 as biomass. It is not clear how the fall in CO_2 levels at this time influenced the global atmospheric temperature, possibly causing a reverse greenhouse effect. It is tempting to speculate that the evolution of wood decay mechanisms by Agaricomycetous fungi had an important role in the loss of coal deposition to help explain why further coal seams have not been formed since this time. Wood decomposition also filled an important gap in the carbon cycle facilitating the release of carbon in woody biomass as CO_2. It is hard to imagine how life would have developed on earth if atmospheric CO_2 had continued to be converted into biomass by photosynthesis without an efficient mechanism to release it once the plant had died.

Comparative genomic analysis (9) suggests that the mechanism to decompose wood emerged once in the Basidiomycota and this event lead to the radiation of the Agaricomycetes into 17 orders and over 20,000 described species. As of writing, at least one member from 15 Agaricomycete orders has a genome sequence publically available (Table 2), allowing researchers to predict gene complements of the last common ancestor between the different orders and identify gene families which may have expanded or reduced in number over time. It is proposed that the ancestor of the Agaricomycetes was a white rot species containing 2 to 7 class II peroxidases capable of attacking lignin, 1 or 2 dye-decolorizing peroxidases and 5 to 8 oxidases implicated in hydrogen peroxide generation. The white rot nutritional form (i.e. depolymerisation and mineralisation of all the components of lignocellulose, including lignin) continued to be the dominant nutritional form during early Agaricomycete evolution, the number of class II peroxidases is predicted to have increased to 3 - 16 copies, and subsequent parallel expansions were then observed in species from the Auriculariales, Corticales, Hymenochaetales, Polyporales and Russulales.

Fungal class II peroxidase enzymes are employed by Agaricomycete white rot species to depolymerise lignin, permitting access to cellulose in wood. The number of class II peroxidases is generally expanded in white rot species compared to other nutritional modes, averaging 14 gene copies (range 5 to 26 copies), underlining their importance in ligninolysis. Peroxidases are key oxidative enzymes found in all cells, yet an alteration to a single fungal class II peroxidase enzyme present in the Basidiomycota common ancestor resulted in the ability to depolymerise lignin (the mechanism of peroxidase enzymes will not be discussed here in any depth). This enzyme was lost in the lineages leading to the other Basidiomycota subphyla (Pucciniomycotina and Ustilaginomycotina) and in species from the Dacrymycetales and Tremellales. There are 4 main groups of fungal class II peroxidases, namely manganese peroxidase, lignin peroxidase, versatile peroxidase and non-ligninolytic low-redox potential peroxidases (sometimes referred to as generic peroxidases). Comparative genome analysis suggests that the first ligninolytic peroxidase enzyme emerged approximately 295 million years ago from an adaptation of a non-ligninolytic peroxidase that allowed the binding of manganese, i.e. a manganese peroxidase. A subsequent gain of a tryptophan residue at position 171 and loss of manganese binding led to the emergence of lignin peroxidase, where the tryptophan residue is responsible for direct lignin oxidation.

Versatile peroxidases possess both a tryptophan residue at position 171 and bind manganese. Interestingly 3 independent origins of versatile peroxidases have been proposed, two in the Polyporales leading to *Trametes versicolor* and *Dichomitus squaliens*, and in *Pleurotus ostreatus* of the Agaricales. This is an example of convergent evolution, where a gene or trait is acquired without being passed down from a common ancestor. Where such events occur, it is usually assumed that the gene or trait provides a strong evolutionary advantage in order to have been selected for independently more than once. A more common example of convergent evolution is the eye of animals, where the eyes of the arthropoda, chordata and cephalopoda serve the same function (having eyes provides a very clear selective advantage), yet they are sometimes strikingly different in appearance due to an independent evolution of the structures.

Evolution of Agaricomycete Nutritional Modes

The majority of Agaricomycete species exhibit a white rot nutritional mode where all the components of lignocellulose are depolymerised and mineralised. Other nutritional modes include brown rot, where cellulose and hemicelluloses in wood are targeted and the lignin is modified, but is left as a polymeric residue, ectomycorrhizal symbiosis, and species exhibiting a range of weak to lethal parasitic interactions. Other species have adapted to particular niches such as leaf litter and humic horizons or coprophilic lifestyles. Ascribing a nutritional mode to a fungus recognises a broad definition of the lifestyle of the fungus while appreciating that individual species may at different times alter their behaviour or not fit securely in a rigidly defined box. For example, some saprotrophic species might form transient associations with plant roots or cause disease on certain host species, or ectomycorrhizal fungi may exhibit a saprotrophic lifestyle for part of their life cycle. Likewise, the decomposition of a complex structure such as lignocellulose almost always involves many overlapping and cooperating mechanisms and pathways, the complexity of which we still do not fully appreciate. It is probable that behaviours or enzyme systems normally associated with one nutritional mode may also occur in other modes, for example, certain white rot species in addition to employing class II peroxidases might employ a Fenton-style attack on lignin normally associated with brown rots. This section considers nutritional modes in their broadest sense and while specific examples will be described, it will not attempt to address all the enzyme mechanisms involved in wood decay in the Agaricomycetes in detail.

The White Rots

The ancestor of the Agaricomycetes exhibited a white rot nutritional mode that was maintained through the early evolution of the class. The most recent common ancestor to the Polyporales, Agaricales and Boletales approximately 160 million years ago is predicted to have contained 66 to 83 carbohydrate active enzymes (CAZys, i.e., glycoside hydrolases and carbohydrate esterases (*10*)) and 27 to 29 extracellular oxidoreductase enzymes (including class II peroxidises) involved in lignocellulose decomposition. As white rot lineages diversified, the number of CAZy and oxidoreductase genes involved in wood decomposition increased in most lineages (Table 3), with the average number of CAZy genes in 17 gene families increasing to 86 gene copies (range 61 to 137) in the "typical white rot". Similarly, extracellular oxidoreductase enzymes in 10 gene families expanded notably to an average of 57 gene copies (range 44 to 74).

General trends in the expansion of certain gene families are observed in the white rot lineages, for example multiple copies of class II peroxidases, multicopper oxidases, glycoside hydrolase (GH families, GH3 (glucosidase), GH5 (endo- and exoglucanases), and GH28 (galactouronidases) families. Endoglucanase / cellobiohydrolase GH6 and GH7 enzymes are also key to the breakdown of crystalline cellulose. In addition, AA9 (previously GH61) copper-dependant lytic polysaccharide monooxygenase and genes with cellulose binding modules are observed in greater numbers in white rot species than in fungi with different nutritional modes.

Table 3. Overview of Putative Lignocellulose Decomposing Carbohydrate Active and Oxidoreductase Gene Number in Selected Sequenced Agaricomycetes (Values Taken from (11))

Gene function	White rots							Brown rots					ECM	NWS	
	Ad	Ds	Fm	Hi	Pc	Ps	Sh	Tv	Cp	Fp	Gt	Pp	Sl	Wc	Lb	Ab	Cc
CAZy	137	72	75	61	69	87	104	79	69	59	49	33	39	37	28	80	103
Oxred	74	53	54	44	47	58	63	60	37	32	26	27	24	30	34	56	41

Abbreviations: ECM, Ectomycorrhiza; NWS, Non-wood degrading saprotroph; CAZy, carbohydrate active enzymes; OxRed, oxidoreductases. Species names: Ad, *Aricularia delicata*; Ds, *Dichomitus squalens*; Fm, *Formitiporia mediterranea*; Hi, *Heterobasidion irregulare*; Pc, *Phanerochaete chrysosporium*; Ps, *Punctularia strigosozonata*; Sh, *Stereum hirsutum*; Tv, *Trametes vesicolor*; Cp, *Coniophora puteana*; Fp, *Formitopsis pinicola*; Gt, *Gloeophyllum trabeum*; Pp, *Postia placenta*; Sl, *Serpula lacrymans*; Wc, *Wolfiporia cocos*; Lb, *Laccaria bicolour*; Ab, *Agaricus bisporus*; Cc, *Coprinopsis cinerea*.

Despite obvious general trends there will always be exceptions which may help explain peculiar adaptations observed in diverging species. Members of the Agaricales tend to have an absence or reduced number of manganese and lignin peroxidise genes compared with other white rot species. *Schizophyllum commune*, which is often considered a white rot despite little evidence supporting an active depolymerisation of lignin by this fungus, contains no class II peroxidases. *Coprinopsis cinerea* also appears to lack lignolytic peroxidases, while *Agaricus bisporus* has two genes – *Phanerochaete chrysospoium* has 15. *Pleurotus ostreatus* has convergently evolved versatile peroxidases as described earlier and this might explain why this species still exhibits an aggressive form of wood decay. In *Agaricus bisporus*, the notable expansion in the number of heme-thiol peroxidases, e.g. aromatic peroxidases, may be seen as an adaptation to the non-woody, partially decomposed lignocellulosic biomass and humic substances which have enabled the fungus to expand into the leaf litter horizon of forests and grassland soils.

The genome sequencing of the Polyporales white rot fungus *Ceriporiopsis subvermispora* provided the opportunity to investigate adaptation in white rot mechanisms. *C. subvermispora* exhibits a sequential/selective decay of wood where lignin is depolymerised with little initial breakdown of cellulose. This contrasts with the more common simultaneous rot exemplified by the closely related *Phanerochaete chrysosoprium* where lignin and cellulose are decomposed at the same time. Comparative and functional genomic comparisons using the sequenced genomes were combined with transcriptomic and proteomic analyses by Fernandez-Fueyo *et al.*, 2012 (*12*). While *C. subvermispora* has comparable glycoside hydrolase gene numbers to *P. chrysosporium*, it has fewer GH7 cellobiohydrolase genes (3, compared with 6 in *P. chrysosporium*), fewer GH3 β-glucosdiases and 16 genes containing cellulose binding modules where *P. chrysosporium* has 31. Importantly, transcript and protein levels of carbohydrate active genes were lower in *C. subvermispora* relative to *P. chrysposporium* when grown on aspen wood, providing compelling evidence of the relevance of gene regulation to nutritional mode adaptation. Just having a gene is not enough as it must also be expressed under appropriate environmental conditions to provide function.

C. subvermispora has an expanded repertoire of manganese peroxidase enzymes, 13 in total, with one lignin peroxidase and an apparent versatile peroxidase, additionally the fungus has 7 laccase genes. This contrasts with 5 manganese peroxidases, 10 lignin peroxidases and no laccases in *P. chrysosporium*. The tendecy of *C. subvermispora* to preferentially target lignin was supported by transcriptomic analysis where increased transcript expression of oxidoreductase enzymes, particularly manganese peroxidases, was identified. Lignolysis in *C. subvermispora* appears to be linked to an active lipid peroxidation system that putatively attacks the non-phenolic structures in lignin that are unaffected by manganese peroxidase activity. Similar mechanisms were not identified in the simultaneous rot mechanism of *P. chrysosporium* which relied on lignin peroxidase activity and elevated levels of cellulolytic CAZy enzymes.

The Brown Rots

Taxonomically a small proportion of wood decay species are brown rots (6%), never-the-less the brown rots dominate the decomposition of conifer wood in boreal forests where they have a major impact on the relatively poor soil condition. Work by Hibbett and Donoghue, 2001 (*13*), provided a correlation between wood decay mechanisms and substrate preference, suggesting that the loss of exoglucanases and lignin degrading enzymes was an evolutionary advancement. The brown rot mechanism, where cellulose and hemicelluloses are decomposed leaving lignin modified but largely intact, has evolved from a white rot ancestry at least 5 times. This suggests that there is a strong selection pressure for ability to cast off the energetically expensive need to depolymerise lignin in conifer-dominated habitats. Brown rot species in the Boletales, particularly the dry rot fungus *Serpula lacrymans*, Gloeophyllales, and Polyporales, particularly *Postia placenta*, have been well studied. However, understanding of the adaptations that lead to the evolution of the brown rot decay form were elusive until the genome sequencing of brown rot fungi was carried out (*8, 9, 14*). While a non-enzymic attack on lignocellulose by an extracellular hydroxyl radical-generating Fenton's reaction based on the reduction of iron (*15*) was largely accepted, it was not understood whether distantly related brown rot species had evolved largely similar or vastly different mechanisms of generating the Fenton's chemistry and subsequent enzymatic decomposition of cellulose.

Comparative genome analysis initially by Eastwood *et al.*, 2011 (*8*) and on more brown species by Floudas *et al.*, 2012 (*9*), highlighted a common pattern of simplification in the lignocellulose decomposition machinery of brown rot species. The brown rot species generally have a much lower repertoire of carbohydrate active enzymes (CAZys) and oxidoreductases involved in wood decay than the average white rot species (Table 3). Brown rots average 48 CAZy (range 33 to 69) and 29 oxidoreductase (range 24 to 37) gene copies. Perhaps the most striking feature is the lack of class II peroxidase enzymes in all brown rot species studied which underlines the distinction in lignin depolymerization mechanisms between white and brown rot species. Other notable features include a much lower number of AA9 (previously GH61) copper-dependant lytic polysaccharide monooxygenases, presumably a consequence of expansion in white rot lineages rather than loss of gene number in the brown rots. Notably, with the exception of most of the Boletales, all the brown rots lack GH6 and GH7 endoglucanase / cellobiohydrolase enzymes, although *Serpula lacrymans* also lacks a GH6. Cellobiose dehydrogenase enzymes are also absent from the brown rot species in the Polyporales. However, rather than consider the relative reduction in gene number a simplification of decay machinery by the brown rots, the expansion in certain gene lineages, e.g. cellulose and hemicellulose decomposing GH5 and GH28 genes respectively, suggests that brown rots have developed a more targeted arsenal of cellulolytic enzymes while relying on non-enzymatic mechanisms to depolymerise and then repolymerize lignin. The multiple emergence and domination of brown rots in boreal forests might be explained by the efficient targeting of resources and circumventing complete lignin decomposition and metabolism to maximise the release of sugars from cellulose and hemicelluloses.

Despite general similarities between the distantly related brown rot species, there are also clear differences between the species. Fenton chemistry ($Fe^{2+} + H_2O_2 + H^+ \rightarrow Fe^{3+} + \cdot OH + H_2O$) is largely accepted as a mechanism by which fungi generate hydroxyl radicals and is used by brown rots to open the lignocellulose structure to access cellulose. However, it was not known whether the different species had evolved the same mechanisms to mediate the reduction of Fe^{3+} to Fe^{2+} in environments where the trivalent form is energetically favoured.

Transcriptomic and proteomic comparison of wood decay between the Polyporales species *Postia placenta* and white rot *Phanerochaete chrysosporium* showed that *P. chrysosporium* simultaneously attacks cellulose and hemicelluloses, while *P. placenta* initially targets hemicelluloses (*16*). Moreover, evidence was presented that supported a 2,5- dimethoxyhydroquinone(DMHQ)-mediated Fe^{3+} reduction by describing genes upregulated on wood that are involved in iron transport and the synthesis of low molecular weight quinones. Similar experiments on *Serpla lacrymans* growing on pine wood failed to identify similar genes or proteins. However, evidence in the literature suggests that 2,5-DMHQ-mediated iron reduction does occur in *S. lacrymans*. Transcriptomic data for *S. lacrymans* identified an iron reductase enzyme with a cellulose binding domain which appeared to have been derived from a cellobiohydrolase dehydrogenase gene that was greatly upregulated on wood, but was absent in *P. placenta*. It was suggested that this enzyme could enable *S. lacrymans* to target iron reduction and, therefore, hydroxyl radical generation close to the substrate, which might in part explain why *S. lacrymans* can decompose crystalline cellulose in the absence of lignin, unlike some other species. In addition, a nutritionally regulated, secondary metabolite-mediated mechanism of iron reduction was also proposed based on the production of varegatic acid. While this topic is under scientific debate, it is plausible that in any fungus multiple mechanisms could exist that generate the same outcome (i.e. reduction of Fe^{3+}), and that these mechanisms may occur simultaneously or be employed at different times during substrate colonisation or under differing environmental conditions. Genomic and transcriptomic analysis have also demonstrated a glyoxylate shunt in *Postia placenta*, possibly of relevance to oxalic acid accumulation and iron mobilization. It seems apparent that different brown rot fungi employ similar and differing mechanisms to reduce iron to enable mediated-Fenton hydroxyl radical generation. Further experimentation, supported by genomic resources, will hopefully expand our knowledge of Fenton chemistry in fungal-mediated wood decomposition.

Comparative analysis of the available Agaricomycete genomes also identified an intriguing similarity in the targeted rationalisation of the brown rot saprotrophic enzyme arsenal and the gene complement of the ectomycorrhizal fungus *Laccaria bicolor*. Both brown rot and ectomycorrhizal species had fewer lignocelluolytic CAZy and oxidoreductase enzymes compared with white rot species. The *L. bicolor* genome contains 28 and 34 CAZy and oxidoreductase genes respectively. Similar to most brown rot species, the ectomycorrhiza-forming fungus lacked class II peroxidises, GH6 and GH7 endoglucanase / cellobiohydrolases, cellobiose dehydrogenase and had a reduced number of genes with cellulose binding modules. GH5 and GH28 enzymes expanded in the brown rots were also retained in *L. bicolor*. The loss of aggressive wood decay enzymes in *L. bicolor* highlights the adaption by the fungus to a more symbiotic lifestyle where intracellular penetration occurs during root colonisation, but aggressive cell wall decomposition does not. Phylogenetic analysis of the Agaricomycetes indicates that the ectomycorrhizal lifestyle has emerged multiple times from a white rot ancestry. It is not known how a white rot species evolves into a mycorrhiza-forming species, perhaps the Agaricales where *L. bicolor* is found may provide a clue as there is a reduction in the class II peroxidases in the saprotrophic species in this order. However, could ectomycorrhizal formation be a natural progression from a less ligninolytic brown rot mechanism and all white rots pass through a transient brown rot stage on route to an ectomycorrhizal lifestyle? In the Boletales at least there is evidence of ectomycorrhiza-forming species emerging from a brown rot ancestry. Approximately 35 million years ago *Austropaxillus* an ectomycorrhizal genus diverged from a brown rot lineage leading to *Serpula lacrymans*. The similarity between brown rot and ectomycorrhizal nutritional modes in the Boletales is supported further by the transition of the brown rot *Hydnomellius pinastri* from an ectomycorrhizal lineage approximately 85 million years ago. The ancestral nutritional mode of the Boletales is weakly predicted to be a brown rot, but this is by no means certain. The apparent plasticity in the transition between nutritional modes within members of the Boletales is a fascinating topic for future study.

Review and Future Work

This chapter has provided a brief overview of how genome sequencing programmes have informed our understanding of wood decay mechanisms and their evolution in the Basidiomycota. The reports referred to will provide more detailed results and explanation of methodology and are recommended for further reading. Comparative and functional genomic studies are becoming more common and more informative as sequencing technologies and bioinformatic analysis improve, therefore, allowing the field of fungal genomics grow and evolve rapidly.

Researchers are already looking beyond individual species and are sequencing whole communities of wood inhabiting an wood decay fungi and bacteria to investigate how they respond to environmental change. How the decomposition of woody material is brought about in nature requires more study, to what extent species interact and adapt to different niches within the forest habitat has been a challenging topic for researchers to address. We can now use functional genomic approaches to investigate two or more intermingling decay fungi and determine the relative activity of each species. Most fungi live in a competitive environment where nutrition is heterogeneously distributed. Linking genomics to a wider appreciation of fungal ecology will be a fascinating challenge for the future.

Further functional approaches should also be employed to test hypotheses resulting from comparative genomic studies. More in-depth study of gene regulation is often required, including promoter analysis, protein processing and consideration of epigenetic regulation of gene expression. Is it also important to test whether the genes identified as putatively important in decay actually carry out their expected role. Often gene function is ascribed by similarity to another gene using a computer-based algorithm, e.g. BLAST. Sometimes these genes and their subsequent proteins are well characterised and previous functional studies have been performed. In many instances though, information from functional studies is missing or inferred from a distantly related species. Furthermore, a large number of genes have no similarity with other species and there is little information on their function. These genes are often ignored when researchers discuss the implications of functional genomic studies, but the role of these genesin the organism might actually be critical. Proper functional analysis is a major challenge that must be addressed if complex processes, such as wood decomposition, are to be understood.

Increasingly the significance of the chemical environment in regulating the behaviour and functioning of cells is being recognised. Fungi in particular actively produce and secrete secondary metabolites and release gaseous chemicals that alter their environments, affect competitors and mediate genetic and morphological change. Metaboliomics is the study of all the metabolites produced by a cell or group of cells under defined conditions. While currently limited for wood decay fungi, metabolomic analysis of species and community decomposition will provide essential information on the process.

Diversification and adaptation in the plant and fungal kingdoms have been aligned as a consequence of close interactions and sharing of habitats. The evolution of wood in plants and the corresponding mechanisms which have evolved in fungi to break it down is an excellent example of how species adaptation alters the wider environment and generates a selection pressure for further evolutionary change. Furthermore, evidence can be presented of co-evolution between species, such as the evolution of the brown rot nutritional mode and the diversification in conifers that occurred at relatively the same time. The lignin-derived residues bind nitrogen and help to create nutrient poor acidic soils which allowed conifer species co-adapted to these conditions to dominate the boreal forests.

It is important to reflect that without the evolution of Agaricomycotina wood decay species approximately 290 – 300 million years ago, the carbon cycle may have stalled due to the locking up of atmospheric CO_2. The selection pressure to evolve a mechanism to access the rich carbon bounty sequestered within wood would have been strong, but it is with reverence to the recalcitrant structure of lignocellulose that efficient decay mechanisms have emerged rarely. This in turn will have had significant implications in the radiation and diversity seen in the Agaricomycetes.

If wood decomposition combined with the loss of vast anoxic swamps caused the end of coal deposition, then is the seemingly popular solution of planting a tree to mitigate the use of fossil fuels valid? Certainly carbon in biomass is sequestered in humic soils to some degree, but will it ever be comparable to the carbon released from fossil fuel, particularly over longer time periods? Interestingly, the lignin-derived residues produced from brown rot decay are long-lived and might accumulate over geological time. Perhaps planting a conifer in a boreal forest might be the more conscientious option for ecologically minded individuals.

References

1. Martinez, D.; Larrondo, L. F.; Putnam, N.; Gelpke, M. D.; Huang, K.; Chapman, J.; Helfenbein, K. G.; Ramaiya, P.; Detter, J. C.; Larimer, F.; et al. *Nat. Biotechnol.* **2004**, *22*, 695–700.
2. Liu, L.; Li, Y.; Li, S.; Hu, N.; He, Y.; Pong, R.; Lin, D.; Lu, L.; Law, M. *J. Biomed. Biotechnol.* **2012**, *2012*, 251364.
3. Bateman, R. M.; Crane, P. R.; DiMichele, W. A.; Kenrick, P. R.; Rowe, N. P.; Peck, T.; Stein, W. E. *Ann. Rev. Ecol. Syst.* **1998**, *29*, 263–292.
4. Weng, J.-K.; Chappel, C. *New Phytol.* **2010**, *187*, 273–285.
5. Taylor, J. W.; Berbee, M. L. *Mycologia* **2006**, *98* (6), 838–849.
6. James, T. J.; Kauff, F.; Schoch, C. L.; Matheny, P. B.; Hofstetter, V.; Cox, C. J.; Celio, G.; Gueidan, C.; Fraker, E.; Miadlikowska, J.; et al. *Nature* **2006**, *443*, 818–822.
7. Stajich, J. E.; Berbee, M. L.; Blackwell, M.; Hibbett, D. S.; James, T. Y.; Spatafora, J. W.; Taylor, J. W. *Curr. Biol.* **2009**, *19* (18), R840–R845.
8. Eastwood, D. C.; Floudas, D.; Binder, M.; Majcherczyk, A.; Schneider, P.; Aerts, A.; Asiegbu, F. O.; Baker, S. E.; Barry, K.; Bendiksby, M.; et al. *Science* **2011**, *333* (6043), 762–765.
9. Floudas, D.; Binder, M.; Riley, R.; Barry, K.; Blanchette, R. A.; Henrissat, B.; Martínez, A. T.; Ortillar, R.; Spatafora, J. W.; Yadav, J. S.; et al. *Science* **2012**, *336*, 1715–1719.
10. Levasseur, A.; Drula, E.; Lombard, V.; Coutinho, P. M.; Henrissat, B. *Biotechnol. Fuels* **2013**, *6*, 41.
11. Morin, E.; et al. *Proc. Natl. Acad. Sci. U.S.A.* **2012**, *109* (43), 17501–17506.
12. Ferandez-Fueyo, E. *Proc. Natl. Acad. Sci. U.S.A.* **2012**, *109* (14), 5458–5463.
13. Hibbett, D. S.; Donoghue, M. J. *Systems Biol.* **2001**, *50*, 215–242.
14. Martinez, D.; Challacombe, J.; Morgenstern, I.; Hibbett, D.; Schmoll, M.; Kubicek, C. P.; Ferreira, P.; Ruiz-Duenas, F. J.; Martinez, A. T.; Kersten, P.; et al. *Proc. Natl. Acad. Sci. U.S.A.* **2009**, *106* (6), 1954–19599.
15. Goodell, B.; Jellison, J.; Liu, J.; Daniel, G.; Paszczynski, A.; Fekete, F.; Krishnamurthy, S.; Jun, L.; Xu, G. *J. Biotechnol.* **1997**, *53*, 133–162.
16. Vanden Wymelenberg, A.; Gaskell, J.; Mozuch, M.; Sabat, G.; Ralph, J.; Skyba, O.; Mansfield, S. D.; Blanchette, R. A; Martinez, D.; Grigoriev, I.; et al. *Appl. Environ. Microbiol.* **2010**, *76* (11), 3599–3610.

Chapter 6

Above Ground Deterioration of Wood and Wood-Based Materials

Grant Kirker*,[1] and Jerrold Winandy[2]

[1]USDA-FS, Forest Products Laboratory, Madison, Wisconsin 53726
[2]Winandy & Associates, LLC, East Bethel, Minnesota 55011
*E-mail: gkirker@fs.fed.us.

Wood as a material has unique properties that make it ideal for above ground exposure in a wide range of structural and non-strucutral applications. However, no material is without limitations. Wood is a bio-polymer which is subject to degradative processes, both abiotic and biotic. This chapter is a general summary of the abiotic and biotic factors that impact service life of wood in above ground exposures, and briefly discusses test methodologies commonly used in North America to determine the durability of wood and wood based materials in above ground exposure. Current efforts to improve service life estimates for wood and wood based materials are also discussed.

Abiotic Factors That Impact Service Life of Wood and Wood-Based Materials above Ground

The term "abiotic degradation" refers to any non-biological related type of wood degradation. The two modes of abiotic degradation include chemical pathways and mechanical action. Each has an independent effect and each interacts with the other. The classic commonly recognized practical modes of abiotic degradation include weathering, mechano-sorptive relationships, and friction/ erosion/ mechanical-related damage. Each is discussed below, and in these discussions it will become apparent how each of these above ground abiotic modes of degradation interact and among themselves and with agents of above

© 2014 American Chemical Society

ground biotic degradation. Additional information on the abiotic processes the degrade wood can also be found in Phil Evans' chapter from the 2008 ACS book (*1*), and further information on weathering and photostability of modified woods are presented in a separate review, also by Evans (*2*).

Modes of Abiotic Degradation

Weathering

Surface Changes

Weathering is a common term describing the abotic degradation of wood when exposed outdoors to sunlight and the direct "weathering" effects of rain, freeze-thaw, alternating thermal loading and wind (Figure 1). The ultraviolet (UV) portion of sunlight reacts with, and causes, a slow deterioration of the lignin exposed at the wood surface as UV light only penetrates wood a few microns (*3*). As the lignin-rich lamella zones around the wood fibers degrade, the fiber-to-fiber bonding at the wood surface is greatly reduced. This abiotic chemical process of lignin degradation is then followed by a physical/mechanical abiotic process.

Figure 1. Surface of weathered exterior-grade plywood after 10-years exposure (right side covered from sunlight; left side exposed to sunlight & elements). (Reproduced from reference (3).)

While the cellulose and hemicelluloses remaining in the now carbohydrate-rich wood fiber (i.e., after UV-induced lignin degradation) are relatively resistant to UV, there can be significant reduction in fiber-to-fiber bond strength because of the lignin degradation. This then allows the combined physical/mechanical internal stresses resulting from cyclic swell/shrink, freeze-thaw, and diurnal thermal loadings, combined with the external forces from rain and/or wind, to cause an on-going erosion of the wood fibers from the wood surface. Earlywood has thinner cell walls than latewood and thus earlywood is generally more prone to suffer from weathering than is latewood (*3*). Because of this tendency, weathered wood often has a wavy texture to its surface (Figure 2).

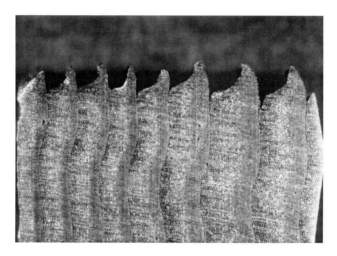

Figure 2. Differential weathering between earlywood and latewood of softwood lumber exposed on radial face. Note how thin-walled earlywood weathers faster than the denser latewood. (Reproduced from reference (3).)

As fresh wood fibers are then exposed to direct sunlight, rain and wind, the process systematically repeats itself on a layer by layer basis (Figure 3).

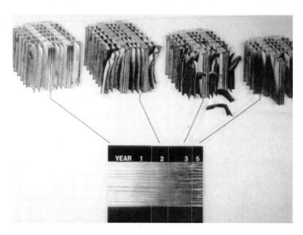

Figure 3. Hypothetical example of the progression of weathering on a softwood surface. Note how the thin-walled earlywood initially weathered faster than the denser latewood, but the latewood fibers also eventually weather and erode away. (Courtesy of R.S. Williams with permission.)

From the previous figure it is apparent that UV weathering is a slow process occurring over many years. When judged on larger wood samples the time required for wood to become fully weathered depends on wood density, species and severity of exposure. In general, weathering is sometimes approximated over a broad range of species/density as 6mm per century (Figure 4).

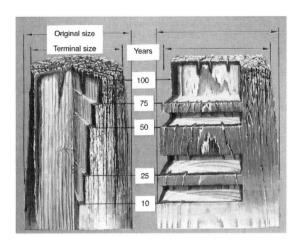

Figure 4. Two hypothetical examples highlighting the progression of the weathering process on the surface of wooden timbers. (Reproduced from reference (3).)

Weathering can take many forms. Often as the lignin is degraded we notice a yellow/brown-to-grey color change because lignin tends to be brownish in color while the carbohydrates appear grey-white under light in our visual spectrum. As lignin degradation develops, fiber-to-fiber bonding is reduced, this in turn also allows wood to develop deeper surface checking. This checking allows deeper penetration of UV light and enhances deeper moisture absorption/desorption. Because the cellulose and hemicelluloses readily absorb moisture in wet or humid conditions, this promotes swelling. During dry conditions, shrinkage occurs so that this cyclic swelling and shrinkage then creates alternating internal stresses that can also lead to still deeper surface checking. As the wood surface(s) shrinks and swells, the external mechanical forces of rain and/or wind can eventually erode these degraded wood fibers from the wood surface.

Extractives often also impart color to the wood and wood surface. These extractives are most often water-soluble and, as the wood surface is progressively degraded and eroded, deeper access to water occurs which accelerates leaching of the extractives. Progressive wet-dry moisture cycling gradually draws more extractives to the wood surface, where rain can leach these extractives from the wood surface. This loss of extractives can further promote color change in wood.

Biological attack of a wood surface by microorganisms is also recognized as a contributing factor to color change or graying of wood. This biological attack, commonly called mildew, does not cause erosion of the surface, but it may cause initial graying or an unsightly dark gray and blotchy appearance. These color changes are caused by dark-colored fungal spores and mycelia on the wood surface, as discussed later in this chapter. In advanced stages of weathering, when the surface has been enriched by cellulose, it may develop a silvery-gray sheen.

Mechancial Damage

Wood changes dimensions as it gains or loses moisture below the fiber saturation point (FSP). Below the FSP, wood shrinks as it loses moisture from the cell walls, and swells when gaining moisture in the cell walls. This shrinking and swelling can result in warping, checking, and splitting. The combined effects of radial and tangential shrinkage can distort the shape of wood pieces because of the difference in shrinkage and the curvature of annual rings. The major types of distortion as result of these effects are illustrated in Figure 5.

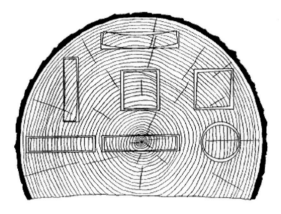

Figure 5. A hypothetical example characterizing how various anatomical orientations can influence the shrinkage in wood with tangential shrinkage being greater than radial shrinkage by an approximate factor of 2x. (Reproduced from reference (3).)

Figure 6. Examples of how once-finished and then unprotected treated-wood decking can experience varying degrees of surface checking and uneven gapping after a 18-year exposure in Wisconsin. (Courtesy of J. Winandy with permission.)

Moisture-related shrinkage and swelling can also create non-uniform gaps along edges of decking or, eventually, surface checking on decking surfaces (Figure 6).

Since wood plastics composites (WPCs) are realitively new to the market long dominated by wood, their long term performance is not fully understood. There is a growing body of work focusing on the thermo- and photstability of wood plastic composites, but the effect of the particular wood species and plastic matrix are still not fully understood and new formulations and blends are still being produced which may have alternative weathering properties. Wood plastic composites are succeptible to abiotic effects of weathering. WPCs vary greatly in the compositon of their plastic matric, wood species and other additives, and all of these factors have been shown to impact photostability and above ground performance. Newer generation WPCs are mostly PVC and HDPE based and contain 50% wood or less, but weathering characteristics have been improved through the addition of UV stabilizers, pigments, and cross-linking agents. In addition, moisture-related issues can also affect WPC decking. If the untreated wood fiber gets wet and stays wet for an extended period of time, abiotic moisture absorption can promote above-ground biotic activity. An example of moisture retention in wood-plastic composite decking is shown in Figure 7. A more in-depth review on the effects of weathering on WPCs is presented by Azwa *et al.* (*4*).

Figure 7. Wood-plastic composite deck boards after a rain shower followed by cloudy conditions. Moisture is retained for days within the decking, especially near joints where end-grain is exposed. (Courtesy of J. Winandy with permission.)

Anatomical/Species Related Issues

As stated earlier, wood density and surface texture can become critical factors in abiotic degradation. Some wood species are denser than others and, within these species, latewood cells are denser than earlywood cells. Aboitic issues such as friction-induced erosion (i.e. erosion on a wearing surface such as flooring or decking) are significantly influenced by wood density and surface texture. Denser woods and tight-grained woods (i.e., wood with many closely spaced growth rings) are more resistant to friction-induced erosion; lower density woods and open-grain woods (i.e. wood with generally faster growth, widely spaced growth rings) are much less resistant. Juvenile wood [see Zobel 1998 (*5*) for an explanation of juvenile wood], used in outdoor decking and exposed to this moisture cycling, greatly increases the effects of surface checking (Figure 8). Species differences in weathering and overall durability of WPCs was investigated by Kim et al. (*6*) and they found enhanced durability of WPCs when incorporating naturally durable woods as feedstock for the WPCs, but also found increased corrosion associated with several species.

Figure 8. A CCA-treated pine deck board with exposed juvenile-wood surface after 5-years (left) and after 18-years (right) of direct outdoor exposurein Wisconsin. (Courtesy of J. Winandy with permission.)

This surface checking is a mechanical issue that promotes deeper moisture access. In some cases this deeper checking enables moisture to reach the untreated core of treated wood decking, promoting greater checking and warping and sometimes leading to fungal staining or decay (Figure 9).

Figure 9. An example of how abiotic and biotic issues can work interactively to promote wood deterioration and destroy wood products in above ground outdoor exposures. Note the presence of an un-dentified fruiting body protruding from crack in the board. (Courtesy of J. Winandy with permission.)

Biotic Factors That Impact Service Life of Wood and Wood-Based Materials above Ground

In this section we will review common causes of biotic degradation in wood. Biotic damage generally refers to damage by organisms which include, but are not limited to, fungi and insects, and we review the role of these biological agents in the biological breakdown of wood and wood-based products in above ground exposure. An important consideration to bear in mind is the effect of climate on rates of biological degradation. The same concept applies above ground as it does in ground contact. Areas with higher mean temperatures and annual rainfall typically have higher decay hazard. A decay hazard map was created by Scheffer (7), and which was recently updated (8, 9) to reflect shifting zones as a consequence of changing climate.

A major factor when considering longevity of above ground structures is thoughtful design and use. Proper choice of fasteners, code adherence, and wood selection are all important decisions to be made when selecting wood for above ground applications and can have a profound impact on the performance of the material. It is imperative to know and understand the structural limitations of the material in its intended exposure and closely follow recommendations. Moisture management is one of the most critical design decisions to be made. By eliminating areas where water collects and causes wood products to remain wet for long periods of time, many of the problems mentioned below can be avoided.

Wood Decay Fungi

Fungi are multicellular microorganisms that spread and propagate via a threadlike mycelium, persisting in soil, wood, and leaf litter. Fungi can form sexual and asexual fruiting bodies that produce spores, allowing the fungi to also disperse by wind or splashing water. Wood decay fungi are a specialized group that can effectively utilize the structural components of wood as a food source and are typically categorized as described below, listed in descending order of relative importance to wood degradation:

Brown-rot fungi – brown-rot fungi belong to the phylum Basidiomycota (true fungi, many in the family Polyporaceae) and have specialized enzyme systems that effectively break down the polysaccharides cellulose and hemicelluloses. Brown-rot decay has a characteristic rusty brown coloration that develops a cracked, cubical appearance in later stages (Figure 10b). Due to the rapid lowering of the cellulose degree of polymerization in the incipient decay stage, brown-rot can cause rapid strength loss of lumber, leading to in-place failures. Brown-rots are a common problem in softwood lumber.

White-rot fungi – white-rot fungi (Basidiomycota) have the ability to break down all of the structural components of wood (cellulose, hemicelluloses, and lignin). Typical indications of white-rot fungi are white, stringy and bleached decayed wood (Figure 10c) that results from lignin removal and the presence of the remaining undigested cellulose. White-rot fungi are typically more common on hardwood species but are sometimes also found on softwoods.

Soft-rot fungi – Soft-rot fungi (Ascomycota) are a problem in areas of higher moisture and available nitrogen (*10*). These fungi cause a characteristic decay pattern that affects cellulose and hemicellulose in the S2 layer of the inner cell wall. Soft-rot does not penetrate deeply into the wood but causes a gradual sloughing off of the outer surface (Figure 10d) that can lead to decreased structural integrity and in-place failures.

Mold Fungi – mold fungi (primarily Ascomycota) are typically asexual fungi that grow and reproduce prolifically through the spread of conidia. Most of the highly successful molds that are encountered on wood produce conidiophores that contain thousands of spores, either singly or in chains, as in the case of *Paecilomyeces* spp., *Penicillium* spp., and *Aspergillis* spp. These spores are transported by wind and can spread rapidly in the environment (*11*). The mold fungi are synonymous to weeds in the plant kingdom - they are biologically pre-disposed to take up space. Most molds are quite aggressive, grow rapidly and many produce secondary metabolites that often repel surrounding micro-organisms (or even insects). Mold fungi are ubiquitous in nature and usually have little impact on the structural integrity of the wood. However, certain molds have been found to degrade biocides, and surface growths of mould do diminish the aesthetic value of the wood (Figure 10a) and can also degrade paints and coatings. The presence of molds is also an indication of moisture problems which can result in subsequent decay.

Sapstains – Sapstain fungi (Ascomycota) are typically introduced by insects before the lumber is cut or can infect lumber as it is being seasoned. Sapstain is caused by pigmented fungi that colonize the sapwood and cause a blue-green discoloration. A very common sapstain is caused by the fungus *Ophiostoma* spp. which is transmitted through feeding by bark beetles. There have been considerable efforts to investigate the effects of sapstain fungi on the strength properties of wood, and it has been found that any strength loss is be dependent on wood and fungus types, particularly in tropical and hardwood species. Although the literature remains unclear on the impacts of sapstain on strength properties, early work by Findlay (*12*) found decreases in toughness and hardness but no decrease in bending and compression strength in *Pinus taeda* exposed to blue stain fungi. Wood permeability can also be increased by the presence of sapstain fungi, which can impact coatings and pressure treatment. For more information on blue stain and it's effects on lumber, see MPBI report (*13*) which details much of the work done on blue-stained lumber. Another resource on the effects of sapstains on wood is presented by Mai *et al.* (*14*) and provides an excellent summary of the colonization and effects of sapstain fungi.

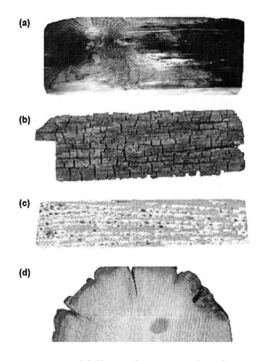

Figure 10. Characteristics of different decay types found on various woods. (a) mold fungi on pine (b) brown rotted pine (c) white rotted maple (d) pole section with soft rot - note the superficial nature of soft rot degradation. (Courtesy of Carol Clausen with permission.)

Yeasts – *Aureobasidum pullulans* is an extremely common fungus that grows on wood. It is currently listed in the ascomycetes, exhibits yeast-like growth and behavior, and is commonly referred to as "black yeast". *Auroeobasidum pullulans* is an important fungus as it can colonize and persist on painted wood surfaces (*15*). The fungus obtains nutrients from the lignin photodegradation products on the surface of the wood and causes major fouling of painted surfaces. This yeast has also been determined to break down water based paints and is a major cause of paint spoilage (*16*). *Aureobasidum pullulans* exhibits resistance to many different chemical compounds and has also been evaluated as a bioremediation tool due to its ability to breakdown organic compounds and withstand metal ions. An image of *A. pullulans* growing on a treated pine piling is shown below (Figure 11).

Figure 11. Aurobasidum pullulans growing on a CCA-treated pine piling in a marina in South Carolina. (Courtesy of Jim Healey, Cox Industries by permission.)

Insects

Insects include a diverse group of organisms capable of causing a wide variety of damage to wood and wood products. Some insects require wood as a major part of their diet while others simply use wood as a substrate for growing fungi to be used as a food source or simply inhabit wood for shelter. Regardless, insects can cause significant damage to wood products in an above ground environment. Beetles are not addressed in this overview as they are typically more common in indoor situations.

Termites – Termites can be a serious problem in above ground exposure. Termites are found nearly worldwide and account for billions of dollars in damage to wooden structures annually, mostly due to subterranean termites. Termites rely on wood as their primary food source, relying on specialized gut microbes that aid in cellulose digestion. Termites are primitive social insects and have an organized hierarchy of castes consisting of workers, soldiers, and reproductives. Workers search, retrieve and bring food back to the colonies to feed the soldiers and reproductives. Caste formation is hormonally controlled so that worker forms can develop into soldiers or reproductive should the colony need more of a given caste (*17*). Economically important termite groups are categorized based on their biology and feeding habits as such:

Drywood Termites (Usually *Incisitermes* spp. or *Cryptotermes* spp.) Drywood termites are typically found along coastal areas and can be a problem in above ground wood products. Drywood termites usually enter structures through attic and foundation vents, fascia boards, and through gaps in windows and doors (*18*). Feeding galleries are smooth and evenly cropped with small pellets in connecting chambers (*19*). These small pellets are a clear indication of drywood termite infestation and are usually found near "kick holes" where the termites eject the frass from their galleries (Figure 12). The ornamentation of the fecal pellets can be so specific that they may be used to identify the species of drywood termite present. Prevention of drywood termite damage is extremely difficult, but proper diagnosis and remedial treatments usually yield good results.

Figure 12. Galleries and frass indicative of western drywood termite, Incisitermes minor. (Courtesy of Whitney Cranshaw, Colorado State. Bugwood.org with permission.)

Subterranean Termites – Subterranean termites are the most common source of insect damage to wood in above and below ground contact. The two most economically important subterranean termite species are:

Reticulitermes flavipes (eastern subterranean termite)- *Reticulitermes flavipes* is native to the US and has a broad distribution. This species causes major damage to wooden structures by excavating the earlywood as it feeds. Members of *R. flavipes* are soft bodied and prone to dessication, so they build shelter tubes out of soil and feces to bridge the soil with their food source (Figure 13). The presence of shelter tubes is often the first indication of an active termite infestation.

Figure 13. Soil tunnels indicative of infestation by subterranean termites. (Courtesy of Rachel Arango, Entomologist, USDA-FS, with permission.)

Common treatments for subterranean termites include soil drenches, barriers, and bait matrices followed by replacement of the damaged wood material. Severe termite damage can cause catastrophic wood failures, especially in cases where the internal damage is extensive and not outwardly visible.

Coptotermes formosanus (Formosan subterranean termite)-*Coptotermes formosanus* is an introduced species that was imported in the 20th century to the southern US mainland and has since spread northward. *C. formosanus* has extremely large colony sizes and can cause catastrophic damage to wooden structures. This species makes "carton material" as a nesting structure from wood/frass/saliva. Carton nests can be found in severely damaged structures and are an adaptation that allows these subterranean termites to carry more soil into higher parts of the structure. This allows them to infest eaves, overheads, and other above ground structures not typically affected by subterranean termites.

Dampwood Termites − Dampwood termites are an occasional problem in wood with excessive moisture. They feed in the earlywood and plug their holes with feces. Dampwood termites typically have rough walled feeding chambers spotted with fecal material. They are an occasional problem in the Pacific Northwest and coastal Florida but are not of major importance on a national scale. Since dampwoods cannot survive in dry wood, eliminating sources of excessive moisture are key to their treatment.

Carpenter Ants – Despite common misconceptions, carpenter ants do not ingest wood but rather excavate it to build their egg chambers, usually in the earlywood bands. They prefer softer wood already degraded by fungi, so the presence of carpenter ants is often indicative of a pre-existing condition brought on by decay or excessive moisture.

Carpenter Bees – Carpenter bees are frequent in above ground wood. They prefer uncoated wood and burrow into the wood to nest. The do not ingest the wood as a food source but simply excavate it to reside in, much like carpenter ants. Carpenter bees lay eggs in these galleries from the inside outward; the brood hatch in reverse order. A diagram of the nesting cavity is shown in Figure 14. The bore holes are not only unsightly, but they also cause disruptions in the surface that can serve as entry points for other deteriorating organisms, such as wood decay fungi.

Figure 14. Cross-section of carpenter bee burrow in wood showing brood chambers. (Courtesy of USDA Forest Service Archive, USDA FS, Bugwood.org with permission.)

Standardized Test Methodologies To Assess above Ground Performance of Treated Wood

In the United States: American Wood Protection Association (AWPA)

Above Ground Tests-AWPA has several standardized test methodologies that can be used to test wood and wood based materials to assess performance of experimental protectants (*20*). They include:

- E9-13: Standard Field Test for Evaluation of Wood Preservatives to be Used Above Ground (UC3A and UC3B); L-Joint Test.
- E16-13: Standard Field Test for Evaluation of Wood Preservatives to be Used Above Ground (UC3B); Horizontal Lap Joint Test.
- E18-13: Standard Field Test for Evaluation of Wood Preservatives to be Used Above Ground (UC3B); Ground Proximity Decay Test.
- E25-13: Standard Field Test for Evaluation of Wood Preservatives to be Used Above Ground (UC3B); Decking Test.
- E27-13: Standard Field Test for Evaluation of Wood Preservatives to be Used Above Ground (UC3B); Accelerated Horizontal Lap Joint Test.

Use Class System-AWPA specifies commodity requirements for all wood products sold in the United States, and assigns the designation of Use Class 3A or 3B to above ground exposed wood products. A listing of approved protectants and target retentions is listed in the Use Class section of the AWPA Book of Standards (20).

Outside of the United States

Internationally, there are additional standards writing organizations that are responsible for developing and maintaining standards for wood products, and they are listed in Table 1 along with sources of more information.

Table 1. Additional International Governing Bodies for Standards for Wood Preservation

Standards Name	Governing Body	Website
EN Standards	CEN	http://www.cen.eu/cen/
Australian Standards	Standards Australia	http://www.standards.org.au/
Japanese Standards	JWPA	http://www.mokuzaihozon.org/english/
Nordic Standards	NWPC	http://www.ntr-nwpc.com/1.0.1.0/2/1/

North American Research Needs for Service Life Prediction in above Ground Use

There has been an international effort to provide more effective methods and models for predicting service life of wood products employed above ground to provide more realistic estimates of the performance of wood and wood based materials in different climates (21–24). Researchers in North America are beginning to make significant advances in this area, but the European research community has a substantial lead in the implementation of models to predict service life.

A major goal of these approaches is to better understand the fungal agents that contribute to the decay process and how they differ in a given region or latitude. When we go back to the Scheffer index (7), we can see that specific hazards exist for a given climatic region but little information is available on the native fungal communities and how they impact wood service life.

Another confounding variable is the incorporation of climate change to the models. It is becoming increasingly clear that climate change is occurring and that increases in mean temperature and carbon dioxide will happen in the near future (*8*, *9*). How these parameters factor into our predictions are currently unknown.

With the increasing volumes of wood plastic composites on the market, a better understanding of the abiotic and biotic factors that effect service life of WPCs needs to be addressed. Lomelia-Romerez et al. (*25*) clearly demonstrated that fungal degradation does occur on WPCs, and several other studies have found similar results. A complete understanding of the fungal specialists that persist on and degrade WPCs are a vital missing component that complicates any real service life estimates of WPCs.

References

1. Evans, P. D. In *Development of Commercial Wood Preservativees: Efficacy, Environmental, and Health Issues*; Schultz, T. P., Militz, H., Freeman, M. H., Goodell, B., Nicholas, D. D., Eds.; ACS Symposium Series 982; American Chemical Society: Washington, DC, 2008; pp 69–117.
2. Evans, P. D. *Wood Mater. Sci. Eng..* **2009**, *4*, 2–13.
3. Williams, R. S. In *Wood Handbook: Wood as an Engineering Material*; U.S. Department of Agriculture: Madison, WI, 1999; Chapter 15, pp 15-1–15-37.
4. Azwa, Z. N.; Yousif, B. F.; Manalo, A. C.; Karunasena, W. *Mater. Design* **2013**, *47*, 424–442.
5. Zobel, B.; Sprague, J. R. In *Juvenile Wood in Forest Trees*; Springer Publishing: Berlin, 1998; pp 300–301.
6. Kim, J. W.; Harper, D. P.; Taylor, A. M. *Wood Fiber Sci.* **2008**, *40*, 519–531.
7. Scheffer, T. C. *For. Prod. J.* **1971**, *21* (10), 25–31.
8. Carll, C. G.; Highley, T. L. *J. Test. Eval.* **1999**, *27*, 150–158.
9. Morris, P. I.; McFarling, S.; Wang, J. *Int. Res. Group Wood Prot.* **2008**, 08–10672.
10. Savory, J. G. *Ann. Appl. Biol.* **1954**, *41* (2), 336–347.
11. Alexopoulos, C. J.; Mims, C. W.; Blackwell, M. In *Introductory Mycology*, 4th ed.; John Wiley and Sons: New York, 1996; 869 pages.
12. Findlay, W. P. K.; Pettifor, C. B. *Forestry* **1937**, *11*, 40–52.
13. Byrne, T.; Stonestreet, C.; Peter, B. In *The Mountain Pine Beetle: A Synthesis of Biology, Management, and Impacts on Lodgepole Pine*; Safranyik, L., Wilson, W. R., Eds.; Natural Resources Canada, Canadian Forest Service, Pacific Forestry Centre: Victoria, British Columbia, 2006; pp 233–253.
14. Mai, C.; Kües, U.; Miltiz, H. *Appl. Microbiol. Biotechnol.* **2004**, *63*, 477–494.
15. Bardage, S. L. *Holz Roh- Werkst.* **1998**, *56*, 359–364.
16. Schoeman, M.; Dickinson, D. *Mycologist* **1997**, *11*, 168–172.
17. Krishna, K.; Weesner, F. M. In *Biology of Termites*; Krishna, K., Weesner, F. M., Eds.; Academic Press: New York, 1969; Vol. I, 600 pages.
18. Gold, R. E.; Glenn, G. J.; Howell Jr., H. N.; Brown, E. Texas A & M Agrilige Extension Pulbication 2005 E-366.

19. Shelton, T. G.; Wheeler, F.; Appel, A. G. Alabama Cooperative Extension System, 2000; ANR 1170, 4 pages.
20. American Wood Protection Association. In *Annual Book of Standards*; American Wood Protection Association: Birmingham, AL, 2013; 647 pages, ISSN 1534-195X.
21. Rapp, A. O.; Peek, R. D.; Sailer, M. *Holzforschung* **2000**, *54*, 111–118.
22. Blom, Å.; Bergström, M. *Wood Sci. Technol.* **2005**, *39* (8), 663–673.
23. Brischke, C.; Bayerbach, R.; Otto Rapp, A. *Wood Mater. Sci. Eng.* **2006**, *1*, 91–107.
24. Råberg, U.; Edlund, M. L.; Terziev, N.; Land, C. J. *J. Wood Sci.* **2005**, *51*, 429–440.
25. Lomelia-Romerez, M. G.; Ochoa-Ruiz, H. G.; Fuentes-Talavar, F. J.; Garcia-Enriquez, S. *Int. Biodeterior. Biodegrad.* **2009**, *63*, 1030–1035.

Chapter 7

Wood Deterioration: Ground Contact Hazards

Robin Wakeling*,[1] and Paul Morris[2]

[1]Beagle Consultancy Limited, 68 Homewood Avenue,
Karori, Wellington 6012, New Zealand
[2]FPInnovations, Durability and Building Enclosure Group, 2665 East Mall,
V6T 1Z4 Vancouver, Canada
*E-mail: robin@beagleconsult.co.nz.

Ground contact poses one of the highest decay hazards that is nonetheless highly variable and not always straightforward to predict. The decay hazard is severe because soil retains damaging moisture for long periods of time, harbours an aggressive preservative depletion hazard and contains a plethora of wood decay microorganisms of a somewhat unpredictable make-up. Relatively few wood products can provide aquate durability in this aggressive environment. Whilst adequate service life for correctly selected preservative-treated wood products is readily achievable, factors other than durability, such as environmental and perceived health profile, increasingly challenge its delivery. This chapter discusses the impact on wood in ground contact of soil-inhabiting wood-rotting basidiomycetes, soft rot fungi, wood degrading bacteria and preservative/extractive detoxifying organisms. It also provides an overview of the ecology of wood decay and other factors that affect product service life in ground contact. It concludes by discussing test methods used to predict the durability of wood in soil and some performance data on treated wood products.

Introduction

Wood in ground contact is subject to a wide range of biodeterioration agents, extractive- or preservative-detoxifying organisms, nutrients and other soil chemicals. Whilst an in-ground decay hazard is almost invariably much greater than an out of ground hazard for any given location, the severity and type of decay

© 2014 American Chemical Society

hazard varies greatly between sites (*1*). Whilst the potential range of in-ground contact fungal decay hazards is very diverse, according to key factors such as soil type, climate, land use and geography, hazards at any one site, or small land area, often comprise a small part of this potential range. Interestingly, the collective diversity of some field test sites at different research centres is also fairly narrow which illustrates the importance of taking account of the effects of the type of site used to test wood preservatives (*2, 3*). Some of these site effects on wood product performance in ground contact are largely predictable and others much less so.

As a general rule the main agents of biodeterioration, in decreasing order of speed of damage, are termites (covered in a separate chapter), disfiguring fungi such as multi-coloured molds and sapstain fungi (no major structural effects), wood-rotting basidiomycetes, soft-rot fungi and wood-degrading bacteria. All of these organisms may colonize a piece of wood in ground contact, but preservative treatment and other influences can affect which one ultimately limits the service life. Untreated wood, or the un-penetrated and poorly penetrated interior of treated wood products, tends to fail to wood-rotting basidiomycetes. However, some hardwoods such as poplar are so susceptible to soft rot, and some soils such as active composts harbour such an aggressive soft rot hazard, that these general rules do not always apply.

Wood treated with less effective preservatives, or low loadings of an effective preservative, tend to fail to preservative-tolerant wood-rotting basidiomycetes. Where wood-rotting basidiomycetes are excluded by more effective preservatives, or other environmental factors such as very high moisture and/or very low oxygen tension, the soft-rot fungi are likely to determine the ultimate life. Where soft-rot fungi are excluded by more extreme environmental conditions, such as water-logging, or bacteria are favoured, such as by alkaline conditions, wood degrading bacteria may limit the service life. For most end use situations untreated wood, even that of naturally durable species (*4*), is either uneconomic, unavailable, or does not last long in ground contact. There are a few exceptions, such as for railway sleepers for which strength and toughness of naturally durable tropical hardwoods is difficult to duplicate with available wood species that can be preservative treated.

Wood-Rotting Basidiomycetes

The initial colonization of wood in ground contact may be via spores, mycelium or mycelial strands, or rhizomorphs, or a mixture thereof. Colonization of wood in ground contact by spores can occur in much the same way as colonization of wood above ground. Whilst spores may be the primary mode of dissemination of an invading basidiomycete decay fungus in some situations such as away from the local effects of a forest, or other vegetation influence, actual colonisation of the wood product may occur via prior mycelial strand formation. In other words, any adjacent woody material acts as a bridgehead for the invading spores where there is a build-up of inoculum potential just before colonisation of the wood product. Garrett (*5*) defined inoculum potential as the "energy of growth of a fungus (or other microorganism) available for colonization of a substrate at the surface of the substrate to be colonized". The potency of this inoculum potential has a profound effect on the success, or failure, of a basidiomycete to decay a wood product (*1*).

Mycelium of wood-rotting basidiomycetes can be found in soil but it typically grows from buried wood (*1*), or other lignocellulosic material and is not evenly distributed over large areas. This can result in rapid initial surface decay which can appear serious but often fails to progress further in wood with sufficient preservative loading as observed by one author (Morris). The larger the volume of the actively decaying adjacent woody material, the greater the likelihood of ongoing and potentially damaging decay to nearby wood products.

Whilst some white rots in grasslands may have a more homogeneous distribution compared to brown rots, they tend not to have such profound effects on performance compared to wood decay fungi, usually brown rot fungi, that originate from more persistent woody material. However, in at least some highly prolific, unmanaged and non-grazed grassland, white rot assumes a much greater significance as a cause of failure of variously preservative treated wood test stakes (*1, 2, 6*). This was in part attributed to the build-up of a white-rot inoculum potential in the soil caused by the prolific deposition of thick woody grass stems. It is likely that a similar phenomenon occurs in other situations wherever there is an abundance of sufficiently woody lignocellulosic plant material.

According to Baldrin (*7*), mycelium of only a few of the well-known wood-rotting basidiomycetes compete well against other microorganisms in soil, and Dowson *et al*. (*8*) recorded that some wood-rotting basidiomycetes form more resistant mycelial strands, not evenly distributed but often found over a much wider area. The largest recorded organism on earth is a soil-inhabiting, strand-forming (rhizomorph-forming) wood-rotting basidiomycete: a colony of *Armillarea* ostoyae which permeates a 10 sq.km.area of forest in Oregon (*9*). A limited number of soil-inhabiting wood-rotting basidiomycetes such as *Wolfiporia cocos* (*10*) and *Leucogyrophana pinastri* (*11*) produce sclerotia, resistant resting structures, that can remain dormant in soil for years and provide a high inoculum potential for colonization if wood is placed into that soil. However, these observations and others probably reflect the specifics of the local environment and may not point clearly to what most likely drives the competitive saprophytic ability of basiodiomycete decay fungi as it affects wood product performance in ground contact.

As a general rule, the competitiveness of a decay fungus mycelium relative to other microorganisms, and its concomitant ability to cause the predominant decay damage in a wood product in ground contact, is largely a function of the frequency of occurrence of nearby rotting woody materials that are of sufficient size to confer a competitive advantage to wood decay fungi that are inherently suited to causing decay in the wood product in question. However, the precise nature of nearby rotting woody materials is likely to have a profound effect. For example, addition of finely divided sawdust would probably not favour a basidiomycete because less specialised deuteromycetes and ascomycetes with shorter life cycles and wider distribution would most likely cause soft rot before basidiomycetes could deploy their specialised biochemistry that would otherwise confer a high competitive saprophytic ability in other situations. In other words, basidiomycetes predominate wherever they are advancing from relatively large centres of actively decaying material, often material that is nearing the end of the decay process, such as dead tree roots, fallen branches and any other suitably-sized woody material.

Copper-tolerant brown rot fungi are consistently found in softwood or gymnosperm forests and at some test sites where a critical mass of soil inoculum potential has built up (*1, 2, 6*) but outside these locations it is difficult to predict where they will show up. Whilst the biochemical apparatus that confers copper-tolerance is clearly a necessary part of the phenomenon of premature failure caused by copper-tolerant fungi, it is likely that it is secondary to the primary requirement of possession of a high inoculum potential in the form of mycelium, or mycelial strands of wood decay fungus, or fungi in the soil, or nearby woody substrate. This is apparent when one considers that there are very few examples of premature failure by copper tolerant fungi in ground contact but where actively growing copper-tolerant decay fungus was not already in the vicinity of the wood product, or out of ground contact. Choi *et al.*, (*12*) and Woo *et al.*, (*13, 14*) found that at least two fungi that are copper-tolerant in the mycelial form do not have copper tolerant basidiospores.

It is likely that possession of a copper-tolerant ability is fairly common amongst brown-rot fungi given the common occurrence of the biochemistry that has been attributed to this phenomenon amongst brown rot fungi, but that the situations where it can be expressed are relatively small because the prerequisite for a high soil inoculum potential is absent from most situations where wood products are used. For example test sites known to harbour copper tolerance (*1, 2, 6, 15–20*) and softwood forests that naturally harbour an abundance of aggressive brown rot fungi (*21*) consistently cause rapid failure of softwood stakes containing ground contact retentions of copper containing wood preservatives whereas sites chosen at random, or for other attributes, only rarely exhibit rapid failure from copper-tolerant wood decay fungi. Their presence is commonly reported when incidents of premature failure of treated wood are brought to the attention of scientists in the field (*22–25*).

A graphic example of the likely consequences of this for wood product performance was observed by Wakeling and Singh (*26*) where large numbers of CCA-treated vineyard trellis posts failed to brown rot fungi within 3–5 years. This was almost certainly the result of a high prevalence of rotting wood in the soil that had been a pine forest immediately prior to introduction of viticulture. As the amount of rotting woody debris declined, the wood preservative tolerance of the site declined, so that in the next five years the site started to behave more like a classic loam soil. Most likely, a soil containing actively rotting softwood will include fungi with sufficient copper tolerance to cause rapid decay of a softwood product containing a ground contact retention of a copper containing preservative. Observations suggest that under such conditions all commonly used copper-containing preservatives are susceptible at ground contact retentions. Preservative retention over and above recommended ground contact retentions and the type of secondary fungicide used has an effect on the severity of brown rot but once the copper has been neutralised it is unlikely that the secondary biocide will be effective by itself.

Copper tolerance amongst white rot fungi is less well understood which may, or may not, reflect a lower incidence compared to brown rot fungi. Certainly, where conditions for brown rot copper tolerance occur, then it is a much more potent and damaging phenomenon compared to white rot copper tolerance. Brown rot fungi are further along the evolutionary pathway (*21*) in that the mechanism of decay is more highly specialised and has thrived because it is a unique and effective way of bypassing the protective effect of recalcitrant lignin on energy rich hemicellulose and cellulose (*26, 27*). However, it comes at a price, in that the fungi that possess it have become more specialised and occupy narrower ecological niches, or microcosms, making them somewhat less adaptable to give them a narrower distribution. In other words, white rot fungi as a whole (*1*), and some white rot fungal species in particular such as *Schizophyllum communae* (*21*), are more tolerant of a wide range of environmental and substrate conditions, giving them a higher competitive saprophytic ability in a much wider range of geographical locations compared to brown rot fungi. This is why brown rot fungi are less numerous than white rot fungi (*21*) and why brown rot decay has a narrower distribution in nature compared to white rot.

Brown rot fungi adopt something of a scorched earth approach to invasion of new woody materials. Typically, they come, conquer and leave (and then die off, sometimes after dissemination via a fruiting body). In contrast, white rots are contained in many types of lignocellulosic debris such as grass which is of course very common, and therefore they tend to be more prevalent in the first place before a wood product is deployed, and they linger for much longer periods because their decay rate is much slower and they can persist in a wider range of more finely divided plant based material. If this were not the case, copper-based wood preservatives would not have enjoyed the success achieved because premature failure to copper-tolerant brown rots would be much more common. In addition to the more well-known copper-tolerant fungi, some wood-rotting basidiomycetes are also tolerant to certain carbon-based preservatives (*13, 14*). Only a few, such as *Tyromyces palustris,* are known to be tolerant to both (*13, 14, 28*).

Whilst basidiomycete brown rots are more typically associated with premature failure, particularly rapid failure of preservative treated softwoods than basidiomycetes that cause white rot, nonetheless white rot is a common cause of more gradual decay in preservative treated wood (*1*). White rot fungi typically do not cause premature failure of correctly preservative treated softwoods but tend to chip away at the outer regions of wood in ground contact as depletion or deactivation of preservative occurs. Presence of more invasive white rot is often a sign that wood was not correctly treated.

Soft Rot-Fungi

The fungi which are capable of causing classical soft-rot cavities, erosion or tunnels are much more widely distributed in most soils than the wood-rotting basidiomycetes. Whilst occurrence of soil-fungi with a soft rot capability are well documented, the significance of the various fungal species cited in terms of their ability to cause degradation of wood products is poorly understood. Nevertheless, it is almost certain that wood introduced into the ground will be colonized to some extent by one or more soft-rot fungi and that at some point in time will likely suffer from soft rot decay. There is a high probability that soft rot will go on to cause significant decay damage and eventually leading to failure, if conditions do not favour basidiomycete decay fungi.

Whilst soft rot fungi typically have a shorter reproductive life cycle than basidiomycetes, are more numerous and probably arrive and proliferate before basidiomycetes, except in forest soils and a few other situations, the subsequent rate of decay caused is typically much slower than brown rot in softwoods but is closer to the rate of decay caused by white rot. This is necessarily a general rule that has some significant exceptions. Because basidiomycetes that cause brown rot and white rot are typically more highly specialised at wood decay than soft rot fungi and are therefore often more competitive, once they have arrived they have the ability to overtake and supplant soft rot fungi. The main exceptions to this, and where soft rot fungi acquire a competitive advantage, include some environmental situations such as very high moisture, or where a substrate factor such as the presence of a wood preservative that is relatively more effective against basidiomycetes, or where a hardwood is especially susceptible to soft rot, or where a particular wood-preservative-environment interaction effect occurs.

Soft-rot fungi grow relatively slowly in the softwoods used for structural applications in many parts of the world, but they can develop rapidly in hardwoods even when preservative treated to loadings that would protect softwoods (*29*). Many soft rot fungi are tolerant of copper and other wood preservative active ingredients (*30–32*). The greater susceptibility of preservative-treated hardwoods to soft rot compared to softwoods is in part due to the difference in lignin chemistry but more importantly in the current context it is related to the different challenges of achieving even penetration of preservative within the wood ultrastructure and across wood cell walls. However, this is a reflection of what has been the status quo with regards the relatively small number of wood-preservative combinations that have dominated the wood products sector. As we introduce new systems, this could change. Soft rot damage typically proceeds slowly inwards from the soil/wood interface, but soft-rot fungi have been known to grow through wood treated to ground-contact preservative loadings and decay the inside of utility poles which have lower loadings (*33*).

Wood Degrading Bacteria

In much the same way that soft rot decay was overlooked, or at least not recognised for what it was and reported for many years, so too was decay caused by bacteria overlooked. In fact this situation has to some extent persisted such that it is a decay type that is far more common than the literature would suggest, largely because only a few workers have been especially inclined towards studying these somewhat elusive microorganisms (*34, 35*). The causative bacteria have not been identified but evidence suggests that they possess some unusual characteristics which in part explains why attempts to identify them have been unsuccessful. In particular, their pleomorphism and mode of movement within wood cell walls sets them apart from the great majority of other bacteria. They most likely occupy a somewhat narrow ecological niche which also accounts for the paucity of knowledge available, few researchers having had cause to encounter them.

In keeping with bacteria that cause biodeterioration in other situations, wood degrading bacteria more commonly occur where pH is close to neutral or is alkaline. In the current context, soil type and preservative type therefore have a profound effect on occurrence (35). They have been found to be prevalent in preservative treated wood exposed in alkaline soils and in wood treated with alkaline copper amine based preservatives (36), although soil alkalinity appeared to have the strongest effect on prevalence (1). There are relatively few in-service situations where wood degrading bacteria assume a significance similar to, or greater than wood decay fungi. Cooling towers and water-logged building supports are two examples were wood degrading bacteria dominate over most fungi, or at least compete favorably. Tolerance of wood degrading bacteria to CCA treated pine in service (1, 37) and to copper amine triazole and ammoniacal copper treated pine in field studies (1) has been suggested, although the rate of decay is slow and has not been linked convincingly to premature failure.

Whilst wood degrading bacteria do indeed occur where water activity is very high this does not necessarily mean that they require such conditions, or that they favour such conditions. Research of Wakeling (1) suggested that tunnelling bacteria decay occurs at moderate to low moisture contents also. These studies suggested that if more rapid and otherwise more competitive wood decay fungi were held in check by some factor such as undesirable pH, moisture content, or preservative, etc., then this allowed wood degrading bacteria to predominate, albeit most likely causing relatively slow rates of decay. Whilst they appear to be relatively tolerant of CCA and other copper-based wood preservatives it is not clear if they possess a real tolerance, or if they simply occur in conditions that promote rapid preservative depletion. Their mode of wood colonisation may contribute to a tolerance of wood preservative. For example, production of copious amounts of extracellular exudates, e.g., polysaccharides, and their tunnelling mechanism may allow them to limit contact with biocides.

It is to be expected that unicellular wood degrading bacteria would predominate over filamentous fungi in an essentially aquatic environment, or where water activity is very high, and evidence does to some extent support this. For example, prevalence in saturated wood in cooling towers alongside soft rot and some types of heart rot white rot basidiomycetes also known for their tolerance of high moisture and/or low oxygen, supports this. Initial rates of bacterial decay in water cooling tower diffuser slats was much higher than subsequent decay rates (1), indicating that the limiting factor was oxygen availability. In other words, once a certain depth of decay had been attained, possibly within preservative depleted wood, low oxygen availability in deeper sound wood kept the decay at bay until such time as distance was reduced by sloughing off of the decayed wood layer. Such a mechanism of wood decay retardation is somewhat analogous to the mechanism of longevity of wooden piles that support the buildings of Venice and other structures.

There are two, or possibly three, types of wood degrading bacteria. Tunnelling bacteria and erosion bacteria are quite different both in morphology and the type of decay caused and to some extent the environment where they occur. Decay caused by erosion bacteria typically occurs where very high moisture most likely causes microaerophilic conditions whereas tunnelling bacteria decay is much more common across a wider range of moisture conditions. A third type called cavitation (*35*) may be a form of erosion (*1*), or at least shares many of the features of erosion bacteria decay.

The Ecology of Wood Decay and Other Factors That Affect Wood Product Service Life in Ground Contact

The ecology of wood decay is a very large, and potentially complex subject which has been dealt with adroitly by Rayner and Boddy (*21*). Whilst is it essentially outside the scope of this chapter to go into any detail, it is nonetheless important to touch on a least some aspects of how the ecology of any given in-service situation affects the longevity of a wood product. Perhaps the most important point is that decay in preservative treated wood in ground contact and the concomitant service life of any given wood product is determined by scenarios of widely varying complexity. The consequence of this is that in some situations the service life of a given wood product is likely to be highly predictable, but in others it is much less so. The greater the reliance on wood products with a long-established track record of many decades, the greater the reliability of predicted performance in varying in-ground situations. However, where greater reliance is placed on newer types of wood preservatives, the reliability of predicted performance becomes more complicated and this is especially the case for diverse in-ground contact situations that are difficult to simulate using rapid testing methodology and any method that employs testing timeframes less than 5–10 years of suitable field testing (*1*).

Since copper-tolerance clearly has a profound effect on the life expectancy of copper-treated wood in ground contact, and since copper continues to be the most important active for protection of softwood in ground contact, it is well worth considering what determines copper tolerance in a little more detail and how this may change as copper-based wood preservatives evolve. It is important to point out that copper-tolerance is not necessarily an all or nothing phenomenon. A ground contact retention of a copper chrome arsenate (CCA) preservative contained within a suitably treatable softwood, normally an outstandingly durable combination, often giving 50–100 years of service life, will inevitably fail within a couple of years if it is unlucky enough to encounter the rarefied conditions where a copper- and arsenic-tolerant brown rot possesses a sufficient inoculum potential. This is essentially an all or nothing effect with devastating consequences for the wood product in question, but it is rare. Part of this rarity is not just the low incidence of situations where copper tolerant fungi possess sufficient inoculum potential to overcome what is otherwise a robust preservative-wood combination but is in part due to the inherent robustness of the more commonly used traditional copper-based wood preservative systems, largely CCA-based.

Other copper-based preservative systems such as discontinued acid copper chromate and copper borate, and more recent ammoniacal copper plus quaternary ammonium compound combinations and amine copper plus triazole (propiconazole plus tebuconazole) are not as resistant to copper-tolerant decay fungi as CCA (*1*). In other words copper-tolerance can assume a much greater significance if the inherent robustness of the copper-based preservative is substantially weaker than CCA. The tipping point for rapid decay and wood product failure caused by copper-tolerance is therefore likely to assume greater significance if, as is likely, the inherent robustness of modern copper-based preservatives is less than for CCA-treated pine.

For more recently introduced wood preservatives in particular, the inherent robustness of copper against copper-tolerant decay fungi is not necessarily straightforward to predict due to the complexity of the ecological interactions that define longevity across different in-ground exposure situations and this is where further research is likely to be of particular benefit. Any change in preservative formulation that affects the distribution of copper within the ultrastructure of wood and across wood cell walls, and any other factor that affects the chemistry and bioavailability of copper at the fungus-wood interface, is likely to be of particular importance. Over-reliance on accelerated testing, or artificially manipulated field sites designed to increase particular hazards such as from copper tolerant brown rot, or reliance on efficacy data less than 5 years in the making may miss important in-service exposure parameters (*1*).

Copper-tolerance is more typically linked to premature failure of softwood caused by copper-tolerant brown rot, and copper tolerant soft rot fungi have more typically been linked to premature failure of hardwoods, e.g., Australian eucalypt. However, there are situations were softwood can fail prematurely to copper tolerant soft rot fungi such as premature carrot-fracture of CCA-treated pine posts (*unpublished observations of Wakeling and* (*1*)). The mechanism of failure caused by soft rot is more likely to be a reflection of less effective copper distribution within the ultrastructure of woody tissues and within the microstructure of the wood cell walls, rather than possession of unique fungal biochemistry as is the case for copper-tolerant brown rot. One of the main differences between failure due to copper-tolerant soft rot and copper-tolerant brown rot is that failures due to soft rot typically take 1–2 decades to show up, almost ten times longer than is the case for brown rot. Furthermore, soft rot decay is not easy to study under laboratory conditions which means that it is likely to take longer to produce realistic predictive data. Whilst tropical soil bed tests typically include an aggressive soft rot challenge, they do not necessarily include the optimal combinations of microorganisms and soil conditions, or the optimal timing and order of their arrival, to promote failure to copper-tolerant soft rot.

Decay that affects wood in ground contact cannot necessarily be easily divided into brown rot, white rot, soft rot and bacterial decay. Mixtures of decay types caused by several fungal species may come together to determine the service life as determined by the ecology of the situation. Wakeling ((*1*), *page 52*) listed 25 publications which referred to the occurrence of fungal species associated with intermediate decay types, or decay fungi which produced a combination of micromorphological features typically considered to be attributable to two, and in some cases three, types of decay. It appeared that wood species and environmental factors were at least in part responsible for expression of different decay types by the same fungus, although sometimes features of more than one type of decay occurred in the same sample of wood. Rayner and Boddy (*21*) discussed the profound affect that extraneous factors have on the colonisation and decay strategy adopted by decay fungi. The point is, what determines decay type and ultimately the service life of a wood product can be complex and difficult to predict because there are many things about in-ground decay hazards that are incompletely understood.

It is likely that environmental conditions and the physico-chemical properties of the wood substrate affect the degree of expression of the various genes that encode the biochemistry associated with wood decay. Decay that is recorded in wood in ground contact is inevitably a consequence not just of the causative decay fungus, or bacterium, but is the combined result of the wood product type and the environment, and their selective effect and subsequent interaction with the causative fungus or bacterium. Soil physico-chemical properties and water availability are of particular importance.

There have been a limited number of systematic studies and a few publications on specific soil factors causing unexpectedly high rates of decay. An opportunity was missed with the collaborative hardwood field experiment due to the paucity of reporting of useable long-term data to the co-ordinators (*38*). Identically treated material had been installed in 34 test sites across the world but after 10 years only 16 participants had provided 5 years-worth of data. On a more limited geographic basis, Wakeling (*1*) installed test material at 13 test sites within New Zealand and Australia. Thousands of test stakes comprising 20 replicates each of four retentions of four preservative types, for each of two wood species (*Pinus radiata* and *Fagus sylvatica*), were exposed at each of 13 test sites in New Zealand and Australia, chosen on the basis of having widely differing soil and climate types. Site location, climate, soil type and decay type occurrence had very highly significant effects on performance of variously preservative treated radiata pine and European beech. Similar findings were made in an earlier study by Wakeling (*39*) which involved much older and similarly diverse preservative treated test material 11–15 years old. One of the overarching findings was that approximately half a dozen carefully chosen test sites were necessary if the various strengths and weaknesses of different wood species-wood preservative combinations were to be understood correctly, and that it took approximately 5–10 years of exposure for these results to emerge.

This is not to say that lesser degrees of field study cannot provide most of the information needed for robust service life prediction most of the time, but it was clear that there was the potential to overlook important interaction effects (ecological effects) that could cause premature failure if inadequate testing and/or inadequate knowledge and experience was brought to bear. It was however reasonably clear that over-reliance on laboratory test data alone, or in combination with artificially accelerated tropical soil bed tests (fungus cellar tests) would be risky and would, sooner or later cause premature failure if traditional life expectancy expectations are retained.

Soil factors credited with increasing the depletion of wood preservatives in ground contact include organic acids (*40, 41*) and waterlogging (*41*). In some cases widely differing soils can have similar rates of depletion (*42*). Other soil factors found to increase the decay rate of treated wood in ground contact include intensive horticultural amendments (*37*), high water-holding capacity soils (*43*), and ferrous iron mobilized by anaerobic conditions (*44*) which detoxifies the arsenic in CCA (*36, 45*), and iron oxide deposits in drainage ditches and seepage areas. The possible impact of calcium uptake from soil or concrete (around fenceposts) in accelerating decay of treated wood in ground contact has not been conclusively proven (*37, 38, 46, 47*). However, Schultz and Nicholas (*48*) found that the pH of wood increases in basic soils, which may favour bacterial growth.

Soil factors can have a greater influence than climatic factors on decay rate of wood treated with certain preservatives, or naturally durable woods. Morris and Ingram (*15*) reported that a test site at Westham Island British Columbia had a decay rate faster than many tropical test sites, based on data from the collaborative hardwood test (*39*). Morris *et al.*, (*4*) found greater decay rates for end-matched naturally durable stakes in a field site in a forest clearing in Ontario with an updated Scheffer Index of 48 (*49*) than in a field site in a forest clearing in British Columbia with an updated Scheffer Index of 63 (erroneously reported as 55). However, these may be specific preservative/site interactions. When data were averaged from thirteen different preservatives, Wakeling (*6*) found a general trend of more rapid decay at warmer and wetter tropical sites compared to cooler and particularly dryer temperate sites. Month to month weather conditions can also affect populations of basidiomycetes colonizing wood in ground contact (*50*).

Detoxifying Organisms

In addition to the agents of biodeterioration in soil, there are many other organisms capable of growing in wood that do no structural damage but may affect the resistance of that material to decay. Crawford and Clausen (*51*) showed that a common soil bacterium, *Bacillus licheniformis,* was capable of removing copper from treated wood. Choi (*52*) found decay in CCA treated decking was associated with the presence of copper-tolerant non-decay fungi commonly found in soils. Certain *Penicillium* species and other mold species produce organic acids which strongly chelate copper (*32, 53, 54*). Copper can become strongly bound to melanin (*55*) and a wide range of fungi with melanised cell walls can colonize wood treated with copper-based preservatives with carbon-based (organic in the strict sense) co-biocides (*56*). With the advent of this type of preservative and the interest in developing metal-free waterborne preservatives for ground contact, biodegradation of carbon-based biocides becomes even more important. Dubois and Ruddick (*57*) identified fungi capable of detoxifying quaternary ammonium compounds. Obanda and Shupe (*58*) showed biodegradation of tebuzonazole by a mold, a soft-rot fungus and a bacterium and suggested a common mechanism. Cook and Dickinson (*59*) identified a wide range of bacteria capable of degrading carbon-based biocides. Wallace and Dickinson (*60*) suggested that bacteria may have a common single mechanism for such biotransformations. Kirker *et al.*, (*61*) found higher richness and diversity of bacteria in treated stakes than untreated stakes during the first 6 months of ground contact, decreasing around the same time as the preservatives were found to undergo depletion. Just as with wood preservatives, the extractives that confer natural durability are also vulnerable to biodegradation in ground contact (*62*). All of these detoxification processes can make treated or naturally durable wood vulnerable to decay by wood-rotting fungi that are not tolerant of the preservative or extractives present in the wood.

Another agent of biodeterioration that can shorten the service life of wood in ground contact is *Homo sapiens* L. Bolt holes drilled for below-ground cross pieces to prevent frost heave permit access for soil fungi through the treated zone. The damage we do by clearing weeds around fenceposts with a line trimmer removes treated surface zones, exposing more vulnerable wood underneath. By adding layers of mulch, humans raise soil levels to contact such damaged posts or cover over concrete foundations designed to separate wood posts from soil. Sometimes the homeowner is unknowingly the decay fungus' best friend.

Methods of Predicting the Durability of Wood in Soil

Standard field test methods of evaluating the durability of wood in soil in North America include the AWPA E7 stake test (*63*) and the AWPA E8 post test (*64*). The AWPA E14 soil-bed test (*65*) is a method designed to simulate field exposure under more controlled conditions with soil moisture content and temperature constantly conducive to decay. A much smaller-scale laboratory unsterile soil test AWPA E23 (*66*) does not require costly infrastructure and is even more controlled, but perhaps less realistic since the inoculum is confined to the organisms present at the time the test was set up. The AWPA E10 soil-block test (*67*), despite having the word soil in the title, does not provide data predictive of field performance. This is because the soil is sterilized and thus contains no live detoxifiers, wood-degrading bacteria or soft-rot fungi. Furthermore, the test wood block is separated from the soil by a feeder strip reducing the potential influx of nutrients and other soil chemicals. The AWPA E10 test is a pure-culture basidiomycete screening test. None of these tests guarantees exposure of treated wood to chemical and biological detoxification followed by basidiomycete attack. This may occur in a field test if the stakes are in an area of the test site permeated by a soil-inhabiting wood-rotting basidiomycete. Otherwise the probability of infection by basidiomycetes is low because wood stakes are not very receptive to colonization by spores (except where spores are entrapped in cracks above ground and cause decay in the above ground part). Some laboratory work has been done in this area (*68*) and further work is underway at FPInnovations (*unpublished*) and at Mississippi State University (*69*) to address this gap in test methods.

Despite all the hazards encountered by wood in ground contact and the short service life of untreated wood products in these applications, properly treated wood can provide a service life that meets consumer demands.

USA utilities responding to a survey (*70*) estimated southern pine, lodgepole pine and Douglas fir poles will last 30–40 years and treated western red cedar poles will last 51–70 years. However, Morrell (*71*) concluded that replacement rates of 0.6% would place average service life in excess of 80 years in many parts of the country. Stewart (*72*) and Nelson (*73*) estimated that while pole users estimated pole life at 30 to 40 years, replacement rate data suggested average lives could be 135 years or more, if only ground line decay were considered. However, he noted other degradation mechanisms come into play as poles age that can affect their life, such as decay of pole tops and connections, splitting of pole tops (preventable problems), and excessive weathering. To this list we would add climbing damage from spurs. Morris *et al.,* (*19*) found lodgepole pine posts with CCA loadings below the pole standard with an average life well over 60 years at a test site in Petawawa, Ontario with a substantial population of soil-inhabiting, strand-forming, wood-rotting basidiomycetes. Remedial treatment in service by application of preservative-containing bandages can extend the life of wood poles by decades. Data from Osmose Utilities inspection and re-treatment program (*74*) showed that if poles with a normal life of 50 years are regularly inspected and remedially treated they can have a service greater than 100 years. Clearly ground contact decay hazards are not insurmountable given suitable wood material, an effective preservative and an effective treatment process.

References

1. Wakeling, R. Ph.D. Thesis, University of Waikato, Hamilton, New Zealand, 2003.
2. Wakeling, R. N. *Int. Res. Group Wood Preserv.* **2006**, Doc. No. IRG/WP/06-20327.
3. Wakeling, R. N. *Int. Res. Group Wood Preserv.* **2006**, Doc. No. IRG/WP/06-10587.
4. Morris, P. I.; Ingram, J. K.; Larkin, G.; Laks, P. *For. Prod. J.* **2011**, *61* (5), 344–351.
5. Garrett, S. D. *Biology of Root-Infecting Fungi*; Cambridge University Press: New York, 1956; pp 293.
6. Wakeling, R. *Int. Res. Group Wood Preserv.* **2001**, Doc. No. IRG/WP/01-20231.
7. Baldrin, P. *Fungal Ecol.* **2008**, *1* (1), 4–12.
8. Dowson, C. G.; Rayner, A. D. M.; Boddy, L. *J. Gen. Microbiol.* **1986**, *132*, 203–211.
9. Ferguson, B. A.; Dreisbach, T. A.; Parks, C. G.; Filip, G. M.; Schmitt, C. L. *Can. J. For. Res.* **2003**, *33*, 612–623.
10. DeGroot, R.; Woodward, B. *Mater. Org.* **1998**, *32* (3), 195–215.
11. Ginns, J. *Can. J. Bot.* **1978**, *56*, 1953–1973.
12. Choi, S. M.; Ruddick, J. N. R.; Morris, P. I. *Int. Res. Group Wood Preserv.* **2002**, Doc. No. IRG/WP/02-10422.
13. Woo, C.; Daniels, C. R.; Stirling, R.; Morris, P. I. *Int. Biodeterior. Biodegrad.* **2010**, *64* (5), 403–408.
14. Woo, C. S.; Morris P. I. *Int. Res. Group Wood Preserv.* **2010**, Doc. No. IRG/WP/10-10707.

15. Morris, P. I.; Ingram, J. K. *Proc. Can. Wood Preserv. Assoc.* **1991**, *12*, 54–78.
16. Preston, A.; Jin, L.; Nicholas, D.; Zahora, A.; Walcheski, P.; Archer, K.; Schultz T. *Int. Res. Group Wood Preserv.* **2008**, Doc. No. IRG/WP/08-30459.
17. Nicholas, D. D.; Schultz, T. P. *Int. Res. Group Wood Preserv.* **2003**, Doc. No. IRG/WP/03-30305.
18. Clausen, C.; Jenkins, K. M. *Chronicles of Fibroporia radiculosa (=Antrodia radiculosa) TFFH 294*. USDA Forest Products Laboratory General Technical Report FPL-GTR-204; U.S. Department of Agriculture, Forest Service, Forest Products Laboratory: Madison, WI, 2011; 5 pages.
19. Morris, P. I.; Ingram, J. K.; Stirling, R. *Proc. Am. Wood-Preserv. Assoc.* **2012**, *108*, 171–181.
20. Raberg, U.; Terziev, N.; Daniel, G. *Int. Biodeterior. Biodegrad.* **2013**, *79* (4), 20–27.
21. Rayner, A. D. M.; Boddy, L. *Fungal Decomposition of Wood: Its Biology and Ecology*; Wiley Interscience Publication: New York, 1988.
22. Levi, M. P. In *Biological Transformation of Wood by Microorganisms*; Liese, W., Ed.; Springer-Verlag: Berlin, 1975.
23. DeGroot, R.; Woodward, B. *Mater. Org.* **1998**, *32* (3), 195–215.
24. Raberg, U.; Daniel, G. *Int. Biodeterior. Biodegrad.* **2009**, *63* (7), 906–912, 27.
25. Bollmus, S.; Rangno, N.; Militz, H.; Gellerich, A. *Int. Res. Group Wood Preserv.* **2012**, Doc. No. IRG/WP/12-40584.
26. Wakeling, R. N.; Singh, A. P. *Int. Res. Group Wood Preserv.* **1993**, Doc. IRG/WP/93-10016.
27. Eastwood, D. C.; Floudas, D.; Binder, M.; Majcherczyk, A.; Schneider, P.; Aerts, A.; Asiegbu, F. O.; Baker, S. E.; Barry, K.; Bendiksby, M.; et al.; S., C. *Science* **2011**, *333* (6043), 762–765.
28. Green, F.; Arango, R. A.; Clausen, C. A. Unique preservative tolerance in Tyromyces palustris. IAWPS 2005: 50th Anniversary of the Japan Wood Research Society: International Symposium on Wood Science and Technology, Pacifico Yokohama, Japan, November 27–30, 2005; Japan Wood Research Society, 2005; p 304–305.
29. Greaves, H.; Nilsson, T. *Holzforschung* **1982**, *36* (4), 207–213.
30. Lundström, H. *Stud. For. Suec.* **1974**, 115.
31. Daniel, G.; Nilsson, T. *Int. Biodeterior. Biodegrad.* **1988**, *24* (4–5), 327–335.
32. Bridžiuviene, D.; Levinskaite, L. *Biologija* **2007**, *53* (4), 54–61.
33. Friis-Hansen, H.; Lundström, H. *Int. Res. Group Wood Preserv.* **1989**, Doc. IRG/WP 1398.
34. Nilsson, T.; Daniel, G. *Int. Res. Group Wood Preserv.* **1983**, Doc. No. IRG/WP/1186.
35. Singh, A. P.; Butcher, J. A. *J. Inst. Wood Sci.* **1991**, *12* (3), 143–157.
36. Morris, P. I.; Ingram, J. K. *Proc. Can. Wood Preserv. Assoc.* **1998**, *19*, 27–46.
37. Butcher, J. *Int. Res. Group Wood Preserv.* **1984**, Doc. No. IRG/WP/1241.

38. Gray, S. M.; Dickinson, D. J. *Int. Res. Group Wood Preserv.* **1989**, Doc. No. IRG/WP/3560.
39. Wakeling, R. N. *Int. Res. Group Wood Preserv.* **1991**, Doc. No. IRG/WP/2034.
40. Cooper, P. *For. Prod. J.* **2001**, *51* (9), 73–77.
41. Wakeling, R. *Int. Res. Group Wood Preserv.* **2008**, Doc. No. IRG/WP/08-30460.
42. Schultz, T. P.; Nicholas, D. D.; Pettry, D. E. *Holzforschung* **2002**, *56* (2), 125–129.
43. Hedley, M. E.; Drysdale, J. A. *Int. Res. Group Wood Preserv.* **1984**, Doc. No. IRG/WP/1225.
44. Ruddick, J. N. R.; Morris, P. I. *Wood Protection* **1991**, *1* (1), 23–29.
45. Morris, P. I. *Material und Organismen* **1994**, *28* (1), 47–54.
46. Murphy, R. J. *Int. Res. Group Wood Preserv.* **1983**, Doc. No. IRG/WP/3221.
47. Hastrup, A. C. S.; Jensen, B.; Clausen, C.; Green, F. *Holzforschung* **2006**, *60*, 339–345.
48. Schultz, T. P.; Nicholas, D. D. *Wood Fiber Sci.* **2010**, *42* (3), 412–416.
49. Morris, P. I.; Wang, J. *Int. Res. Group Wood Preserv.* **2008**, Doc. No. IRG/WP/10672.
50. Kirker, G. T.; Prewitt, M. L.; Diehl, S. V. *Int. Res. Group Wood Preserv.* **2007**, Doc. No. IRG/WP/07-30429.
51. Crawford, D. M.; Clausen, C. A. *Int. Res. Group Wood Preserv.* **1999**, Doc. No. IRG/WP 99-20155.
52. Choi, S. Ph.D. Thesis, University of British Columbia, 2004.
53. Murphy, R. J.; Levy, J. F. *Trans. Br. Mycol. Soc.* **1983**, *81* (1), 165–168.
54. Kartal, N.; Katsumata, N.; Imamura, Y. *For. Prod. J.* **2006**, *56* (9), 33–37.
55. Gadd, G. M. *New Phytol.* **1993**, *124*, 25–60.
56. Bech-Andersen, J.; Elborne, S. A. *Int. Res. Group Wood Preserv.* **1994**, Doc. No. IRG/WP/94-10078.
57. Dubois, J.; Ruddick, J. N. R. *Int. Res. Group Wood Preserv.* **1998** Doc. No. IRG/WP/98-10263. 10p.
58. Obanda, D. N.; Shupe, T. F. *Wood Fiber Sci.* **2009**, *41* (2), 157–167.
59. Cook, S. R.; Dickinson, D. J. *Int. Res. Group Wood Preserv.* **2004**, Doc. No. IRG/WP/04-10544.
60. Wallace, D. F.; Dickinson, D. J. *Int. Res. Group Wood Preserv.* **2006**, Doc. No. IRG/WP/06-10585.
61. Kirker, G. T.; Diehl, S. V.; Prewitt, M. L.; Diehl, S. V. *Int. Res. Group Wood Preserv.* **2008**, Doc. No. IRG/WP/07-20377.
62. Stirling, R.; Morris, P. I. *Proc. Can. Wood. Preserv. Assoc.* **2011**, 32.
63. American Wood Protection Association. *AWPA E7*; AWPA: Birmingham AL, 2013a; 9 pages.
64. American Wood Protection Association. *AWPA E8*; AWPA: Birmingham AL, 2013b; 3 pages.
65. American Wood Protection Association. *AWPA E14*; AWPA: Birmingham AL, 2013d; 4 pages.
66. American Wood Protection Association. *AWPA E23*; AWPA: Birmingham AL, 2013e; 5 pages.

67. American Wood Protection Association. *AWPA E10*; AWPA: Birmingham AL, 2013c; 12 pages.
68. Molnar, S.; Dickinson, D. J. *Int. Res. Group Wood Preserv.* **2000**, Doc. No. IRG/WP/00-20185.
69. Schultz, T. P.; Nicholas, D. D. *Holzforschung* **2010**, *64* (5), 673–679.
70. Mankowski, M.; Hansen, E.; Morrell, J. J. *For. Prod. J.* **2002**, *52* (11/12), 43–50.
71. Morrell, J. J. *Technical Bulletin*; North American Wood Pole Council, 2008, 6 pages; http://www.woodpoles.org/PoleLifeAndLifeCycle.htm
72. Stewart, A. H. *Wood Pole Newsletter 20: 2-5* Western Wood Preservers Institute, 1996.
73. Nelson, R. F. *Proceedings of the Wood Pole Conference*, October 21-22, 1999, Reno, NV; Northwest Public Power Association: Vancouver, WA, 1999; pp 61–82
74. Pope, T. *Proc. Am. Wood-Preserv. Assoc.* **2004**, *100*, 255–262.

Chapter 8

Thermal Degradation and Conversion of Plant Biomass into High Value Carbon Products

Xinfeng Xie[*,1] and Barry Goodell[2]

[1]Division of Forestry & Natural Resources, West Virginia University,
Morgantown, West Virginia 26506
[2]Department of Sustainable Biomaterials,
Virginia Polytechnic Institute and State University,
Blacksburg, Virginia 24061
*E-mail: Xinfeng.Xie@mail.wvu.edu.

Plant biomass materials are thermally degradable due to their polymeric nature. Treatment of plant biomass at high temperatures removes all carbohydrates with the resulting carbon-rich material highly durable and is not subject to biological degradation due to the lack of nutrient sources for microorganism and insects. Considerable interest has developed in recent years in the use of plant biomass as an inexpensive and renewable feedstock to produce advanced carbon materials for engineering and energy applications. Plant biomass derived carbon has shown great potential for production of carbon-polymer composites, carbon-carbon composites, carbide ceramics, and carbon fiber. Both non-graphitic and graphitic nanostructures have been produced from plant biomass, and they are promising alternatives to petroleum-based carbon nanomaterials. New studies have improved our understanding on the evolution of the carbon structure in carbonized plant materials. Recent studies demonstrated the possibility to produce carbon nanotubes (CNTs) and mesoporous carbon with regularly arranged channels directly from plant biomass materials.

© 2014 American Chemical Society

Thermal Decomposition of Plant Biomass in Inert Atmosphere

Plant biomass materials, such as wood, decompose on heating due to their polymeric nature. There are four basic types of products generated during a thermal degradation process under ambient pressure (*1*):

1) Non-condensable gases, including carbon monoxide, carbon dioxide, hydrogen and methane. They are produced at temperatures between 200°C and 450°C with a maximum production at about 350°C to 400°C.
2) Condensable pyroligneous material, containing about 50% moisture. The production rate of pyroligneous material reaches a maximum from 250°C to 300°C and ceases at about 350°C.
3) Tar with no moisture; formed from 300°C to 450°C.
4) Solid carbon-rich residue.

One previous study (*2*) indicated that only gases were produced as wood was pyrolyzed up to 170°C. At these temperatures, except during long-term heat treatment, and other than the products of wood decomposition, water vapor predominates among the gases. Baileys *et al.* (*3*) found that extreme mass loss during wood thermal degradation occurs between 300°C and 350°C due to the rapid decomposition of cellulose. When higher temperatures are used, the carbon content of the solid residue increases further while the hydrogen and oxygen content decreases (*4*). The process of increasing the carbon content of an organic polymer material by pyrolysis is also called carbonization.

The thermal decomposition of wood is a superpositioning of the thermal degradation of its three major polymer components, i.e. cellulose, lignin and hemicelluloses (*1, 5*). Zeriouh and Belkbir (*6*) investigated the thermal decomposition of a Moroccan wood under a nitrogen atmosphere and found that decomposition of hemicellulose, cellulose and lignin occurred discretely during wood thermal degradation. A study by Beall (*7*) indicated that hemicelluloses were thermally the least stable wood component. Their decomposition is almost completed before cellulose starts to decompose. The thermal stability of lignin is greater than hemicellulose and less than that of cellulose. The decomposition of lignin starts at about 200°C and does not complete until about 650°C (*8*).

Studies by Shafzadeh (*9*) indicated that cellulose decomposes upon heating via two pathways. The first pathway dominates at temperatures below 300°C and it involves reduction in the degree of polymerization (DP). The major decomposition products of this pathway are CO, CO_2, H_2O, and solid carbon-rich residue. The second pathway, which dominates at temperatures greater than 300°C, involves cleavage of molecules and disproportionation reactions to produce a mixture of anhydro tar sugars and low molecular weight volatiles. Intensive oxidation of the solid carbonaceous residue gives glowing combustion, while violent oxidation of the combustible volatiles gives flaming combustion. The decomposition of the polymers in plant biomass is generally complete at about 800°C, because there is no significant mass loss after the material is carbonized at a temperature greater than 800°C (*10*).

The ability to obtain higher carbon yields by chemically modifying lignocellulosic precursors has been studied, initially with the intention of suppressing the flammability of the materials. In the 1960's, cellulose fiber was carbonized in an hydrogen chloride atmosphere, resulting in a carbon yield 14% greater than that would obtained when nitrogen was used (*11*). It is well known that alkali and alkaline earth metals, including potassium, sodium, calcium, and magnesium, are strong catalysts for decomposition of lignocellulosic polymers. It has been reported that these metals can increase the carbon yield in pyrolysis of biomass materials at temperatures lower than 500°C (*12, 13*). In addition, phosphorus and boron containing compounds also can catalyze the formation of solid carbon during pyrolysis of lignocellulosic materials by promoting the dehydration reactions in the materials at temperatures lower than 300°C (*14, 15*). Because of their ability to increase solid carbon yield and reduce the production of flammable volatiles at relatively low temperatures, many phosphorus and boron compounds have been extensively used as fire retardant chemicals for lignocelluloses-based materials and composites.

Evolution of the Carbon Structure during Carbonization of Plant Biomass Materials

The properties of a material are generally determined by its structure. In the development of advanced carbon materials from plant biomass, it is of great importance to understanding the evolution of the carbon structure during the carbonization process. There are two critical factors governing the carbonization process and the material properties: carbonization temperature and heating speed.

Table 1. Mass Yield at Different Carbonization Temperatures

Carbonization temperature (°C)	Slow heating rate 3°C / hour	Fast heating rate 60°C / hour
600	31.26% (0.0039)	29.71% (0.0023)
800	30.61% (0.0029)	28.18% (0.0028)
1000	30.44% (0.0033)	27.93% (0.0017)

Data are from reference (*10*). Values in the brackets are standard deviation.

A previous study (*10*) on carbonization of solid wood indicated that slower heating rates lead to significantly higher mass yield when the material is carbonized to the same temperature (Table 1). Carbonized wood exhibited an anisotropic shrinkage behavior when the processing temperature was higher than 600°C, with the least shrinkage in the longitudinal direction and the greatest the tangential direction. The difference was more prominent when a slower heating rate was used (Table 2).

Table 2. Shrinkage at Different Carbonization Temperatures

Carbonization temperature (°C)	Slow heating rate 3°C / hour		Fast heating rate 60°C / hour	
	Longitudinal	Tangential	Longitudinal	Tangential
600	17.7% (0.0009)	36.1% (0.0067)	17.5% (0.0027)	33.9% (0.0041)
800	20.4% (0.0021)	37.8% (0.0029)	21.4% (0.0039)	37.3% (0.0116)
1000	20.9% (0.0017)	38.4% (0.0037)	21.7% (0.0006)	37.9% (0.0024)

Data are from reference (*10*). Values in the brackets are standard deviation.

At atomic level, carbonized plant biomass has a turbostratic crystallite carbon structure, which is non-graphitic and exhibits 2-dimensional peaks in x-ray diffraction profiles. Plant biomass materials undergoing carbonization between 300°C and 1000°C are characterized by a continuous increase in the amount of turbostratic crystallites and a continuous growth in the dimension of graphene sheets, while the number of graphene layers in the turbostratic crystallites does not significantly change (*10, 16, 17*).

The graphene sheets in the turbostratic crystallites are preferentially oriented parallel to the longitudinal direction of plant biomass cells (*10, 16, 18, 19*) with a certain angle to the circumference of the cell walls in the cross-sectional plane of the plant cell (*10*). The orientation of the graphene sheets was believed to be attributed to the original orientation of cellulose microfibrils in the plant cell wall structure. However, recent studies have shown that the phenyl-propane units of lignin follow the cellulose microfibril arrangement, and also have a preferred orientation along the fiber axis (*20–22*). Given that lignin generates a higher carbon yield compared to cellulose, the preferred orientation of lignin may also contribute to the orientation of the graphene sheets. The preferred orientation of the turbostratic crystallites primarily contributes to the anisotropic properties of carbonized plant biomass materials.

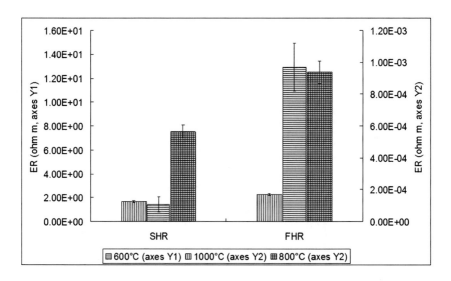

Figure 1. ER of samples prepared at different carbonization temperatures. SHR: slow heating rate at 3°C per hour; FHR: fast heating rate at 60°C per hour. Each point is an average of 24 measurements of 3 samples, error bar: standard deviation. Data are from reference (10).

Carbonized plant biomass can be considered to be a two-phase system including disordered carbon and turbostratic crystallites with large graphene layers. Turbostratic crystallites possess higher electric conductivity and mechanic strength compared to the disordered carbon. The electric resistivity (ER) of carbonized wood along the fiber longitudinal direction decreases 3-4 levels of magnitude when carbonization temperatures increase from 600°C to 800°C (Figure 1). This dramatic decrease in electric resistivity is closely related to a significant increase in the size of graphene layers in the same temperature range (*10, 23*). Both high carbonization temperature and slow heating rate can promote the formation and growth of graphene sheets in the turbostratic crystallites (*10*). Therefore, important properties, such as ER and Young's modulus, which are developed at high carbonization temperatures, can also be obtained at lower temperatures if a slow heating rate is used (Figure 1 and Table 3).

Table 3. Young's Modulus of Samples Prepared at Different Temperatures

Carbonization temperature (°C)	Slow Heating Rate 3°C / Hour (Gpa)	Fast Heating Rate 60°C / Hour (Gpa)
600	8.74 (0.4022)	6.67 (0.3385)
800	14.45 (1.0448)	12.49 (1.1699)
1000	19.66 (0.5630)	15.34 (0.2072)

Data are from reference (*10*). Values in the brackets are standard deviation.

Large crack-free monolithic carbon blocks have been produced from thick solid wood and medium density fiberboard (MDF) using slow heating rates, which reduces shrinkage stresses associated with the decomposition differential between the exterior and interior parts of the material. Thermally conductive materials, such as graphite powder, sand or granular silica, and graphite plates, usually were used to surround the wood during the heating process to reduce uneven heating of the samples. The carbon obtained has excellent machinability, high reactivity, and outstanding dimensional stability, making the material an excellent net-shape preform for producing carbide ceramics, carbon-polymer composites, and carbon-carbon composites. Figure 2 shows a sample of 1 inch thick crack-free, carbonized oak (the original thickness was 1.5 inches), two pieces of carbonized MDF-polymer composites, a cylinder-shaped wood carbon, and three pieces of carbon/carbon composite made from phenolic resin infused carbonized MDF. Because MDF is more homogeneous compared to solid wood at the millimeter scale, large deformation-free monolithic carbon panels were produced from MDF (Figure 3) using heating rates slightly faster than those for carbonization of solid wood.

Figure 2. From back to front: 1) a large monolithic carbon block (1 inch think) produced from oak using slow heating rates; 2) carbonized MDF-polymer composite; 3) carbonized solid wood machined into a cylinder; 4) carbonized MDF based carbon/carbon composite.

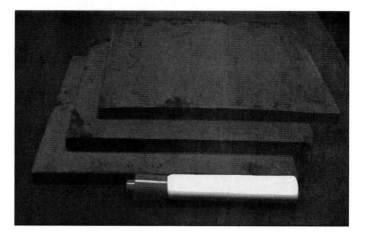

Figure 3. Large monolithic carbon panels produced from MDF using slow heating rates.

Thermal Conversion of Plant Biomass into Carbon Nanostructures

Carbon nanostructures including graphene, nanotubes, nanofibers, nanohorns, nanocapsulates, nanocages, and ordered meso- and microporous carbon have drawn much attention recently because of their great potential for applications in electrochemical energy storage systems, such as rechargeable batteries and supercapacitors, and in energy conversion systems, such as solar cells and artificial leaf devices (24). Plant biomass provides a renewable, abundant and inexpensive carbon source for the production of carbon nanostructures. Certain types of biomass also provide a promising alternative to petroleum-based carbon precursors. Although previous studies have shown some promising results in transferring plant biomass carbon into high performance carbon nanostructures, much fundamental research and technical development is needed before large scale production and applications of biomass-derived carbon nanostructures.

The carbon from plant biomass is non-graphitizable, which means that it cannot be transformed from non-graphitic into graphitic carbon solely by heating the material to 3000°C at atmospheric or lower pressure (25). However, most of the carbon nanostructures aforementioned are in graphitic carbon forms. In order to promote the transformation of plant biomass carbon from non-graphitic to graphitic at relatively low carbonization temperatures ranging up to 1300°C, catalyzed heat treatments using transition metals, such as nickel and iron, have been developed (26–28). Some aluminum compounds were also found to be effective in the catalytic graphitization of wood charcoal as well, but at a much higher temperature of 2200°C (29). The use of other catalysts, including cobalt and copper in the catalytic graphitization of plant biomass has not been reported, although they were found effective in graphitizing non-graphitic carbons from synthetic polymers.

There are two widely accepted mechanisms for catalyzing the graphitization of non-graphitic carbon (30, 31): 1) dissolving of non-graphitic carbon into metals or metallic compounds, and subsequent precipitation of graphitic carbon; and 2) formation and subsequent decomposition of intermediate carbide into metal and graphite. The difference in free energy between non-graphitic carbon and graphitic carbon is the driving force behind the mechanisms (32).

During catalytic graphitization only the non-graphitic carbon in contact with the metal catalyst will be converted into graphitic carbon. Therefore the graphitization is rather localized. Catalytic graphitization of solid carbons derived from lignocellulosic biomass produces an inhomogeneous material with localized regions of graphitic carbon surrounded by non-graphitic carbon. However, the range of order of catalytically graphitized lignocellulosic materials at the nanometer level can be as good as traditional petroleum-based graphite. One study (28) reported that the ordering of the graphitic carbon was comparable to that of the pitch-derived graphite when samples of hardwood species were soaked in nickel nitrate solution under vacuum for 120 hours and then treated at 1600°C for 6 hours.

Using catalytic techniques, graphitic carbon nanostructures have been fabricated from lignocellulosic biomass. The process generally includes five important steps (26, 27): 1) oven drying of lignocellulosic materials; 2) impregnation of the dried materials with a solution of metallic salt catalyst; 3) heat treatment of the catalyst-loaded material to enable catalytic graphitization; 4) removal of the metal catalyst using acids; and 5) removal of the non-graphitic carbon by oxidation.

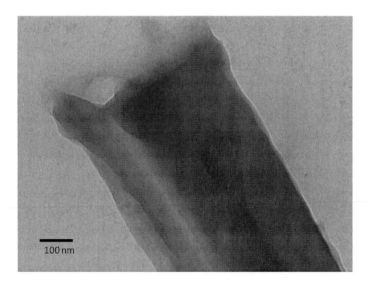

Figure 4. Homogeneous morphology of carbonized plant material without pretreating at 250°C.

In addition to catalytic graphitization, non-catalytic technologies have been developed as well to produce carbon nanostructures from lignocellulosic materials.

Traditional production of charcoal involves heating woody biomass continuously to high temperatures with limited or no oxygen supply. The carbon produced retains the structures of the original biomass material at cellular or fiber level, but within the carbonized cell wall, at nanometer level, the material is homogeneous (Figure 4). New studies have demonstrated that the original arrangement of the cellulose microfibril and lignin-containing matrix can be retained when a step-wise oxidative carbonization process at defined temperatures is employed (33, 34). A typical step-wise oxidation process includes treating plant materials in air at about 250°C followed by oxidation at temperatures higher than 400°C. Carbon nanotubes (CNTs) (33) and mesoporous carbon with nanochannels (Figure 5) have been observed in carbonized plant materials produced using step-wise oxidation processes.

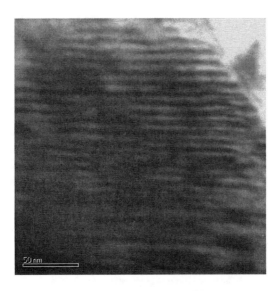

Figure 5. Nanochannels in carbonized plant fibers prepared using step-wise oxidative carbonization method.

It is believed that the production of CNTs and nanochannels within the plant cell wall is attributed to preferential ablation of cellulose microfibrils within the lignin-containing matrix of the intact secondary plant cell walls (*33*). It is hypothesized that the ablation of cellulose microfibrils results in the formation of nanochannels in the carbonized plant cell wall, while the nanochannels, formed from carbonized lignin residues, may act as a template that facilitates the formation of CNTs from the volatilized cellulose carbon gases.

Table 4. Apparent Kinetic Parameters of Cellulose Carbon and Lignin Carbon

Carbonization temperature	Cellulose Carbon		Lignin Carbon	
	n	E (kJ mol^{-1})	n	E (kJ mol^{-1})
400°C	1.05	89.8	0.65	98.4
500°C	0.75	101.2	0.50	109.7
700°C	0.65	143.4	0.55	141.1
1000°C	0.55	167.3	0.50	165.8

Data are from reference (*35*).

A study (*35*) focused on a comparison of the oxidation behavior of cellulose and lignin carbons prepared at different temperatures reported the discovery that cellulose carbon had a higher reaction order (n) and lower activation energy (E) than lignin carbon when oxidized in air if they were prepared under identical conditions, and at temperatures lower than 500°C. The difference in oxidation decreased dramatically when the initial carbonization temperature was at 700°C or greater (Table 4).

Table 5. Pore Volume and Surface Area of Cellulose Carbon and Lignin Carbon

Carbonization temperature	Total pore volume ($cm^3\ g^{-1}$)		BET surface area ($m^2\ g^{-1}$)	
	Cellulose carbon	Lignin carbon	Cellulose carbon	Lignin carbon
400°C	0.0092	0.0011	2.602	1.321
500°C	0.0108	0.0087	5.244	0.846
700°C	0.2238	0.2221	437.63	448.10
1000°C	0.2256	0.2130	449.06	432.33

Data are from reference (*35*).

Experimental data from the Fourier transform infrared (FTIR) absorption studies verified that the oxidative differences observed in the cellulose carbon and lignin carbon were influenced primarily by the chemical structure of the carbonized materials (*35*). Cellulose carbon contained more paraffinic carbon structures than lignin carbon when these carbons were formed at lower temperatures. However, the chemical structures were similar from the perspective of carbonization, when higher temperature carbons were compared. The results from nitrogen adsorption at 77 K comparing the pore volume and Brunauer-Emmett-Teller (BET) surface area of both materials indicated that the surface and porosity properties played only a minor role in the oxidation of cellulose carbon and lignin carbon (Table 5).

References

1. Beall, F. C.; Eickner, H. W.; Forest Products Lab. *Thermal degradation of wood components: a review of the literature*; U.S.D.A. Forest Service: Madison, WI, 1970; pp 2–3.
2. Beall, F. C. *Wood Sci.* **1972**, *5*, 102–108.
3. Baileys, R. T.; Blankenhorn, P. R. *Wood Sci.* **1982**, *15*, 19–28.
4. Setton, R. In *Carbon Molecules and Materials*, 1st ed.; Setton, R, Bernier, P., Lefrant, S., Ed.; Taylor & Francis Inc.: New York, 2002; pp 1–50.
5. Beall, F. C. *Wood Sci. Technol.* **1971**, *5*, 159–175.
6. Zeriouh, A.; Belkbir, L. *Thermochim. Acta* **1995**, *258*, 243–248.
7. Beall, F. C. *Wood Fiber* **1969**, *1*, 215–226.
8. Fierro, V.; Torné-Fernández, V.; Montané, D.; Celzard, A. *Themochim. Acta* **2005**, *433*, 142–148.

9. Shafizadeh, F. In *The Chemistry of Solid Wood*; Rowell, R., Ed.; American Chemical Society: Washington, DC, 1984; pp 489–529.
10. Xie, X.; Goodell, B.; Qian, Y.; Peterson, M.; Jellison, J. *Holzforschung* **2008**, *62*, 591–596.
11. Shindo, A.; Nkanishi, Y.; Soma, I. In *High Temperature Resistant Fibers from Organic Polymers: American Chemical Society Symposium*; Preston, J., Ed.; Applied Polymer Symposia; Interscience Publishers: New York, 1969; Vol. 9, pp 271–284.
12. Raveendran, K.; Ganesh, A.; Khilar, K. C. *Fuel* **1995**, *74*, 1812–1822.
13. Wang, Z.; Wang, F.; Cao, J.; Wang, J. *Fuel Process. Technol.* **2010**, *91*, 942–950.
14. Grexa, O.; Horvathova, E.; Besinova, O.; Lehocky, P. *Polym. Degrad. Stab.* **1999**, *64*, 529–533.
15. Wang, Q.; Li, J.; Winandy, J. E. *Wood Sci. Technol.* **2004**, *38*, 375–389.
16. Paris, O.; Zollfrank, C. *Carbon* **2005**, *43*, 53–66.
17. Kercher, A. K.; Nagle, D. C. *Carbon* **2003**, *41*, 15–27.
18. Byrne, C. E.; Nagle, D. C. *Carbon* **1997**, *35*, 267–273.
19. Greil, P.; Lifka, T.; Kaindl, A. *J. Eur. Ceram. Soc.* **1998**, *18*, 1961–1873.
20. Fromm, J.; Rockel, B.; Lautner, S.; Windeisen, E.; Wanner, G. *J. Struct. Biol.* **2003**, *143*, 77–84.
21. Åkerholm, M.; Salmén, L. *Holzforschung* **2003**, *57*, 459–465.
22. Deng, Y.; Feng, X.; Yang, D.; Yi, C.; Qiu, X. *BioResources* **2012**, *7*, 1145–1156.
23. Nishimiya, K.; Hata, T.; Imamura, Y.; Ishihara, S. *J. Wood. Sci.* **1998**, *44*, 56–61.
24. Su, D. S.; Centi, G. *J. Energy Chem.* **2013**, *22*, 151–173.
25. Edwards, I. A. S. In *Introduction to Carbon Science*; Marsh, H., Ed.; Butterworth & Co. (publishers) Ltd: London, 1989; pp 1–36.
26. Kodama, Y.; Sato, K.; Suzuki, K.; Saito, Y.; Suzuki, T.; Konno, T. *Carbon* **2012**, *50*, 3486–3496.
27. Sevilla, M.; Sanchís, C.; Valdés-Solís, T.; Morallón, E.; Fuertes, A. B. *J. Phys. Chem. C* **2007**, *111*, 9749–9756.
28. Johnson, M. T.; Faber, K. T. *J. Mater. Res.* **2011**, *26*, 18–25.
29. Bronsveld, P.; Hata, T.; Vystavel, T.; DeHosson, J.; Kikuchi, H.; Nishimiya, K.; Imamura, Y. *J. Eur. Ceram. Soc.* **2006**, *26*, 719–723.
30. Marsh, H.; Warburton, A. P. *J. Appl. Chem.* **1970**, *20*, 133–142.
31. Ōya, A.; Marsh, H. *J. Mater. Sci.* **1982**, *17*, 309–322.
32. Fitzer, E.; Kegel, B. *Carbon* **1968**, *6*, 433–436.
33. Goodell, B.; Xie, X.; Qian, Y.; Daniel, G.; Peterson, M.; Jellison, J. *J. Nanosci. Nanotechnol.* **2008**, *8*, 2472–2474.
34. Xie, X.; Goodell, B.; Qian, Y.; Daniel, G.; Zhang, D.; Nagle, D.; Peterson, M.; Jellison, J. *For. Prod. J.* **2009**, *59*, 26–28.
35. Xie, X.; Goodell, B.; Zhang, D.; Nagle, D. C.; Qian, Y.; Peterson, M.; Jellison, J. *Bioresour. Technol.* **2009**, *100*, 1797–1802.

Chapter 9

Termites and Timber

Don Ewart*,[1] and Laurie J. Cookson[2]

[1]Consulting Entomologist, P.O. Box 1044, Research, Victoria 3095, Australia
[2]School of Biological Sciences, Monash University,
Clayton, Victoria 3800, Australia
*E-mail: dme@drdons.net.

The management of termite risk for the protection of timber structures mainly from subterranean termites, are discussed. Inedible wood may be produced from naturally durable timber, impregnation using extractives from naturally durable wood, wood preservatives such as preservative oils, copper-based preservatives, pyrethroids and other organic insecticides, modified wood and preserved wood composites. These measures should be used in conjunction with a whole-of-structure approach that protects buildings using barriers such as graded particles that termites cannot penetrate, resistant sheet materials, or soil treatments. Traditional and electronic detection methods are available for detecting termite activity in a structure, which then allows additional control methods such as baiting and dusting to be introduced. An understanding of the termite hazard and their habits will allow for improved building design and maintenance regimes that will mitigate the risk of termite attack.

Introduction

Termites are a diverse group stemming from social cockroaches (Blattodea: Termitoidae) (*1*, *2*) with a cellulose diet ranging from humus through grasses to wood fiber in all its forms. Their colonies may comprise a few hundred to over 2,500,000 individuals (*3*, *4*), with the majority being workers followed by soldiers and reproductives. Termites may forage over distances of 100 metres (*5*) or more (*6*), usually less. While workers and soldiers lack eyes, they maintain contact and communication with the colony through grooming, olfaction and pheromones, and

© 2014 American Chemical Society

the vibrations produced by head-banging and feeding (*7*). The latter feature can be used as the basis for acoustic detection (*8*). Wood-feeding species also tend to confine their activities within defined galleries, while grass-feeders often leave the protection of their galleries to forage over the ground's surface.

The interactions of termites and people may be complex (*9*) and termites provide valuable ecosystem services (*10*). Termites recycle plant material, and their mounds and galleries can become important nutrient sources (*11*). Termites often alter the landscape by changing soil structure (*12, 13*) and have recently been identified as the cause of 'fairy rings' in the Namib desert (*14*). They degrade plant sugars such as cellulose and hemicellulose, and cause minor modification to lignin, with the aid of a rich gut fauna of up to 10^{11} cells/mL (*15*). The symbiotic fauna includes archaea, bacteria, and eukaryote yeasts and flagellates, except in the Termitidae, which do not use flagellates (*16–18*). The composition of the gut community varies with diet (*19*). Insect cellulolysis has been reviewed by Watanabe and Tokua (*20*). The involvement of termites in the production of the greenhouse gases methane, carbon dioxide (*21*) and nitrous oxide (*22*) have also been studied.

Many termite species are a major threat to structural timbers around the world, declining in importance only in the coldest climates. They also damage non-wood targets, such as buried cables (*23*). The negative impacts of termites are usually expressed according to the economic costs of damage, repair and prevention measures (*24*). Damage in the USA easily exceeds $1.5 billion per annum (*25*) while in Australia cost has been estimated at $780 million per annum (*26*). Termite risk management is often required by law or building statutes. The risks vary greatly with factors such as latitude, elevation, rainfall, evapotranspiration, temperature, humidity, vegetation, construction and land use history (*27*), so that rates of attack may vary greatly across a single continent (*28*). The attacking termites may be native wildlife or introduced pests. The dampwood termites (Termopsidae) are regularly controlled through moisture management which leaves the timbers too dry for them to exploit. The drywood termites (*e.g.* Kalotermitidae) are pests of the tropics and sub tropics and may be managed through the use of inedible timber or by control actions such as regular fumigation (*29*). Infestations of drywood termites begin with a pair of flying termites initiating a nest in a piece of wood. A structure may contain hundreds of separate drywood termite colonies. By far the bulk of timber damage comes from the diverse group called subterranean termites because of their propensity to travel through and attack, from the soil. Their use of soil enables them to enter structures using hidden pathways so that their actions may not be discovered before considerable structural damage has been done.

Managing Subterranean Termites

Management of the subterranean termite at construction comes in two basic approaches: making the structure from timber that is inedible or stopping the termites from accessing a structure made of edible timber. The former approach encompasses the use of naturally resistant timbers, preserved timbers, modified timber products or substitution with non-timber structural members such as concrete and steel. This approach does not prevent the incidental damage that termites may cause to fittings, furniture, floor coverings and contents. By contrast the latter approach tends to provide a whole-of-structure system which aims to exclude all concealed access by termites from the ground. The usual approach to limit access from the ground is to apply barriers to entry, either physical or chemical. In some areas, notably the USA, it is permissible to protect a newly constructed structure by maintaining a baiting system with the intention of intercepting termite attacks (30) however there is a risk of termites reinfesting the gallery systems of dead colonies (31). Regardless of the approach taken, the management of termite risk still requires regular, competent inspections and the growing acceptance of integrated pest management principles (32) has meant that construction prescriptions are now better described as 'termite management systems' than they were with the 'barrier' term (33). Morris (34) discussed the range of termite management options in relation to risk.

In rare circumstances, a subterranean termite colony may attack a building where the alate termites have flown in and found suitably large, moist timbers because of poor building conditions. Such terrestrial attacks are most often in high-rise structures (35) although the incidence of subterranean termites in boats can be quite high (36, 37) where their water is obtained from rainwater in the bilge. Subterranean termites may also be found in the tops of marine piles when they obtain sufficient rain moisture, even though they are repelled from the water-line by seawater. Most infestations of marine piles are by rhinotermitid subterranean termites but in New Zealand, the introduced Australian termopsid dampwood, *Porotermes adamsoni*, is successful around Christchurch (38, 39).

Resistant Framing

Natural Durability

Timbers vary in their natural durability to termites (40, 41). In most cases natural durability refers to outer heartwood, as the sapwood is nearly always non-durable and the inner heartwood is often less durable than outer heartwood due to the presence of juvenile heartwood (42) and pith, and the gradual degradation of extractives in what was mature heartwood. This later feature is noticed in older trees where the inner heartwood may be decayed by fungi and/or hollowed out by termites (piping), providing habitat for many vertebrate species (43). However, this differentiation can be obscured in some tropical 'sapwood' trees, and unusually, the sapwood of some species such as belian (*Eusideroxylon zwageri*) can be relatively durable (44).

Natural durability arises due to a wide range of extractives (45, 46), although for termites simple hardness can also reduce consumption (47–49). The extractives involved are often polyphenols and terpenoids (50). Some of the specific extractives responsible for greatest repellency or control have been identified, and there is the potential that these could be isolated and synthesised as protective treatments for non-durable wood (*51–57*). Conversely, the attractive components in susceptible or modified timbers could be removed by hot water extraction to make them less palatable (*58*).

The resistance of teak (*Tectona grandis*) to termites has been largely attributed to quinones (*59*) which can also be found in ebony (*Diospyros* spp.). Native cypress pine (*Callitris glaucophylla*) in Australia has termite resistance partly attributed to 1-citonellic acid and β-eudesmol (*60–62*). Termite resistance by Alaska yellow cedar (*Chamaecyparis nootkatensis* (Lamb) Spach) is due to nootkatone and for western red cedar (*Thuja plicata*) it is due to thujaplicins although many other compounds in the extractive mix also contribute (*63, 64*). The shrub/tree *Eremophila mitchelli* Benth contains oil that confers termite resistance (*65*).

Repellency can also vary according to the termite genera involved. For example, the heartwood and sapwood of *Pinus radiata* is highly susceptible to *Coptotermes acinaciformis* but usually resists *Nasutitermes exitiosus* (*66*) due to the presence of α-pinene, a compound that is repellent and also found in the defensive secretions of *Nasutitermes* soldiers (*67*). As well as 'higher' termites, the heartwood of some *Pinus* species such as maritime pine (*P. pinaster*), slash pine (*P. elliottii*), and Caribbean pine (*Pinus caribaea*) can resist the 'lower' termites *Coptotermes acinaciformis* and *Mastotermes darwiniensis*, making it easier to meet standard treatment requirements as the heartwood does not require treatment in indoor above-ground (H2) applications (*68*). This feature is especially important for sawn timber as the exposed heartwood is difficult to penetrate with preservatives.

Modified Wood

Termite resistance to modified wood is a more recent field of investigation, and mixed results have been obtained according to the specific kinds of modification involved. Acetylated wood with weight gain of 20% in beech (*Fagus sylvatica*) failed to provide protection against termites (*69*); whereas, weight gains above 20% in *Pinus radiata* showed good resistance to *Coptotermes* species (*70*). Trials on furfurylated wood have varied from no control (*69*) to some control (*71*) and good control against *Crypotermes cynocephalus* and *Macrotermes gilvus* when furfurylation weight gains were 43% or more (*72*). In the few trials that have been published, heat treated wood has not increased resistance to termites (*73*), and further trials with *Fagus sylvatica* (beech) and *Pinus sylvestris* confirmed this failure (*69*). Indeed, heat treatment often increases termite susceptibility. Simply oven drying wood at 105°C for 24 h has been shown to enhance termite damage (*74*), as did heating *P. sylvestris* at 210°C for 15 minutes (*75*). Steam-heating Japanese beech (*Fagus crenata* Blume) also made the wood more attractive to termites due to the production of certain water soluble compounds (*76*). On the other hand, treatment of *Pinus elliottii* with dimethyloldihydroxy-ethyleneurea (DMDHEU) gave good resistance to termites (*69*), as did an amide wax treatment in *F. sylvatica* (*77*).

Preservatives

There are a range of preservatives that are effective against termites. Copper based preservatives are generally effective when sufficient retentions are used, and include CCA (*78*), ACQ (*75*, *79*), Copper azole (*79*, *80*), Cu-HDO (*81*), CDDC (*82*), and copper naphthenate (*83*). This efficacy does not seem to be affected when the copper is applied in micronized/particulate form rather than solubilised form (*84*, *85*).

Creosote is another effective preservative (*86–88*). Additional insecticides could be added to pigment emulsified creosote if greater protection was required (*89*). Arsenical creosote was developed for vertical retort creosotes in Australia for additional control against the voracious *Mastotermes darwiniensis* which can breach the thin-depth treatments obtained in envelope treated eucalypt sleepers and sapwood-thin natural rounds (*90*). However, the arsenic could not be solubilised in modern horizontal retort creosotes so is no longer used. Other oil-borne preservatives such as chlorothalonil and PCP could have insecticides added to boost termite efficacy (*88*, *91*, *92*).

Boron has long been used for the prevention of *Lyctus* beetle damage in the sapwood of hardwood species (*93*), but its use for termite control is limited. As boron can leach from treated wood, it is mainly used in house-framing that is protected from the weather. Since around 1991 borate-treated timber has been widely utilised in Hawaii for the control of *Coptotermes formosanus* (*94*). Borates are non-repellent slow acting toxicants (*95*, *96*), and termites can recover from sublethal doses (*97*). Therefore, control can take time to develop allowing some minor surface feeding to occur, until the foraging termites die, thereby reducing the strength of the trail pheromone leading to the treated wood (*95*). The effectiveness of boron-treated wood can also depend on the availability of nearby untreated wood that can dilute toxicity or recruit greater termite activity (*98*). In laboratory bioassays boric-acid-equivalent (BAE) retentions below 0.5% m/m often control termites (*99*, *100*). However, several field tests have shown that this retention level can fail (*68*, *101*), although Peters and Allen (*102*) obtained control in a field test of borate-treated hoop pine (*Araucaria cunninghamii*) against two species of *Coptotermes*. The discrepancy arises according to the volume of untreated wood available near boron-treated wood, so that tests with reduced choice give greater control (*98*). In Australia, borates are registered for H2 (house framing) use at 2% m/m BAE; however, if all or most of the house frame was boron-treated, control may still occur at 0.5% m/m or 1.0% m/m BAE depending on the amount of untreated wood in the vicinity (*98*). Some termites such as *Nasutitermes* sp. (*103*) and *M. darwiniensis* are relatively borate-tolerant (*104*). Boron is known to affect some species of protozoa in the gut of lower termites (*105*), and also inhibits a number of metabolic processes although the main mode of action remains unclear (*106*, *107*).

There are a number of organic insecticides that can be impregnated into timber for protection from termites, and they are mostly restricted to above-ground applications either outdoors (H3) or indoors (H2). These compounds include (*108*) imidacloprid (*109*) and the pyrethroids permethrin (*110*, *111*), cypermethrin (*112*), deltamethrin (*113*), and bifenthrin (*114*). When used for H3 applications the formulations would also contain a fungicide, along with waxes and resins for water-repellency. For H2 applications the treatment can exclude such additives. The solvent used can vary, and for permethrin can be white spirit or a microemulsion in water (*112*, *115*), or even supercritical carbon dioxide (*116*). *C. acinaciformis* is more sensitive to pyrethroids than *C. formosanus* (*117*). A number of organic insecticides can gradually degrade due to volatilisation and photodegradation, especially in the outer layers of wood (*118*, *119*), so that pyrethroids tend to be used mainly for indoor applications (*115*).

A relatively recent development in Australia has been H2F (H2 framing) envelope treatments for softwoods that can be used south of the Tropic of Capricorn where *M. darwiniensis* does not occur (*120–122*). In this treatment, both the sapwood and heartwood of softwoods only require 5 mm (permethrin) or 2 mm (bifenthrin) penetration rather than full sapwood penetration (*108*). These treatments can be applied economically by spray and in-line dip treatments, and a blue dye is added to the formulation for easy recognition of treated timber on the building site. Retention may be checked in-line using near infrared (NIR) spectroscopy (*123*). Blue pine (blue-dyed) framing has grown to comprise some 30% of the total structural pine framing market in Australia (*124*). Performance is mainly attributed to the repellency of these insecticides. The majority of ends cut after treatment are still protected as normally they will abut the repellent face of a similarly treated timber, while the few cut ends exposed (in corners) have also found protection in simulated wall cavity tests (*125*, *126*). This surface repellency was tested for up to 12 months in these trials. Longer-term testing would be useful.

Composites

Wood composites can often be protected using conventional treatment processes, although glue-bonds can present difficulties for penetration. A more efficient treatment method for some composites is to add insecticide to the glue or furnish during manufacture, thereby obtaining deep and relatively even treatment. However, the insecticide must survive the often harsh chemical environment and the heat and pressing stages employed to produce boards (*127*). For example, while permethrin can be used in urea-formaldehyde bonded composites, it usually degrades in the alkaline conditions found in phenol-formaldehyde glues. Successful glueline additives for plywood and laminated veneer lumber (LVL) include bifenthrin, deltamethrin, imidacloprid (*128*) and thiacloprid (*129*). Zinc borate can also be added to composites during manufacture for termite resistance, especially particleboard products and oriented strand board (OSB) (*130*, *131*). These means of protection rely on insecticide in the glueline, along with some seepage of glue into veneers through peeler checks. Therefore, there are limits to the thickness of veneer that a glueline can protect.

Whole-of-Structure Approaches

For a whole-of-house termite management system, the components are usually applied so as to prevent termites from gaining unseen access to the structure from the building footprint. Barrier components may be for example, applied to the soil beneath a floor, to a concrete flooring slab or to pillars and stumps that support a suspended floor. Regular inspection is still required as termites may bypass systems and there is always a risk that systems may degrade or be rendered incomplete during renovations or landscaping changes so that concealed termite entry occurs.

Particle Barriers

The use of a layer of coarse sand to retard termites was independently discovered by Ebeling (*132*) and Tamashiro et al. (*133*) and has found widespread acceptance in Hawaii and Australia (*33*). The particle size distribution must be of a mix so that termites cannot easily move or pass between the particles and the stone must be durable and hard enough that the termites cannot damage it. In the volcanic Hawaiian islands where the pests are two species of *Coptotermes* (*134*), the particles are made from crushed and graded basalt whereas in Australia, where the particles must work against a diverse termite threat (*135*) and despite abundant basalts, the dominant Granitgard system uses only a single source of granite (*136*). Although much work has been done to accurately define the required particle characteristics beyond simple sieve ranges (*137*), testing against termites is the best measure (*33, 138*). Granitgard also developed a 'tropical grade' product capable of simultaneously excluding the largest and smallest termites (*138*). Other mineral particles such as volcanic ash (*139*) and crushed waste glass (*140*) have been tried with varying success. Impregnated plastic pellets as Termigranuls® with permethrin by Cecil were first used in France and followed in Australia with Homeguard GT® a bifenthrin-impregnated plastic pellet from FMC.

Planar Physical Barriers

Metal sheeting across building cavities to prevent concealed termite entry has a history at least back to the 1920s (*141*) and is still widely used (*24, 142, 143*). Perhaps the most widely used is the cap placed on top of sub-floor stumps or piers. Oxidation and galvanic corrosion may reduce service life, particularly in older installations where soldered joints were used (*33*). Woven stainless steel mesh, as Termi-Mesh, is used on several continents. Rigid uPVC sheet is also used in some systems (*144*).

Pipe Collars

Properly formed concrete floor slabs are termite-resistant, but termites may still gain access through cutouts, joints and the gaps around penetrating services. Pipe collars may take the form of rigid metal, metal mesh, rigid plastic, plastics impregnated with termiticide and particle barriers.

Termiticide Impregnated Plastic Sheet

The first termiticide impregnated plastic sheet product was Termifilm® with permethrin by Cecil in France. This product was followed by those based on termiticide-treated non-woven fabric sandwiched between polythene layers, mostly using deltamethrin (Kordon® by Bayer, and several clones *e.g.* Trithor® by Ensystex) and later by another polythene film product HomeGuard® by FMC which contains bifenthrin. In each of these products there is the expectation that the termiticide will have a service life well in excess of its half-life in soil. The products used in Queensland are claimed by the manufacturer's, as a requirement of the Building Code, as providing service for a building's 50 year design life (*145*). Termiticides used in extruded LDPE films must be capable of tolerating at least 105 °C (*146*). Termiticides applied to a non-woven fabric matrix must bind evenly and permanently.

Termiticide Applied to Soil

The application of termiticide emulsion to soil during various stages of a building's construction is still widely practised and has been reviewed by Hu (*147*) and Wiltz (*148*). The termiticide residues obtained are seldom uniform, with a tendency for the chemical to remain near the soil surface (*149*) and the response of termites is also not constant (*150*). Modern soil termiticides fall broadly into two categories (*151*), repellent formulations (pyrethroids) and non-repellent formulations (chlorantraniliprole, fipronil, imidacloprid etc.). Termites are able to detect and avoid barriers of repellent pyrethroids whereas, as they enter zones treated with non-repellent termiticides, they may collect and share a lethal dose before symptoms appear (*e.g.* (*152*)).

Reticulation Systems

Reticulation systems are used on several continents but have scant mention in scientific literature. They are primarily employed to distribute termiticide into soil although they may also be used to spray termiticide into building cavities. Reticulation, compared with one-off hand spraying of chemical, allows a lower-dose or shorter service life chemical to be used as the barrier can be replenished at will. Reticulation systems are capable of delivering a range of termiticide formulations depending upon need or preference. The Termguard® system uses mostly bifenthrin in Australia, fipronil in Japan and the USA and imidacloprid in China, Indonesia, Malaysia and the Philippines (*153*). If the termiticide formulation is incompatible with the piping used, the termiticide can cause damage (*154*). Another way to manage the risk of termiticide causing pipe damage is to flush the system with water after injection, thus reducing prolonged termiticide-pipe contact. Flushing is specified for the Altis® system which uses a non-rigid pipe with emitters similar to drip irrigation systems (*155*).

Termite Detection

The presence of termites in buildings may be deduced by tapping timbers to detect the reduced resonance associated with hollowing, probing wood with a knife especially in any damp or discoloured areas, and searching for mud tubes. More recently, a range of other methods have been developed. Termites may be detected by the noise they produce while chewing (*156*), the heat generated by their metabolism especially from surfaces behind which they have congregated (*157*), microwave radar movement detection (Termatrac®), odours and metabolic gases that are detected electronically (*158*) or by sniffer dogs (*159*), increases in relative humidity (*160*) or baits placed within walls or around the grounds that can be inspected regularly for termite activity (*161*). Live termites are seldom seen and apart from Termatrac radar, evidence of activity is usually indirect. Pest managers typically rely on a mixture of visual inspection, tapping and moisture detection with the radar, detector dogs and thermal imaging cameras reserved for second level or more expensive inspection. Probing timbers with a knife has limited acceptance due to the damage caused so that it is limited to confirmation of infestation or to limited use on non-cosmetic timbers (*162*).

Colony Control

Colony control can be used as a prophylactic measure before construction or as a response to infestation post-construction. Colony control has the advantage of removing the immediate threat, but does not provide protection from reinfestation and so other management actions are usually required .

Baits

The use of baits has been reviewed by Lenz and Evans (*163*), Quarles (*164*) and Dhang (*165*). Baits provide opportunity for colony control where no central nest can be found, where observed termite activity is too low for direct termiticide application and where building 'features' prevent the installation of a sub-floor termite barrier. In addition, baiting has become the method of choice for many pest managers, replacing other options. Successful bait toxicants fall roughly into two classes: insect growth regulators and inhibitors of energy production and all to date are slow-acting and non-repellent (at the levels employed). The major commercial toxicants include chlorfluazuron, hexaflumuron, noviflumuron, and bistrifluron (*166*).

Dusts

Termiticidal dust formulations have been used commercially since the 1930s (*167*) but have recently found a resurgence (*168, 169*). Most dust formulations require skilled application so that a small amount of dust is introduced into the termites' galleries and as much as possible, directly onto termites. Even the most acceptable dust formulations can repel termite activity if applied in excess or allowed to block galleries. For this reason, there is a risk of assuming successful colony elimination where the termite activity is reduced at the observable points but not overall. It is wise then, to not dust all accessible galleries but to leave at least one for the purpose of observing any ongoing activity. Commercial non-repellent dust formulations for colony elimination include compounds of arsenic, triflumuron, fipronil, and the repellent permethrin which is used for injection into nests.

Other Colony Controls

Killing the queen does not necessarily kill a colony (*170*) as colonies may recover from a loss of primary reproductives and colonies may be polycalic (*171, 172*) so that loss of a single nest does not result in overall control. Nevertheless, direct application of termiticide emulsion to nests located in buildings and trees is generally successful. If the nest cannot be found, slow-acting, non-repellent termiticides may be picked up by contact with treated substrates, or by contact with other affected termites through grooming and trophallaxis. Such termiticides, applied to soil or structures, may result in colony deaths (*173–175*) but the value of the approach is controversial (*147, 176*), particularly as termiticide transfer before death may be retarded by the high concentrations specified for long-term barrier products (*177*).

In manner similar to dusts, non-repellent termiticides such as fipronil and imidacloprid may be applied as liquid or foam directly into termite workings (*178*). Another method of control for subterranean termites is with fumigants, more widely used in the USA (*179*), while in Australia Dazomet® (3,5-dimethyl-1,3,5-thiadiazinane-2-thione) was effective against termites infesting eucalypt poles (*180*).

Risk Reduction Strategies

The most effective forms of risk management involve the elimination of risk factors (*181*). We have discussed the use of treated timbers to render the food source unavailable, but for whole-of-structure risk management the other factors which attract termites need to be addressed. Before construction, the site should be assessed in terms of the known level of local risk (*182, 183*), including the types and species of termites likely to be encountered. This knowledge can be used to drive the choice of appropriate structural design and termite management system. Outside of areas of permanently high humidity (e.g. the tropics), the most common action is to prevent the accumulation of moisture in timbers. This can be achieved by wall and sub-floor venting, eaves over exterior walls and drainage that keeps the soil under and around the structure drier than the surroundings. The greater the distance for a foraging termite between moisture source and food source, the lower the chance of sustained feeding.

The cryptic nature of termites means that they may be present in a structure for some time before there is sufficient evidence to allow effective detection. This lack of certainty is managed through the scheduling of regular inspections, so that the severity of infestation may be contained. While termites normally conceal their activities, subterranean termites almost exclusively attack buildings from the ground. Therefore, the ability to inspect is enhanced when design and construction allows ample access at ground level for the termite inspector. The space required also helps to reduce humidity. It is preferable to have sub-floor access than to construct any sort of flooring at ground level. Similarly it is useful to have full visual access to the building perimeter as this ensures that no external shelter tubes are concealed, and prevents local favourable microclimates from developing over time (e.g. from garden mulch or firewood stacked against a wall, attaching a shed).

Termite risk management has, since the early part of last century, involved barriers to entry, timber treatments, inspection and detection, colony elimination and design to reduce risk factors. Today, there are many tools at our disposal for the control of termites, and with an understanding of their habits and preferences their impact can be contained or prevented. However, one of their most prevailing habits is to find and exploit any weaknesses or oversights that we might leave in those defences.

References

1. Inward, D.; Beccaloni, G.; Eggleton, P. Death of an order: a comprehensive molecular phylogenetic study confirms that termites are eusocial cockroaches. *Biol. Lett.* **2007**, *3*, 331–335.
2. Lo, N.; Engel, M. S.; Cameron, S.; Nalepa, C. A.; Tokuda, G.; Grimaldi, D.; Kitade, O.; Krishna, K.; Klass, K-D.; Maekawa, K.; Miura, T.; Thompson, G. J. Save Isoptera: A comment on Inward et al. *Biol. Lett.* **2007**, *3*, 562–563.
3. Gay, F. J.; Wetherly, A. H. The population of a large mound of *Nasutitermes exitiosus* (Hill) (Isoptera: Termitidae). *J. Aust. Entomol. Soc.* **1970**, *9*, 27–30.
4. Evans, T. A.; Lenz, M.; Gleeson, P. V. Estimating population size and forager movement in a tropical subterranean termite (Isoptera: Rhinotermitidae). *Environ. Entomol.* **1999**, *28*, 823–830.
5. Su, N. Y.; Scheffrahn, R. H. Foraging population and territory of the Formosan subterranean termite (Isoptera: Rhinotermitidae) in an urban environment. *Sociobiology* **1988**, *14*, 353–359.
6. Evans, T. A. Termites. Presented at AEPMA National Conference 2009, Caloundra, Australia, July 15–17, 2009; Australian Environmental Pest Managers Association: Sydney, 2009.
7. Hager, F. A.; Kirchner, W. H. Vibrational long-distance communication in the termites *Macrotermes natalensis* and *Odontotermes* sp. *J. Exp. Biol.* **2013**, *216*, 3249–3256.

8. Mankin, R. W.; Benshemesh, J. Geophone detection of subterranean termite and ant activity. *J. Econ. Entomol.* **2006**, *99*, 244–250.
9. Sileshi, G. W.; Nyeko, P.; Nkunika, P. O.; Sekematte, B. M.; Akinnifesi, F. K.; Ajayi, O. C. Integrating ethno-ecological and scientific knowledge of termites for sustainable termite management and human welfare in Africa. *Ecol. Soc.* **2009**, *14*, 48–69.
10. Jouquet, P.; Traoré, S.; Choosai, C.; Hartmann, C.; Bignell, D. Influence of termites on ecosystem functioning. Ecosystem services provided by termites. *Eur. J. Soil Biol.* **2011**, *47*, 215–222.
11. Park, H. C.; Majer, J. D.; Hobbs, R. J. Contribution of the Western Australian wheatbelt termite, *Drepanotermes tamminensis* (Hill), to the soil nutrient budget. *Ecol. Res.* **1994**, *9*, 351–356.
12. Abe, S. S.; Kotegawa, T.; Onishi, T.; Watanabe, Y.; Wakatsuki, T. Soil particle accumulation in termite (Macrotermes bellicosus) mounds and the implications for soil particle dynamics in a tropical savanna Ultisol. *Ecol. Res.* **2012**, *27*, 219–227.
13. Rückamp, D.; Martius, C.; Bornemann, L.; Kurzatkowski, D.; Naval, L. P.; Amelung, W. Soil genesis and heterogeneity of phosphorus forms and carbon below mounds inhabited by primary and secondary termites. *Geoderma* **2012**, *170*, 239–250.
14. Juergens, N. The biological underpinnings of Namib desert fairy circles. *Science* **2013**, *339*, 1618–1621.
15. Ohkuma, M.; Brune, A. Diversity, structure, and evolution of the termite gut microbial community. In *Biology of Termites: A Modern Synthesis;* Bignell, D. E., Roisin, Y., Lo, N., Eds.; Springer: Dordrecht, The Netherlands, 2011; pp 439−475.
16. Brune, A.; Ohkuma, M. Role of the termite gut microbiota in symbiotic digestion. In *Biology of Termites: A Modern Synthesis;* Bignell, D. E., Roisin, Y., Lo, N., Eds.; Springer: Dordrecht, The Netherlands, 2011; pp 439−475.
17. Hongoh, Y. Toward the functional analysis of uncultivable, symbiotic microorganisms in the termite gut. *Cell Mol. Life Sci.* **2011**, *68*, 1311–1325.
18. König, H.; Li, L.; Fröhlich, J. The cellulolytic system of the termite gut. *Appl. Microbiol. Biotechnol.* **2013**, *97*, 7943–7962.
19. Huang, X. F.; Bakker, M. G.; Judd, T. M.; Reardon, K. F.; Vivanco, J. M. Variations in diversity and richness of gut bacterial communities of termites (*Reticulitermes flavipes*) fed with grassy and woody plant substrates. *Microb. Ecol.* **2013**, *65*, 531–536.
20. Watanabe, H.; Tokuda, G. Cellulolytic systems in insects. *Ann. Rev. Entomol.* **2010**, *55*, 609–632.
21. Jamali, H.; Livesley, S. J.; Butley, L. B.; Fest, B.; Arndt, S. K. The relationships between termite mound CH_4/CO_2 emissions and internal concentration ratios are species specific. *Biogeosciences* **2013**, *10*, 2229–2240.
22. Brümmer, C.; Brüggemann, N. Greenhouse gas exchange in West African savanna ecosystems-how important are emissions from termite mounds? *EGU General Assembly Conference Abstracts* **2012**, *14*, 13436.

23. Fuscaldi, R.; Silveira, F. H.; De Conti, A.; Visacro, S. Power quality problems caused by termites on 138-kV underground cables. *Ing. Invest.* **2011**, *31*, 148–152.
24. Saw, S. S.; Sia, M. K. A review of methods used to control termites in buildings in Malaysia. In *The Sheffield Hallam University Built Environment Research Transactions*; **2011**, *3* (Special International Edition), 23−37; http://research.shu.ac.uk/ds/bert/V03_0Special%20BERT%20TARC.pdf#page=23.
25. Woodrow, R. J.; Grace, J. K. Termite control from the perspective of the termite: A 21st century approach. In *Development of Commercial Wood Preservatives*; Schultz, T., Militz, H., Freeman, M. H., Goodell, B., Nicholas, D. D. Eds, ACS Symposium Series 982; American Chemical Society: Washington, DC, 2008; pp 256−271.
26. Ewart, D. Managing termite risks – An Australian perspective and a cautionary tale. *The International Research Group on Wood Protection*, 2012; Document No. IRG/WP 12-20482.
27. Dambros, C. D. S.; da Silva, V. N. V.; Azevedo, R.; de Morais, J. W. Road-associated edge effects in Amazonia change termite community composition by modifying environmental conditions. *J. Nature Conserv.* **2013**, *21*, 279–285.
28. Cookson, L. J.; Trajstman, A. C. Termite survey and hazard mapping. *CSIRO Forestry and Forest Products*; 2002, Technical Report No. 137; http://www.csiro.au/files/files/plum.pdf.
29. Scheffrahn, R. H.; Su, N. Y.; Busey, P. Laboratory and field evaluations of selected chemical treatments for control of drywood termites (Isoptera: Kalotermitidae). *J. Econ. Entomol.* **1997**, *90*, 492–502.
30. Grace, J. K.; Su, N. Y. Evidence supporting the use of termite baiting systems for long-term structural protection (Isoptera). *Sociobiology* **2001**, *37*, 301–310.
31. Husseneder, C.; Simms, D. M.; Riegel, C. Evaluation of treatment success and patterns of reinfestation of the Formosan subterranean termite (Isoptera: Rhinotermitidae). *J. Econ. Entomol.* **2007**, *100*, 1370–1380.
32. Su, N. Y.; Scheffrahn, R. H. A review of subterranean termite control practices and prospects for integrated pest management programmes. *Integr. Pest Manage. Rev.* **1998**, *3*, 1–13.
33. Ewart, D. M. Termite barriers for new construction in Australia (Isoptera). *Sociobiology* **2001**, *37*, 379–388.
34. Morris, P. I. Integrated control of subterranean termites: the 6s approach. *Proc. Am. Wood-Preserv. Assoc.* **2000**, *96*, 93–106.
35. Rust, M. K.; Su, N. Y. Managing social insects of urban importance. *Annu. Rev. Entomol.* **2012**, *57*, 355–375.
36. Gui-Xiang, L.; Zi-Rong, D.; Biao, Y. Introduction to termite research in China. *J. Appl. Entomol.* **1994**, *117*, 360–369.
37. Scheffrahn, R. H.; Crowe, W. Ship-Borne Termite (Isoptera) Border Interceptions in Australia and Onboard Infestations in Florida, 1986-2009. *Fla. Entomol.* **2011**, *94*, 57–63.

38. Ewart, D. *Management options for the Australian dampwood termite Porotermes adamsoni in New Zealand*; Report to Ministry of Agriculture and Forestry Biosecurity New Zealand, Wellington, New Zealand; 2009, 27 pp.
39. Pearson, H. G.; Bennett, S. J.; Philip, B. A.; Jones, D. C.; Zydenbos, S. M. The Australian dampwood termite, *Porotermes adamsoni,* in New Zealand. *N. Z. Plant Prot.* **2010**, *63*, 241–247.
40. Morales-Ramos, J. A.; Rojas, M. G. Nutritional ecology of the Formosan subterranean termite (Isoptera: Rhinotermitidae): Feeding response to commercial wood species. *J. Econ. Entomol.* **2001**, *94*, 516–523.
41. Arango, R. A.; Green, F., III; Hintz, K.; Lebow, P. K.; Miller, R. B. Natural durability of tropical and native woods against termite damage by *Reticulitermes flavipes* (Kollar). *Int. Biodeterior. Biodegrad.* **2006**, *57*, 146–150.
42. Dünisch, O.; Richter, H.-G.; Koch, G. Wood properties of juvenile and mature heartwood in *Robinia pseudoacacia* L. *Wood Sci. Technol.* **2010**, *44*, 301–313.
43. Adkins, M. F. A burning issue: using fire to accelerate tree hollow formation in *Eucalyptus* species. *Aust. For.* **2006**, *69*, 107–113.
44. Wong, A. H. H.; Singh, A. P. The high decay resistance in the sapwood of the naturally durable Malaysian hardwood belian (*Eusideroxylon zwageri*). *The International Research Group on Wood Protection*, 2001; Document No. IRG/WP/10-10410.
45. Scheffer, T. C.; Cowling, E. B. Natural resistance of wood to microbial deterioration. *Annu. Rev. Phytopathol.* **1966**, *4*, 147–168.
46. Kirker, G. T.; Blodgett, A. B.; Arango, R. A.; Lebow, P. K.; Clausen, C. A. The role of extractives in naturally durable wood species. *Int. Biodeterior. Biodeg.* **2013**, *82*, 53–58.
47. Behr, E. A.; Behr, C. T.; Wilson, L. F. Influence of wood hardness on feeding by the eastern subterranean termite, *Reticulitermes flavipes* (Isoptera: Rhinotermitidae). *Ann. Entomol. Soc. Am.* **1972**, *65*, 457–460.
48. Morales-Ramos, J. A.; Rojas, M. G. Wood consumption rates of *Coptotermes formosanus* (Isoptera: Rhinotermitidae): a three-year study using groups of workers and soldiers. *Sociobiology* **2005**, *45*, 707–719.
49. Scholz, G.; Militz, H.; Gasćon-Garrido, P.; Ibiza-Palacios, M. S.; Oliver-Villanueva, J. V.; Peters, B. C.; Fitzgerald, C. J. Improved termite resistance of wood by wax impregnation. *Int. Biodeterior. Biodegrad.* **2010**, *64*, 688–693.
50. Scheffrahn, R. H. Allelochemical resistance of wood to termites. *Sociobiology* **1991**, *19*, 257–281.
51. Carter, F. L.; Smythe, R. V. Extractives of bald cypress, black walnut, and redwood and survival of the eastern subterranean termite, *Reticulitermes flavipes*. *Ann. Entomol. Soc. Am.* **1972**, *65*, 686–689.
52. Carter, F. L.; Beal, R. H. Termite responses to susceptible pine wood treated with antitermitic extracts. *Int. J Wood Preserv.* **1982**, *2*, 185–191.
53. Grace, J. K.; Yamamoto, R. T. Natural resistance of Alaska-cedar, redwood, and teak to Formosan subterranean termites. *For. Prod. J.* **1994**, *44*, 41–45.

54. Grace, J. K.; Ewart, D. M.; Tome, C. H. M. Termite resistance of wood species grown in Hawaii. *For. Prod. J.* **1996**, *46*, 57–60.
55. Kennedy, M. J.; Powell, M. A. Methodology challenges in developing a transfer of natural durability from sawmill residues, illustrated by experiences with white cypress (*Callitris glaucophylla*). *The International Research Group on Wood Protection*, 2000; Document No. IRG 00-20203.
56. Singh, T.; Singh, A. P. A review on natural products as wood protectants. *Wood Sci. Technol.* **2012**, *46*, 851–870.
57. Syofuna, A.; Banana, A. Y.; Nakabonge, G. Efficiency of natural wood extractives as wood preservatives against termite attack. *Maderas: Cienc. Tecnol.* **2012**, *14*, 155–163.
58. Doi, S.; Takahashi, M.; Yoshimura, T.; Kubota, M.; Adachi, A. Attraction of steamed Japanese larch (*Larix leptolepis* (Sieb.et Zucc.) Gord.) heartwood to the subterranean termites *Coptotermes formosanus* Shiraki (Isoptera: Rhinotermitidae). *Holzforschung* **1998**, *52*, 7–12.
59. Rudi, D.; Bhat, I. H.; Khalil, H. P. S. A.; Naif, A.; Hermawan, D. (2012). Evaluation of antitermitic activity of different extracts obtained from Indonesian teakwood (*Tectona grandis* L.f). *Bioresources* **2012**, *7*, 1452–1461.
60. Rudman, P. The causes of natural durability in timber XVII. The cause of decay and termite resistance in *Callitris columellaris* F. Muell. *Holzforschung* **1965**, *19*, 52–57.
61. French, J. R. J.; Robinson, P. J.; Yazaki, Y.; Hillis, W. E. Bioassays of extracts from white cypress pine (*Callitris columellaris* F. Muell.) against subterranean termites. *Holzforschung* **1979**, *33*, 144–148.
62. Yazaki, Y. Volatility of extractive components in white cypress pine (*Callitris columellaris* F. Muell.). *Holzforschung* **1983**, *37*, 231–235.
63. Taylor, A. M.; Gartner, B. L.; Morrell, J. J.; Tsunoda, K. Effects of heartwood extractive fractions of *Thuja plicata* and *Chamaecyparis nootkatensis* on wood degradation by termites and fungi. *J. Wood Sci.* **2006**, *52*, 147–153.
64. Verma, M.; Sharma, S.; Prasad, R. Biological alternatives for termite control: A review. *Int. Biodeterior. Biodegrad.* **2009**, *63*, 959–972.
65. Scown, D. K.; Creffield, J. M.; Hart, R. S. Laboratory study on the termiticidal efficacy of Eremophilone oil. *The International Research Group on Wood Protection*, 2009; Document No. IRG/WP 00-30497.
66. Gay, F. J. Termite attack on radiata pine timber. *Aust. For.* **1957**, *21*, 86–91.
67. Everaerts, C.; Bonnard, O.; Pasteels, J. M.; Roisin, Y.; König, W. A. (+)-α-Pinene in the defensive secretion of *Nasutitermes princeps* (Isoptera, Termitidae). *Experientia* **1990**, *46*, 227–230.
68. Kennedy, M. J.; Creffield, J. W.; Eldridge, R. H.; Peters, B. C. Field evaluation of the above-ground susceptibility of Pinus heartwood and untreated or treated sapwood of two species of Australian subterranean termites. *The International Research Group on Wood Protection*, 1996; Document No. IRG/WP/96-10147.
69. Militz, H.; Peters, B. C.; Fitzgerald, C. J. Termite resistance of some modified wood species. *The International Research Group on Wood Protection*, 2009; Document No. IRG/WP/09-40449.

70. Bongers, F.; Hague, J.; Alexander, J.; Roberts, M.; Imamura, Y.; Suttie, E. The resistance of high performance acetylated wood to attack by wood-destroying fungi and termites. *The International Research Group on Wood Protection*, 2013; Document No. IRG/WP/13-40621.
71. Lande, S.; Eikenes, M.; Westin, M.; Schneider, M. C. Furfurylation of wood: Chemistry, properties and commercialization. In *Development of Commercial Wood Preservatives*; Schultz, T., Nicholas, D., Militz, H., Freeman, M. H., Goodell, B., Eds.; ACS Symposium Series 982; American Chemical Society: Washington DC, 2008; pp 337−355.
72. Hadi, Y. S.; Westin, M.; Rasyid, E. Resistance of furfurylated wood to termite attack. *For. Prod. J.* **2005**, *55*, 85–88.
73. Militz, H. Processes and properties of thermally modified wood manufactured in Europe. In *Development of Commercial Wood Preservatives*; Schultz, T., Nicholas, D., Militz, H., Freeman, M. H., Goodell, B., Eds.; ACS Symposium Series 982; American Chemical Society: Washington DC, 2008; pp 372−388.
74. Peters, B. C.; Allen, P. J. Susceptibility of conditioned softwood baits to *Coptotermes* spp. (Isoptera: Rhinotermitidae). *Mater. Org.* **1995**, *29*, 47–65.
75. Shi, J. L.; Kocaefe, D.; Amburgey, T.; Zhang, J. A comparative study on brown-rot fungus decay and subterranean termite resistance of thermally-modified and ACQ-C-treated wood. *Holz Roh- Werkst.* **2007**, *65*, 353–358.
76. Doi, S.; Kurimoto, Y.; Ohmura, W.; Ohara, S.; Aoyama, M.; Yoshimura, T. Effects of heat treatments of wood on the feeding behaviour of two subterranean termites. *Holzforschung* **1999**, *53*, 225–229.
77. Scholz, G.; Militz, H.; Cascón-Garrido, P.; Ibiza-Palacios, M. S.; Oliver-Villanueva, J. V.; Peters, B. C.; Fitzgerald, C. J. Improved termite resistance of wood by wax impregnation. *Int. Biodeterior. Biodegrad.* **2010**, *64*, 688–693.
78. Johnson, G. C.; Thornton, J. D. An Australian test of wood preservatives II. The condition after 25 years' exposure of stakes treated with waterborne preservatives. *Mater. Org.* **1991**, *26*, 303–315.
79. Lin, L.-D.; Chen, Y.-F.; Wang, S.-Y.; Tsai, M.-J. Leachability, metal corrosion, and termite resistance of wood treated with copper-based preservatives. *Int. Biodeterior. Biodegrad.* **2009**, *63*, 533–538.
80. Creffield, J. W.; Drysdale, J. A.; Chew, N. In-ground evaluation of a copper azole wood preservative (Tanalith® E) at a tropical Australian test site. *The International Research Group on Wood Protection*, 1996; Document No. IRG/WP 96-30100.
81. Kim, G. H.; Hwang, W.-J.; Yoshimura, T.; Imamura, Y. Laboratory evaluation of the termiticidal efficacy of copper HDO. *J. Wood Sci.* **2010**, *56*, 166–168.
82. Freeman, M. H. Copper dimethyldithiocarbamate: An effective wood preservative. *Proc. Canadian Wood Preserv. Assoc.* **1993** (14), 77–97.
83. Grace, J. K.; Yamamoto, R. T.; Laks, P. E. Evaluation of the termite resistance of wood pressure treated with copper naphthenate. *For. Prod. J.* **1993b**, *43*, 72–76.

84. Freeman, M. H.; McIntyre, C. R. Copper-based wood preservatives. *For. Prod. J.* **2008**, *58*, 6–27.
85. Cookson, L. J.; Creffield, J. W.; McCarthy, K. J.; Scown, D. K. Trials on the efficacy of micronized copper in Australia. *For. Prod. J.* **2010**, *60*, 6–12.
86. Loseby, P. J. A.; Krogh, P. M. D. The persistence and termite resistance of creosote and its constituent fractions. *J. S. Afr. For. Assoc.* **1944**, *11*, 26–32.
87. Creffield, J. W.; Greaves, H.; Chew, N.; Nguyen, N.-K. A field trial of pigment-emulsified creosotes: 11 year data. *For. Prod. J.* **2000**, *50*, 77–82.
88. Cookson, L. J. 40 year results from the largest preservative in-ground stake trial conducted in Australia. *The International Research Group on Wood Protection*, 2013; Document No. IRG/WP 13-30624.
89. Greaves, H.; Chin, C. W.; Watkins, J. B. Improved PEC preservatives with added biocides. *The International Research Group on Wood Protection*, 1985; Document No. IRG/WP/3322.
90. Johanson, R. Affinity of the phenols in creosote for As_2O_3. *Chem. Ind.* **1966**, *29*, 1269–1270.
91. Grace, J. K.; Laks, P. E.; Yamamoto, R. T. Efficacy of chlorothalonil as a wood preservative against the Formosan subterranean termite. *For. Prod. J.* **1993a**, *43*, 21–24.
92. Creffield, J. W.; Chew, N. Efficacy of chlorothalonil and chlorothalonil plus chlorpyrifos against termite attack. *For. Prod. J.* **1995**, *45*, 46–50.
93. Cummins, J. E. The preservation of timber against the attacks of the powder post beetle (*Lyctus brunneus* Stephens) by impregnation with boric acid. *Aust. J. CSIR* **1939**, *12*, 30–49.
94. Grace, J. K. Hawaiian experience with treated building components. In *Enhancing the Durability of Lumber & Engineered Wood Products*; Forest Products Society: Madison, WI, 2002; pp 305–309.
95. Campora, C. E.; Grace, J. K. Foraging behaviour of the Formosan subterranean termite (Isoptera: Rhinotermitidae) in response to borate treated wood. *The International Research Group on Wood Protection*, 2007; Document No. IRG/WP/07-10605.
96. Gentz, M. C.; Grace, J. K.; Mankowski, M. E. Horizontal transfer of boron by the Formosan subterranean termite (*Coptotermes formosanus* Shiraki) after feeding on treated wood. *Holzforschung* **2009**, *63*, 113–117.
97. Gentz, M. C.; Grace, J. K. The response and recovery of the Formosan subterranean termites (*Coptotermes formosanus* Shiraki) from sublethal boron exposures. *Int. J. Pest Manage.* **2009**, *55*, 63–67.
98. Peters, B. C.; Fitzgerald, C. J. Borate protection of softwood from *Coptotermes acinaciformis* (Isoptera: Rhinotermitidae) damage: Variation in protection thresholds explained. *J. Econ. Entomol.* **2006**, *99*, 1749–1756.
99. Mauldin, J. K.; Kard, B. M. Disodium octaborate tetrahydrate treatments to slash pine for protection against Formosan subterranean termite and eastern subterranean termite (Isoptera: Rhinotermitidae). *J. Econ. Entomol.* **1996**, *89*, 682–688.
100. Ahmed, B. M.; French, J. R. J.; Vinden, P. Evaluation of borate formulations as wood preservatives to control subterranean termites in Australia. *Holzforschung* **2004**, *58*, 446–454.

101. Preston, A. F.; McKaig, P. A.; Walcheski, P. J. Termite resistance of treated wood in an above ground field test. The International Research Group on Wood Protection, 1986; Document No. IRG/WP/1300.
102. Peters, B. C.; Allen, P. J. Borate protection of hoop pine (*Araucaria cunninghamii*) sapwood from attack by subterranean termites (*Coptotermes* spp., Isoptera: Rhinotermitidae). *Aust. For.* **1993**, *56*, 249–256.
103. Gay, F. J.; Harrow, K. M.; Wetherly, A. H. Laboratory studies of termite resistance III. A comparative study of the anti-termitic value of boric acid, zinc chloride, and "Tanalith U". Division of Entomology, CSIRO, Melbourne, Australia, 1958; Technical Paper No. 4, .
104. Peters, B. C.; Fitzgerald, C. J. Field exposure of borate-treated softwood to subterranean termites (Isoptera: Rhinotermitidae, Mastotermitidae). *Mater. Org.* **1998**, *32*, 41–66.
105. Nunes, L.; Dickinson, D. J. The effect of boric acid on the protozoan numbers of the subterranean termite, *Reticulitermes lucifugus*. *The International Research Group on Wood Protection*, 1996; Document No. IRG/WP/96-10148.
106. Lloyd, J. D. Borates and their biological applications. *The International Research Group on Wood Protection*, 1998; Document No. IRG/WR 98-30178.
107. Gentz, M. C.; Grace, J. K. A review of boron toxicity in insects with an emphasis on termites. *J. Agric. Urban Entomol.* **2006**, *23*, 201–207.
108. Australian Standard. *Specification for preservative treatment. Part 1: Sawn and round timber.* 2012, Standards Australia, Sydney, NSW, AS 1604.1-2012.
109. Freeman, M. H. The insecticide/termiticide imidacloprid a sound choice for sustainability: Keeping the bugs out of boards. *Proc. Am. Wood-Preserv. Assoc.* **2010**, *106*, 204–218.
110. Creffield, J. W.; Howick, C. D. Comparison of permethrin and fenvalerate as termiticides and their significance to Australian quarantine regulations. *The International Research Group on Wood Protection*, 1984; Document No. IRG/WP/1230.
111. Obanda, D. N.; Shupe, T. F.; Freeman, M. H. Permethrin: An effective wood preservative insecticide. *Proc. Am. Wood-Preserv. Assoc.* **2007**, *103*, 36–47.
112. Read, S. J.; Berry, R. W. An evaluation of the synthetic pyrethroid cypermethrin in organic solvent and emulsion formulations. *The International Research Group on Wood Protection*, 1984; Document No. IRG/WP/3290.
113. Adams, A. J.; Lindars, J. L. A review of the efficacy and uses of deltamethrin for wood preservation. *The International Research Group on Wood Protection*, 1996; Document No. IRG/WP/96-30105.
114. Creffield, J. W.; Watson, K. Correlation between a laboratory bioassay and field trial conducted to determine the termiticidal effectiveness of bifenthrin. *The International Research Group on Wood Protection*, 2002; Document No. IRG/WP 02-20248.

115. Lloyd, J. D.; Schoeman, M. W.; Brownsill, F. Losses of pyrethroids from treated wood due to photodegradation. *The International Research Group on Wood Protection*, 1998; Document No. IRG/WR 98-30177.
116. Cookson, L. J.; Qader, A.; Creffield, J. W.; Scown, D. K. Treatment of timber with permethrin in supercritical carbon dioxide to control termites. *J. Supercrit. Fluids* **2009**, *49*, 203–208.
117. Creffield, J. W.; Lenz, M.; Scown, D. K.; Evans, T. A.; Zhong, J.-H.; Kard, B. M.; Hague, J. R. B.; Brown, K. S.; Freytag, E. D.; Curole, J. P.; Smith, W. R.; Shupe, T. F. International field trials of pyrethroid-treated wood exposed to *Coptotermes acinaciformis* in Australia and *Coptotermes formosanus* (Isoptera: Rhinotermitidae) in China and the United States. *J. Econ. Entomol.* **2013**, *106*, 329–337.
118. Orsler, R. J.; Stone, M. W. S. The permanence of permethrin in wood preservation. *The International Research Group on Wood Protection*, 1984; Document No. IRG/WP/3288.
119. Hunt, A. C.; Humphrey, D. G.; Wearne, R.; Cookson, L. J. Permanence of permethrin and bifenthrin in framing timbers subjected to hazard class 3 exposure. *The International Research Group on Wood Protection*, 2005; Document No. IRG/WP/05-30383.
120. Cobham, P. R.; Snow, J. Tanalith® T a new preservative system for protecting house frames in Australia from termite attack. *The International Research Group on Wood Protection*, 2003; Document No. IRG/WP 03-30306.
121. Kennedy, M. J.; Cobham, P. R. S. Controlled envelope treatments of Pinus sapwood, achieved by modifications to impregnation process and carrier solvents. *The International Research Group on Wood Protection*, 2003; Document No. IRG/WP/03-40258.
122. Sukartana, P.; Creffield, J. W.; Ismanto, A.; Lelana, N. E. Effectiveness of a superficial treatment of bifenthrin to protect softwood framing from damage by subterranean and drywood termites in Indonesia. *The International Research Group on Wood Protection*, 2009; Document No. IRG/WP/09-40443.
123. Meder, R.; Ebdon, N. NIR spectroscopy for rapid determination of permethrin or bifenthrin retention in *P. radiata* sapwood. *The International Research Group on Wood Protection*, 2013; Document No. IRG/WP/13-20507.
124. *AP3. Blue pine framing*. Australian Plantation Products and Paper Industry Council, 2013; http://www.bluepine.com.au/.
125. Lenz, M.; Creffield, J. W.; Runko, S. Is there a need for re-sealing cut ends of envelope-treated softwood framing timber to protect against attack from *Coptotermes* spp. (Isoptera)? *The International Research Group on Wood Protection*, 2004; Document No. IRG/WP/04-10524.
126. Peters, B. C.; Creffield, J. W. Susceptibility of envelope-treated softwood to subterranean termite damage. *For. Prod. J.* **2004**, *54*, 9–14.
127. Norton, J.; Stephens, L. Bifenthrin recovery from glue-line-treated plywood. *The International Research Group on Wood Protection*, 2007; Document No. IRG/WP/07-20355.

128. Doyle, J.; Webb, K.; Rae, W. R.; Siraa, A. F.; Malcolm-Black, J. Permatek IM 30 as a termiticide treatment for veneer-based wood products. *The International Research Group on Wood Protection*, 2003; Document No. IRG/WP/03-30325.
129. Day, K.; Siraa, A.; and Lobb, P. Thiacloprid as a glueline termiticide for veneer based wood products. *The International Research Group on Wood Protection*, 2012; Document No. IRG/WP 12-30592.
130. Laks, P. E.; Manning, M. J. Update on the use of borates as preservatives for wood composites. In *The Second International Conference on Wood Protection with Diffusible Preservatives and Pesticides*; Forest Products Society: Wisconsin, 1997; pp 62–68.
131. Larkin, G. M.; Merrick, P.; Gnatowski, M. J.; Laks, P. E. In-process protection of wood composites: An industry perspective. In *Development of Commercial Wood Preservatives*; Schultz, T., Nicholas, D., Militz, H., Freeman, M. H., Goodell, B., Eds.; ACS Symposium Series 982; American Chemical Society: Washington DC, 2008, pp 458–469.
132. Ebeling, W.; Pence, R. J. Relation of particle size to the penetration of subterranean termites through barriers of sand or cinders. *J. Econ. Entomol.* **1957**, *50*, 690–692.
133. Tamashiro, M.; Yates, J. R.; Yamamoto, R. T.; Ebesu, R. H. Tunneling behavior of the Formosan subterranean termite and basalt barriers. *Sociobiology* **1991**, *19*, 163–170.
134. Uchima, S. Y.; Grace, J. K. Interspecific agonism and foraging competition between *Coptotermes formosanus* and *Coptotermes gestroi* (Blattodea: Rhinotermitidae). *Sociobiology* **2009**, *54*, 765–776.
135. Lee, C.; Vongkaluang, C.; Lenz, M. Challenges to subterranean termite management of multi-genera faunas in Southeast Asia and Australia. *Sociobiology* **2007**, *50*, 213–222.
136. Mawson, J. B.; Carlin, G. *Loads of experience: Mawsons 1912-2012*; E. B. Mawsons & Sons Pty Ltd: Cohuna, Victoria, 2012; 387 pp.
137. Keefer, T. C.; Zollinger, D. G.; Gold, R. E. Evaluation of aggregate particles as a physical barrier to prevent subterranean termite incursion into structures. *Southwest. Entomol.* **2013**, *38*, 447–464.
138. Ewart, D. M.; Rawlinson, E. R.; Tolsma, A. D.; Irvin, G. C.; Wiggins, P. R. Development of a Granitgard (registered trade-mark) particulate termite barrier for use in tropical Australia. *The International Research Group on Wood Protection*, 1997; Document No. IRG/WP/97-10190.
139. Acda, M. N.; Ong, H. B. Use of volcanic debris as physical barrier against the Philippine milk termite (Isoptera: Rhinotermitidae). *Sociobiology* **2005**, *46*, 117–129.
140. United Nations Environment Programme. *Finding Alternatives to Persistent Organic Pollutants (POPs) for Termite Management*; UNEP/FAO/Global IPM Facility Expert Group on Termite Biology and Management, 2000; 47 pp.
141. Snyder, T. E. Termites modify building codes. *J. Econ. Entomol.* **1927**, *20*, 316–321.

142. Myles, T. G. Termite Control in Canada. University of Toronto Centre for Urban and Community Studies. *Res. Bull.* **2004** (23), 4.
143. Rashid, R. Traditional house of Bangladesh: typology of house according to materials and location. Presented at Virtual Conference on Sustainable Architectural Design and Urban Planning, September 15−24, 2007; pp 15−24, http://unaus.eu/pdf/A014.pdf.
144. Australian Standard. *Termite management, Part 1: New building work*; 2000, Standards Australia, Sydney, NSW, AS 3660.1-2000.
145. Building Services Authority. *Termite Management Systems − February 2010*; Building Services Authority: Queensland, 2010; 12 pp, accessed as http://www.bsa.qld.gov.au/SiteCollectionDocuments/Generic/Publications/Termite%20Management%20Systems%20-%20Advisory%20Notes%20for%20Homeowners%20and%20Builders%20Booklet.pdf.
146. Cantor, K. *Blown Film Extrusion*; Hanser Gardner Publications: Cincinnati, 2006; 182 pp.
147. Hu, X. P. Liquid termiticides: their role in subterranean termite management. In *Urban Pest Management: An Environmental Perspective*; Dhang, P., Ed.; CAB: Wallingford, U.K., 2011; Chapter 8, pp 114−132.
148. Wiltz, B. A. Factors Affecting Performance of Soil Termiticides. In *Insecticides–Basic and Other Applications*; Soloneski, S., Ed.; InTech: Rijeka, Croatia, 2012; pp 153−170, http://www.intechopen.com/books/insecticides-basic-and-other-applications/factors-affecting-performance-of-soil-termiticides.
149. Mulrooney, J. E.; Davis, M. K.; Wagner, T. L.; Ingram, R. L. Persistence and efficacy of termiticides used in preconstruction treatments to soil in Mississippi. *J. Econ. Entomol.* **2006**, *99*, 469–475.
150. Osbrink, W. L.; Lax, A. R. Effect of tolerance to insecticides on substrate penetration by Formosan subterranean termites (Isoptera: Rhinotermitidae). *J. Econ. Entomol.* **2002**, *95*, 989–1000.
151. Yeoh, B. H.; Lee, C. Y. Tunneling responses of the Asian subterranean termite, *Coptotermes gestroi* in termiticide-treated sand (Isoptera: Rhinotermitidae). *Sociobiology* **2007**, *50*, 457–468.
152. Gautam, B. K.; Henderson, G. Effect of soil type and exposure duration on mortality and transfer of chlorantraniliprole and fipronil on Formosan subterranean termites (Isoptera: Rhinotermitidae). *J. Econ. Entomol.* **2011**, *104*, 2025–2030.
153. Phua, E. General Manager − Asia and U.S.A., Termguard International Pty Ltd. Personal Communication, 2013.
154. Smith, J. A.; Saunders, J.; Koehler, P. G. Combined effects of termiticides and mechanical stress on chlorinated polyvinyl chloride (CPVC) pipe. *Pest Manage. Sci.* **2008**, *64*, 147–155.
155. CSIRO. ALTIS Standard Partial Anti-Termite Irrigation System. Technical Assessment 222, CSIRO Appraisals, Commonwealth Scientific and Industrial Research Organisation, Melbourne 1997, 9 pp.
156. Fujii, Y.; Noguchi, M.; Imamura, Y.; Tokoro, M. Using acoustic emission monitoring to detect termite activity in wood. *For. Prod. J.* **1990**, *40*, 34–36.

157. Evans, T. A. Assessing efficacy of Termatrac™; a new microwave based technology for non-destructive detection of termites (Isoptera). *Sociobiology* **2002**, *40*, 575–583.
158. Yanase, Y.; Fujii, Y.; Okumura, S.; Yoshimura, T. Detection of metabolic gas emitted by termites using semiconductor gas sensors. *For. Prod. J.* **2012**, *62*, 579–583.
159. Brooks, S. E.; Oi, F. M.; Koehler, P. G. Ability of canine termite detectors to locate live termites and discriminate them from nontermite material. *J. Econ. Entomol.* **2003**, *96*, 1259–1265.
160. Oliver-Villanueva, J. V.; Abián-Pérez, M. A. Advanced wireless sensors for termite detection in wood constructions. *Wood Sci. Technol.* **2013**, *47*, 269–280.
161. Broadbent, S. A Stand-alone Termite Management Technology in Australia. In *Urban Pest Management: An Environmental Perspective*; Dhang, P., Ed.; CAB: Wallingford, U.K., 2011; Chapter 10, pp 145–155
162. Clarke, S.; Ewart, D.; Farrow, K. *A Code of Practice For Prior to Purchase Timber Pest Inspections*; Australian Environmental Pest Managers' Association, Sydney, 2011; 29 pp, available at www.aepma.com.au/Codes-of-Practice.
163. Lenz, M.; Evans, T. A. Termite bait technology: perspectives from Australia. In *Proceeding of the 4th International Conference on Urban Pests*; July 7–10, 2002, Charleston, South Carolina; Jones, S. C., Zhai, J., Robinson, W. H., Eds.; pp 27–36.
164. Quarles, W. IPM for termites–termite baits. *IPM Pract.* **2003**, *25*, 1–18.
165. Dhang, P. Insect baits and baiting. In *Urban Pest Management: An Environmental Perspective*; Dhang, P., Ed.; CAB: Wallingford, U.K., 2011; Chapter 13, pp 187–206.
166. Evans, T. A. Rapid elimination of field colonies of subterranean termites (Isoptera: Rhinotermitidae) using bistrifluron solid bait pellets. *J. Econ. Entomol.* **2010**, *103*, 423–432.
167. Randall, M.; Herms, W. B.; Doody, T. C. The toxicity of chemicals to termites. In *Termites and Termite Control*; Kofoid, C. A., Ed.; University of California Press: Berkeley, 1934; pp 368–384.
168. Kleinschmidt, S. Dust Composition for Combating Insects. U.S. Patent Application 12/678,413, 2007.
169. Barwary, Z.; Hu, X. P.; Hickman, R. B. The Effect of Foraging Tunnel Treatment with Termidor® DRY on 119231973 *Reticulitermes flavipes* (Isoptera: Rhinotermitidae). *J. Agric. Urban Entomol.* **2013**, *29*, 25–34.
170. Tasisa, J.; Gobena, T. Evaluation of Chemical, Botanical and Cultural Managements of Termites Control. *World Appl. Sci. J.* **2013**, *22*, 583–588.
171. Vargo, E. L.; Husseneder, C. Biology of subterranean termites: insights from molecular studies of *Reticulitermes* and *Coptotermes*. *Annu. Rev. Entomol.* **2009**, *54*, 379–403.
172. Raffoul, M.; Hecnar, S. J.; Prezioso, S.; Hecnar, D. R.; Thompson, G. J. Trap response and genetic structure of eastern subterranean termites (Isoptera: Rhinotermitidae) in Point Pelee National Park, Ontario, Canada. *Canadian Entomologist* **2011**, *143*, 263–271.

173. Parman, V.; Vargo, E. L. Colony-level effects of imidacloprid in subterranean termites (Isoptera: Rhinotermitidae). *J. Econ. Entomol.* **2010**, *103*, 791–798.
174. Neoh, K. B.; Hu, J.; Yeoh, B. H.; Lee, C. Y. Toxicity and horizontal transfer of chlorantraniliprole against the Asian subterranean termite *Coptotermes gestroi* (Wasmann): effects of donor:recipient ratio, exposure duration and soil type. *Pest Manage. Sci.* **2012**, *68*, 749–756.
175. Vargo, E. L.; Parman, V. Effect of fipronil on subterranean termite colonies (Isoptera: Rhinotermitidae) in the field. *J. Econ. Entomol.* **2012**, *105*, 523–532.
176. Tsunoda, K. Improved management of termites to protect Japanese homes. In Proceedings of the Fifth International Conference on Urban Pests, Singapore; Lee, C.-Y., Robinson, W. H.; Eds.; Perniagaan: Malaysia, 2005; pp 10–13.
177. Saran, R. K.; Rust, M. K. Toxicity, uptake, and transfer efficiency of fipronil in western subterranean termite (Isoptera: Rhinotermitidae). *J. Econ. Entomol.* **2007**, *100*, 495–508.
178. Hickman, R.; Forschler, B. T. Evaluation of a localized treatment technique using three ready-to-use products against the drywood termite *Incisitermes snyderi* (Kalotermitidae) in naturally infested lumber. *Insects* **2012**, *3*, 25–40.
179. La Fage, J. P.; Jones, M.; Lawrence, T. A laboratory evaluation of the fumigant, sulfuryl fluoride (Vikane®), against the Formosan termite Coptotermes formosanus Shiraki. *The International Research Group on Wood Protection*, 1982; Document No. IRG/WP/1164.
180. Horwood, M. A.; Westlake, T.; Kathuria, A. Control of subterranean termites (Isoptera: Rhinotermitidae) infesting power poles. *J. Econ. Entomol.* **2010**, *103*, 2140–2146.
181. Covello, V. T.; Mumpower, J. Risk analysis and risk management: an historical perspective. *Risk Anal.* **1985**, *5*, 103–120.
182. Foliente, G.; Leicester, R. H.; Chi-Hsiang, W.; Mackenzie, C.; Cole, I. Durability design for wood construction. *For. Prod. J.* **2002**, *52*, 10–19.
183. Kirton, L. G. The importance of accurate termite taxonomy in the broader perspective of termite management. In Proceedings of the Fifth International Conference on Urban Pests, Singapore; Lee, C.-Y., Robinson, W. H.; Eds.; Perniagaan: Malaysia, 2005; pp 1–7.

Biocides

Chapter 10

Fungicides and Insecticides Used in Wood Preservation

Rod Stirling[*,1] and Ali Temiz[2]

[1]FPInnovations, Durability and Sustainability Group, 2665 East Mall, V6T 1Z4 Vancouver, Canada
[2]Karadeniz Technical University, Forestry Faculty, 61080 Trabzon, Turkey
*E-mail: rod.stirling@fpinnovations.ca.

Chemicals used commercially or which have been studied or previously employed to protect wood from degradation by fungi and insects are reviewed. Basic information on their properties, mechanisms and typical uses is provided. High cost limits the development of new fungicides and insecticides specifically for wood preservation. Instead innovations use different actives combinations or adjuvants to increase toxicity to target organisms, improve penetration, provide protection against oxidation or biodegradation, reduce physical depletion, or improve ancillary properties such as water repellency.

Introduction

In order to be commercially viable, wood products need to have a service life that meets consumer expectations. Moreover, to be sustainable wood products should last long enough to grow replacement fiber. Building design can and should be used to maximize durability (*1*, *2*), but in many cases this is not enough, particularly for structures completely exposed to the elements. Naturally durable species may also be used, but their supply is limited, and their durability may be insufficient for demanding exposures. Preservatives are

© 2014 American Chemical Society

needed to provide adequate service life for non-durable species placed at risk of fungal or insect attack. Fungicides are used to control mold, sapstain, and decay by basidiomycetes and soft-rots. Insecticides are largely used to control termites, but may also be used to protect wood against other insects such as carpenter ants and wood-boring beetles. A few active ingredients are effective against both fungi and insects. To be selected for use in a wood preservative formulation, compounds must be effective, cost competitive and meet health and environmental regulations. Requirements for the ideal wood protection biocide have been outlined by Leightley (*3*).

Fungicides are chemical compounds that kill fungi. This is different than a fungistat, which inhibits the growth of fungi but does not kill the organism. Compounds with both modes of action are used in wood preservation. The biodegradation process of wood-rotting fungi is based on free-radical reactions catalyzed by a variety of wood cell wall degrading enzymes (e.g. ligninolytic, cellulolytic and hemicellulolytic) (*4*, *5*). Some fungicides may work by inhibiting these extracellular processes but this mode of action is not well studied. Other common modes of action include cell membrane disruption (e.g. quaternary ammonium compounds), enzyme inactivation (e.g. borates), cell division inhibition (e.g. carbendazim), respiratory inhibition (e.g. copper), and lipid synthesis inhibition (e.g. triazoles) (*6*).

Formulations used to control insects may have repellent and/or insecticidal activity. Repellents are typically volatile compounds that insects sense and avoid. This effect is observed in many oilborne preservative systems, as heavy oil solvent alone can deter termites (*7*). Insecticides may be grouped by their target system, including water balance, the nervous system, endocrine system, energy production and cuticle production (*8*). Most of the insecticides used in wood preservation affect either water balance (e.g. borates) or the nervous system (e.g. permethrin).

Government regulations aim to ensure safe and appropriate use of pesticides. Pesticides are regulated in the United States by the Environmental Protection Agency (EPA) under the Federal Insecticide, Fungicide and Rodenticide Act (FIFRA). In Canada, Health Canada's Pest Management Regulatory Agency (PMRA) regulates the use of pesticides under the Pest Control Products Act. In Europe, the European Chemical Agency (ECHA) manages the Restriction, Evaluation, Authorisation and Restriction of Chemicals (REACH). Actives used in Europe are regulated by the new Biocidal Products Regulation (BPR 528/2012) adopted in May 2012. New text was added in September 2013, with a transitional period for certain provisions. Other countries have their own regulatory systems.

Developing a new fungicide or insecticide and bringing it to market requires years of evaluation and is extremely expensive. As a result, there are relatively few new chemicals being developed for wood preservation. Those that are developed usually come from agriculture, which is a much larger market and can bear much of the development cost. Nevertheless, wood preservation is a major use for pesticides. In the EU alone 18 million m^3 of wood is treated annually for hazard classes 1-4 (*9*). Similar amounts are produced in the United States (*10*). This chapter reviews the major fungicides and insecticides used or examined to protect wood. More historical information can be found in earlier reviews (*6*, *11–15*).

Oilborne Preservatives

Oilborne preservatives use hydrocarbon solvents to carry fungicides and insecticides into wood. Heavy oils are typically used for industrial preservatives such as creosote and penta. Such formulations are most often used in industrial applications (e.g. poles, ties) where durability is paramount and an oily surface is not a detriment. Light organic solvents are typically used to solubilize carbon-based preservatives, copper naphthenate, zinc naphthenate or oxine copper. These are termed light organic solvent preservatives (LOSP) and are most commonly used in Australia and New Zealand for the treatment of framing lumber and millwork.

The use of creosote in wood preservation was patented by Moll in 1836 and Bethell in 1838, and it has been used ever since (*15*). It is a tar oil distillate derived from the carbonization of bituminous coal (*16*). Consequently, creosote contains hundreds of chemicals, including polycyclic aromatic hydrocarbons (PAHs), alkyl-PAHs, heteronuclear aromatics, and tar acids and bases. Some of these compounds, including benzo-a-pyrene, have been classified as carcinogen category 1B (*17*). Typical concentrations of the more abundant compounds in creosote are listed by Nicholas (*12*). Creosote for wood preservation is produced as a by-product of coke production for steel making. It is applied by pressure to wood neat, or as part of a petroleum solution. Creosote is effective against a wide range of wood destroying fungi and insects, though due to its complexity, its mode of action is not known (*18*). Long-term field tests have shown excellent performance in ground contact for more than 50 years (*19–21*). Creosote is used primarily for treatment of railway ties, utility poles and cross arms, bridge timbers, and marine pilings (*22*). Environmental and health concerns have led to restrictions on creosote (*23*). Despite these concerns, after many years of intense scrutiny, regulators in the United States and Canada recently reregistered creosote for industrial uses (*23, 24*). In the EU, creosote is similarly limited to industrial uses, and it must also contain benzo-a-pyrene at a concentration of less than 0.005 % by mass and water extractable phenols at a concentration of less than 3 % by mass (*25*).

Penta (2,3,4,5,6-Pentachlorophenol) is a synthetic compound that is dissolved in hydrocarbon solvents and used to treat utility poles, railway ties, and large timbers. While penta can be solubilized in water or other solvents, its primary use today is as an oilborne preservative for industrial applications. It works primarily by inhibiting oxidative phosphorylation (*26*). Long-term field tests have shown penta to be highly effective against fungi and insects in ground contact exposure (*21, 27*). Environmental and health concerns about penta, and some of the contaminants found in technical grade penta (polychlorinated dibenzo-p-dioxins and dibenzofurans), have led to outright bans in some countries and restrictions on its use in others (*28*). It remains registered for industrial use in the United States and Canada following recent review and reregistration (*24, 29*). In Europe, pentachlorophenol is subject to restrictions under the Directive of 1999/51/EC (*30*).

Inorganic Biocides

Inorganic compounds have long been used to protect wood (mercuric chloride and zinc chloride were patented as wood preservatives in the 1800s). Modern wood preservation still uses many inorganic compounds (Table 1). In general, they are valued for their proven performance, broad spectra of activity, and relative resistance to degradation. The inorganics used in wood preservation today all have at least some activity against both fungi and insects. Cobiocides are added to enhance activity against specific groups of organisms that are resistant to the main active ingredient (e.g. copper-tolerant fungi).

Pentavalent arsenic is a highly effective co-biocide used in chromated copper arsenate (CCA) and ammoniacal copper zinc arsenate (ACZA) to control decay by copper-tolerant basidiomycetes and insects. Arsenic is a broad spectrum biocide that is a competitive inhibitor for phosphorus in adenosine triphosphate (ATP) synthesis (*18*). Human health concerns have led to the restriction of arsenic-containing preservatives to primarily industrial applications in many parts of the world. However, with proper handling these risks can be minimized, and these preservatives have recently been reregistered in the United States and Canada for industrial uses (*24, 31*).

Boron is typically found as boric acid salts or esters (*32*). Borates used in wood preservation include sodium tetraborate, sodium pentaborate, sodium octaborate and boric acid, disodium octaborate tetrahydrate (DOT), and zinc borate. Borates have low mammalian toxicity and are effective against a wide range of insects, including Formosan termites, and fungi, though less effective against molds (*32, 33*). Borates' mode of action has been suggested to come from the borate anion complexing with biologically important polyols, specifically the oxidized coenzymes NAD^+, NMN^+ and $NADP^+$ (*34*). Against termites, borates are associated with reduced numbers of symbiotic gut flora, though activity at the cellular level is also likely occurring (*35*). Borate efficacy against fungi and insects has been reviewed by Drysdale (*36*) and Freeman et al. (*32*). While organisms vary in their susceptibility to borates, none seem to be resistant to borates in the way copper-tolerant fungi are relatively insensitive to copper.

Borates are diffusible and highly leachable. Diffusion is beneficial as borates will migrate within the wet parts of wood that are most likely to be attacked by decay fungi. Diffusion also allows borates to penetrate refractory species (*37*). However, their leachability limits their use in exposed exterior applications. To decrease boron leachability, a number of systems have been proposed including surface coating, water repellants, organo-boron compounds, metallo-borates, stabilized boron esters, chelators including protein borates, *in situ* polymerization and boron-silicate combinations (*38–50*). None of these has yet found widespread commercial success.

Borates are used to protect against decay and insects in protected exposures, in composites as zinc borate (ZB) which has low water solubility, in remediation products (e.g. borate glycols and fused borate rods), in antisapstain formulations, and as cobiocides in some copper-based systems such as copper boron azole type A (CBA), and chromated copper borate (CCB) (*32, 40*). Recently they have been used with creosote in dual treatments of railways ties (*51*). At higher concentrations borates may also contribute to reduced flame spread, and as such are used in fire retardant treatments (*52*).

Copper is still the workhorse of modern wood preservation. It is widely used in many forms and formulations (*53*). The cupric ion is the primary active form of copper. It has a wide spectrum of activity derived from its ability to oxidize proteins, enzymes and lipids (*53*). The role of copper (Cu^{2+}) as a fungicide and termiticide can be attributed to generation of highly active free radicals which damage proteins and DNA (*54*). Carboxylic acid groups in hemicellulose, phenolic groups in lignin and alcoholic hydroxyl groups in cellulose are potential binding sites for copper (*54*). The form in which copper is put into treated wood and the structure of precipitates, complexes or other reaction products affect efficacy and resistance to leaching (*55*, *56*). It is effective against a wide range of fungi, including soft rots, but has minimal effect on some copper-tolerant basidiomycetes. These tolerant fungi detoxify copper by complexation with oxalic acid produced as part of the decay process (*57*). Spores of copper tolerant fungi arriving at the wood surface do not produce oxalic acid and are therefore susceptible to copper (*58*, *59*). The main roles of the cobiocides in copper-containing formulations is to control copper tolerant fungi (*59*). Copper is also effective in preventing degradation by termites (*60*).

Copper as metal ions or present in particles is mostly used in waterborne preservative systems; however it is also used in oilborne copper-organic complexes, specifically copper naphthenate (CuN) and oxine copper (copper quinolin-8-olate).

Copper is the primary biocide in the industrial preservatives CCA and ACZA. Copper in CCA is dissolved in acid, while in ACZA copper is dissolved in an ammoniacal solution (*16*). CCA and ACZA are restricted use pesticides, though this is largely due to the presence of the arsenic, not the copper.

Copper is also the primary biocide in many formulations currently used for the residential market. Several manufacturers solubilize copper in ammonia or amine solutions and formulate with cobiocides to control copper tolerant fungi. These formulations include alkaline copper quaternary (ACQ), copper azole (CA), copper-HDO (N-cyclohexyldizeniumdioxide) (CX), and alkaline copper betaine (KDS). Variants of these systems include using different solvents, actives ratios, or specific actives. For example, ACQ-A has a 1:1 copper:quat ratio, while most other formulations have a 2:1 ratio, ACQ-B uses ammonia to solubilize the copper rather than ethanolamine, and ACQ-C and ACQ-D use different quats as the cobiocide. Alternatively, copper may be present as small particles of basic copper carbonate. Preservatives of this type include micronized copper azole (MCA/μCA) and micronized copper quaternary (MCQ) (*53*). Leaching into aquatic environments is the primary environmental concern for copper-based preservatives, especially arsenic- and chromium-free wood preservatives, due to the high levels of copper content in the new systems and the absence of the oxidant chromium. The effect of copper and other preservatives in the aquatic environment has been extensively reviewed by Brooks (*61*).

Zinc has fungicidal and insecticidal activity (*62*). It is used in the end-cut preservative zinc naphthenate (ZnN). ZnN is less effective than copper naphthenate in this role, but may be preferable when a colorless formulation is desired (*18*). Zinc is used as a cobiocide in the industrial preservative ACZA, which is largely used as an alternative to CCA to treated Douglas-fir and other refractory species (*63*). Zinc is also used in ZB, which is used to protect wood composites from decay and termite attack (*64–66*). Particulate zinc oxide has been evaluated against fungi and termites. It had variable efficacy against decay fungi, but was effective against termites (*67*), though less so than particulate copper (*60*).

Other metals have also been investigated for use in wood preservation. Tin compounds, including bis(tributyltin) oxide (TBTO) which was commercialized as a millwork preservative, have been researched (*68–72*) Silver has also been recently researched as a potential wood preservative (*73, 74*); however, it has not been commercialized, largely due to cost.

Preservative retentions vary depending on intended use, and also between jurisdictions. In general, higher retentions are required for ground contact exposures, and for critical infrastructure. Legal ranges of target retentions are defined on preservative labels. More precise requirements are given in building codes and wood preservation standards (*16, 75, 76*).

Table 1. Inorganic Actives in Wood Preservation

Preservative	Arsenic	Boron	Copper	Zinc
ACQ		+[2]	+	
ACZA	+		+	+
CA		+[2]	+	
CCA[1]	+		+	
CCB[1]		+	+	
CuN			+	
Oxine Cu			+	
CX		+	+	
KDS		+	+	
MCA/µCA			+	
MCQ			+	
SBX		+		
ZB		+		+
ZnN				+

[1] Contains chromium to improve leach resistance. [2] Boron only present in some formulations.

Organic Biocides

Organic biocides used in wood preservation are "organic" only in the strict chemical sense of the word since special interest groups have adopted the word to mean pesticide-free in an agricultural context. The term "carbon-based" is sometimes used to avoid confusion. Organic biocides are used as cobiocides in all of the major residential copper systems. Opportunities to avoid issues around copper leaching, corrosivity, and disposal have also led to the development of totally organic preservative systems. Carbon-based actives can be formulated in solvents (LOSP), or in waterbased systems where they are either dissolved, emulsified, or suspended.

Most of the organic biocides used for wood preservation were developed for agriculture to control a limited number of fungi or insects (Table 2). However, when used in wood preservation they are expected to control a much wider variety of fungi and/or insects (*77*). The spectrum of activity of organic biocides is often less than that of creosote, penta, or metal-based systems, so combinations of organic biocides are often used to ensure efficacy against a broader range of fungi and insects. Moreover, in addition to leaching, organic biocides can be depleted by biodegradation, UV-light, evaporation and chemical degradation (*13*). It has been suggested that organic biocides are typically not active against soft-rot (*14*); however, good performance in pure culture and poor performance in soil (*78*) suggests they are more likely degraded by bacteria or other microorganisms (*79*). Either or both of these issues currently limit the use of wholly organic systems in ground contact.

Triazoles are generally highly effective against basidiomycete decay, but less effective against soft-rot (*80*). They work by inhibiting the synthesis of ergosterol. Triazoles are not effective against insects. Propiconazole and tebuconazole were introduced to wood preservation in the late 1980s (*81–83*) and have become frequently used actives in anti-sapstain (*84*), millwork (*85*) and wood preservative formulations (e.g. CA, PTI) (*86*). To enhance efficacy against specific fungi and broaden the spectrum of activity, triazoles are often formulated with other biocides, such as quats or IPBC, and with adjuvants such as amine oxides (*87, 88*). Propiconazole and tebuconazole have been suggested to work synergistically against basidiomycete decay fungi (*89*), but it is not clear whether these compounds would offer much cross-protection against detoxifying fungi.

Thiabendazole (4-(1H-1,3-benzodiazol-2-yl)-1,3-thiazole) is a thiazole effective against mold fungi (*90*).

TCMTB (2-(Thiocyanomethylthio) benzothiazole) is primarily used to control mold and sapstain. However, it also has activity against decay fungi (*91*) and termites (*92*).

Fenpropimorph ((+/-)-cis-4-[3-(4-tert-butylphenyl)-2-methylpropyl]-2,6-Dimethylmorpholine) is used to control blue stain and decay fungi. It is used in antisapstain treatments and is formulated with other carbon-based biocides in wood preservatives for above ground use.

Table 2. Organic Actives in Wood Preservation

Class	Active	Primary Uses		
		Decay	Mold/ Sapstain	Insects
Aromatics	Chlorothalonil	+	+	+
Benzothiazoles	TCMTB	+	+	+
Benzoylurea	Flufenoxuron			+
Carbamates	IPBC	+	+	
	Carbendazim		+	
	Fenoxycarb			+
Isothiazolones	CMIT		+	
	DCOI	+	+	+
	MIT		+	
	OIT		+	
Morpholines	Fenpropimorph	+	+	
Neonicotinoids	Clothianidin			+
	Imidacloprid			+
	Thiacloprid			+
	Thiamethoxam			+
Organophosphates	Chlorpyrifos			+
Pyrethroids	Bifenthrin			+
	Cypermethrin			+
	Deltamethrin			+
	Ethofenprox			+
	Permethrin			+
Quaternary ammonium compounds	BAC	+	+	+
	DDAC	+	+	+
	DMAP	+	+	+
	DPAB	+	+	+
Sulfamides	Dichlofluanid		+	
Thiazole	Thiabendazole		+	

Continued on next page.

Table 2. (Continued). Organic Actives in Wood Preservation

Class	Active	Primary Uses		
		Decay	Mold/Sapstain	Insects
Triazoles	Cyproconazole	+	+	
	Propiconazole	+	+	
	Tebuconazole	+	+	

Dichlofluanid (N-(Dichlorofluoromethylthio)-N',N'-dimethyl-Nphenylsulfamide) is used to control blue stain fungi. It is used in above ground, exterior applications often in factory-applied or user-applied primers. Dichlofluanid is used in solventborne formulations and is not recommended for interior use.

Carbamate fungicides used in wood preservation include IPBC (3-Iodo-2-propynyl butylcarbamate) and carbendazim (methyl benzimidazol-2-yl carbamate). Though carbamates are a well known class of insecticides, these specific compounds are not noted for insecticidal activity. IPBC was created as a mildewcide for paints and was introduced to wood protection in the 1980s (*93–95*). It has been widely used in anti-sapstain treatments (*84*) and in millwork and joinery systems (*96*). IPBC's mode of fungal activity is not known (*97*). Fenoxycarb (ethyl N-[2-(4-phenoxyphenoxy)ethyl]carbamate) is a carbamate insecticide that acts as an insect growth regulator and is used against wood boring insects in Europe (*98*).

Chlorothalonil (2,4,5,6-tetrachloroisophthalonitrile) is primarily used to control mold and sapstain. It works by reacting with glutathione, which inhibits enzyme function and leads to cell death. Though it is also effective against decay (*99*) and termites (*7*), it is poorly soluble in common solvents and difficult to formulate (*13*), and is only used in minor amounts for a few wood composites.

Several quaternary ammonium compounds (quats) are used in wood preservation, including didecyldimethylammonium chloride (DDAC) and its bicarbonate/carbonate analog (DDACarb), benzalkonium chloride (BAC), Didecylpolyoxyethylammonium borate (DPAB), and didecylmethylpoly(oxyethyl)ammonium propionate (DMPAP). They are commonly used in antisapstain formulations, as well as in wood preservatives (e.g. ACQ, MCQ, KDS). Quats are active against insects, fungi and bacteria, but often require relatively high concentrations and typically do not have sufficient activity to be used alone. Failures of quats as standalone preservatives have been reported in ground contact field tests (*100*), as well as above-ground in service (*101*), so they should be combined with with other biocides.

Isothiazolones are effective against a wide range of fungi and work by binding to thiol groups in fungal enzmyes, which leads to a loss of cell viability (*102*). DCOI (4,5-dichloro-*N*-octylisothiazolin-3-one) is effective against decay fungi, molds and termites (*80*). It is the primary biocide in EL2, where it is combined with the insecticide imidacloprid and a moisture control stabilizer *(15)*. Other isothiazolones include MIT (2-Methylisothiazol-3(2H)-one), CMIT (5-Chlor-2-methyl-4-isothiazolin-3-one) and OIT (2-n-Octyl-4-isothiazolin-3-one). These latter compounds are used primarily for mold and sapstain control.

Pyrethroids are synthetic insecticides modeled on naturally occurring insecticides found in chrysanthemums. Pyrethroids examined for use in wood preservation include permethrin (3-Phenoxybenzyl (1RS)-cis,trans-3-(2,2-dichlorovinyl) -2,2-dimethylcyclopropanecarboxylate), bifenthrin (2-Methyl-3-phenylphenyl)methyl (1S,3S)-3-[(Z)-2-chloro-3,3,3-trifluoroprop-1-enyl]- 2,2-dimethylcyclopropane-1-carboxylate), cypermethrin ([[Cyano-(3-phenoxyphenyl)methyl]3-(2,2-dichloroethenyl)-2,2-dimethylcyclopropane-1-carboxylate), deltamethrin ([(S)-cyano-(3-phenoxyphenyl)-methyl] (1R,3R)-3-(2,2-dibromoethenyl)-2,2-dimethyl-cyclopropane-1-carboxylate), and ethofenprox (1-ethoxy-4-[2-methyl-1-([3-(phenoxy)phenyl]methoxy)propan-2-yl]benzene). Pyrethroids are active against the insect's nervous system. They bind to proteins that open and close sodium channels, which prevents the insect from coordinating movement *(8)*. Pyrethroids combined with fungicides are employed for wood protection in Europe, Australia and New Zealand.

Neonicotinoid incesticides used or being examined for wood preservation include imidacloprid ((E)-1-(6-chloro-3-pyridylmethyl)-N-nitroimidazolidin-2-ylideneamine), thiacloprid ({(2Z)-3-[(6-Chloropyridin-3-yl)methyl]-1,3-thiazolidin-2-ylidene}cyanamide), thiamethoxam (3-[(2-Chloro-1,3-thiazol-5-yl)methyl]-5-methyl-N-nitro-1,3,5-oxadiazinan-4-imine) and clothianidin ((E)-1-(2-Chloro-1,3-thiazol-5-ylmethyl)-3-methyl-2-nitroguanidine). These compounds affect the insect's nervous system as an acetocholine receptor agonist *(8)*. Imidacloprid is used with triazoles in the above-ground carbon-based preservatives PTI and EL2 *(16)*.

Chlorpyrifos (O,O-diethyl O-3,5,6-trichloro-2-pyridyl phosphorothioate) is an organophosphate incesticide that has been used in wood preservation *(103, 104)*. It also affects the nervous system as an acetylcholinesterase inhibitor *(8)*.

Flufenoxuron (1-[4-(2-chloro-α,α,α-trifluoro-p-tolyloxy)-2-fluorophenyl]-3-(2,6-difluorobenzoyl)urea)) is an insect growth regulator used against wood boring insects in Europe *(87)*.

Future Outlook

Given the cost for approving new actives for the relatively small wood preservation market ($0.6 billion US annually compared with $36 billion US for agrochemicals) *(77)*, it is unlikely that any new active will be specifically developed for wood preservation in the near future. Innovations are more likely to result in new combinations of actives and non-biocidal adjuvants. The latter may be designed to increase toxicity to target organisms *(105)*, increase penetration *(88)*, provide protection against oxidation *(106)* or biodegradation *(79)*, reduce physical depletion, or improve ancillary properties such as water repellancy *(107)*, or enhance environmental performance *(108)*.

Regulatory trends suggest that further restrictions on existing actives are likely. Despite recent reregistration in the US and Canada (*23, 24, 29, 31*), creosote, penta and many metal-based systems remain vulnerable to future restrictions. End of service life disposal remains an issue for all wood preservatives. Although new technologies show promise in recovering metals from treated wood (*109*), it is currently uneconomic to do so. Wood treated with organic systems at the end of service life have proven no easier to recycle (*110, 111*).

Modified wood (e.g. acetylation, furfurylation and thermal modification) offers enhanced durability without the use of biocides. These systems have had some commercial success, largely in Europe, and will likely continue to grow in some segments of the market for durable wood products. However, the cost of these treatments remains high, giving preservative treated wood a cost advantage.

Natural biocides such as wood and plant extracts, chitosan, and essential oils continue to be investigated for use in wood preservation (*112–115*). Despite some promising efficacy of natural biocides, little commercialization has taken place. Ironically, the stringent regulations that govern the use of biocides in wood protection also form a huge barrier to commercialization of potentially more benign systems. Studies on naturally durable woods have also identified new mechanisms through which fungi and insects are controlled (*116*). This has led to the development and evaluation of wood preservative formulations containing antioxidant and metal chelating compounds (*117–119*).

The once promising field of exploration which targeted the enzymic and non-enzymic wood breakdown mechanisms of the fungi for disruption (*120–124*) has not yet borne fruit in terms of commercial implementation and seems to have received little attention recently. However, the potential for development of biostats is probably still worth pursuing.

Nanotechnologies are also expected to lead to new wood innovations in wood preservation. The unique properties of nanomaterials may enable the development of formulations with improved penetration and distribution of biocide within the treated wood. However, special regulatory requirements for nanomaterials in many jurisdictions may slow this trend.

Recently published life cycle analyses of treated wood products compared to alternative products show that treated wood has a very strong environmental profile (*125–127*). This recognition may increase demand for treated wood products as environmentally preferable materials. Thus while specific biocides employed may change, the future commercial viability for treated wood remains strong.

Notes

Fungicides and insecticides discussed in this chapter may not be approved for use in all jurisdictions, and may be harmful to human health and the environment if used improperly. Always follow the label.

Acknowledgments

This project was financially supported by the Canadian Forest Service under the Contribution Agreement existing between the Government of Canada and FPInnovations.

References

1. Hazleden, D. G.; Morris, P. I. 8th International Conference On Durability of Building Materials and Components, Vancouver, May 30 to June 3, 1999.
2. Morris, P. I. *Proc. Am. Wood-Preserv. Assoc.* **2000**, *96*, 93–106.
3. Leightley, L. E. In *Wood Deterioration and Preservation: Advances in Our Changing World*; Goodell, B., Nicholas, D. D., Schultz, T. P., Eds.; ACS Symposium Series 845; American Chemical Society: Washington DC, 2003; pp 390–398.
4. Daniel, G. In *Wood Deterioration and Preservation: Advances in Our Changing World*; Goodell, B., Nicholas, D. D., Schultz, T. P., Eds.; ACS Symposium Series 845; American Chemical Society: Washington DC, 2003; pp 34–72.
5. Reading, N. S.; Welch, K. D.; Aust, S. D. In *Wood Deterioration and Preservation: Advances in Our Changing World*; Goodell, B., Nicholas, D. D., Schultz, T. P., Eds.; ACS Symposium Series 845; American Chemical Society: Washington DC, 2003; pp 16–33.
6. Reinprecht, L. In *Fungicides*; Carlisse, O., Ed.; InTech: Rijeka, Croatia, 2010; pp 95–122.
7. Grace, J. K.; Laks, P. E.; Yamamoto, R. T. *For. Prod. J.* **1993**, *43*, 21–24.
8. Valles, S. M.; Koehler, P. G. *Insecticides used in the urban environment: Mode of action*; Entomology and Nematology Department, Florida Cooperative Extension Service, Institute of Food and Agricultural Sciences, University of Florida; ENY282, 2011.
9. Humar, M.; Peek, R. D.; Jermer, J. In *Environmental Impacts of Treated Wood*; Townsend, T. G., Solo-Gabriele, H., Eds.; Taylor & Francis Group: Boca Raton, FL, 2006; pp 37–58.
10. Smith, S. T. *Economic analysis of regulating treated wood waste as hazardous waste in California.* Report to Western Wood Preservers' Institute, 2003.
11. Levi, M. P. In *Wood Deterioration and Its Prevention by Preservative Treatments. Vol. I. Degradation and Protection of Wood*; Nicholas, D. D., Ed.; Syracuse University Press: Syracuse NY, 1973; pp 183–216.
12. Nicholas, D. D. In *Wood and Cellulosic Chemistry*; Hon, D. N.-S., Shiraishi, N., Eds.; Marcel Dekker Inc: New York, 2000; pp 795–806.
13. Freeman, M. H.; Nicholas, D. D.; Schultz, T. P. In *Environmental Impacts of Treated Wood*; Townsend, T. G, Solo-Gabriele, H., Eds.; Taylor & Francis Gruop: Boca Raton, FL, 2006; pp 19–36.
14. Mai, C.; Militz, H. In *Wood Production, Wood Technology and Biotechnological Impacts*; Kües, U., Ed.; Universitätsverlag Göttingen: Göttingen, Germany, 2007; pp 259–272.

15. Ruddick, J. N. R. In *Uhlig's Corrosion Handbook*; Revie, W., Ed.; John Wiley and Sons: Hoboken, NJ, 2011; pp 469−477.
16. AWPA. *AWPA Book of Standards*; American Wood Protection Association: Birmingham, AL, 2012; 644 pages.
17. Commission Directive (1272/2008). *Off. J. Eur. Communities* 2008; L 353/1.
18. Zabel, R. A.; Morrell, J. J. *Wood Microbiology – Decay and its Prevention*; Academic Press: San Diego, CA, 1992; 476 pages.
19. Webb, D. A.; Fox, R. F.; Pfeiffer, R. G. *Proc. Am. Wood-Preserv. Assoc.* **2009**, *105*, 182–187.
20. Morris, P. I.; Ingram, J. K.; Stirling, R. *Proc. Am. Wood-Preserv. Assoc.* **2012**, *108*, 12p..
21. Woodward, B. M.; Hatfield, C. A.; Lebow, S. T. *Research Note FPL-RN-02*; U.S. Department of Agriculture, Forest Service, Forest Products Laboratory: Madison, WI, 2011.
22. Webb, D. A.; Harward, D. *Proc. Am. Wood-Preserv. Assoc.* **2007**, *103*, 178–182.
23. EPA. *Reregistration Evaluation Decision (RED) Document for Creosote*; United States Environmental Protection Agency. Case 0139, 2008.
24. PMRA. *Heavy Duty Wood Preservatives: Creosote, Pentachlorophenol, Chromated Copper Arsenate (CCA) and Ammoniacal Copper Zinc Arsenate (ACZA)*; Re-evaluation Decision RVD2011-06, 2011.
25. Commission Directive (2001/60/EC). *Off. J. Eur. Communities* 2001; L 283/41.
26. Ozanne, G. *Int. Res. Group Wood Preserv.* **1995** Doc. No. IRG/WP/95-50040-07.
27. Nicholas, D. D.; Freeman, M. H. *Int.. Res. Group Wood Preserv.* **2000** Doc. No. IRG/WP/00-30243.
28. Freeman, M. H.: Wilkinson, J. *Proceedings of the 2010 Southeastern Wood Pole Conference*; Forest Products Society: Madison, WI, 2010.
29. EPA. *Reregistration Evaluation Decision for Pentachlorophenol*; United States Environmental Protection Agency, Case 2505, 2008.
30. Commission Directive (1999/51/EC), *Off. J. Eur. Communities* **1999**, L 142/22.
31. EPA. *Reregistration Evaluation Decision for Chromated Arsenicals*; United States Environmental Protection Agency, Case 0132, 2008.
32. Freeman, M. H.; McIntyre, C. R.; Jackson, D. *Proc. Am. Wood Prot. Assoc.* **2009**, *104*, 279–294.
33. Jin, L.; Walcheski, P.; Preston, A. F. *Int. Res. Group Wood Prot.* **2012** Doc. No. IRG/WP/12-30588.
34. Lloyd, J. D. *Int. Res. Group Wood Preserv.* **1998** Doc. No. IRG/WP/98-30178.
35. Grace, J. K. *Second International Conference on Wood Protection with Diffusible Preservatives and Pesticides*; Forest Products Society: Madison, WI, 1997.
36. Drysdale, J. A. *Int. Res. Group Wood Preserv.* **1994** Doc. No. IRG/WP/94-30037.

37. Byrne, A.; Morris, P. I. *Second International Conference on Wood Protection with Diffusible Preservatives and Pesticides*; Forest Products Society: Madison, WI, 1997.
38. Vinden, P.; Romero, R. *Second International Conference on Wood Protection with Diffusible Preservatives and Pesticides*; Forest Products Society: Madison, WI, 1997.
39. Sites, W. H.; Williams, L. H. *Second International Conference on Wood Protection with Diffusible Preservatives and Pesticides*; Forest Products Society: Madison, WI, 1997.
40. Kartal, S. N. In *Handbook on Borates*; Chung, M. P., Ed.; Nova Science: Hauppauge, NY, 2010.
41. Mohareb, A.; Van Acker, J.; Stevens, M. *Int. Res. Group Wood Preserv.* **2002** Doc. No. IRG/WP 02-30290.
42. Yalinkilic, M. K.; Yoshimura, T.; Takahashi, M. *J. Wood Sci.* **1998**, *44*, 152–157.
43. Yalinkilic, M. K.; Yusuf, S.; Yoshimura, T.; Su, W.; Tsunoda, K.; Takamashi, M. *Int. Res. Group Wood Preserv.* **1997** Doc. No. IRG/WP 97-40083.
44. Kartal, S. N.; Green, F., III *Holz Roh- Werkst.* **2003**, *61*, 388–389.
45. Amburgey, T. L.; Freeman, M. H. *Proc. Int. Conf. Utility Line Structure* **2000**.
46. Lloyd, J. D.; Fogel, J. L. U.S. Patent 6896908, 2005.
47. Thevenon, M. F.; Pizzi, A. *Holz Roh- Werkst.* **2003**, *61*, 457–464.
48. Cui, W.; Kamdem, D. P. *Int. Res. Group Wood Preserv.* **1999** Doc. No. IRG/WP/ 99-30202.
49. Kartal, S. N.; Yoshimura, T.; Imamura, Y. *Int. Biodeterior. Biodegrad.* **2009**, *63*, 187–190.
50. Temiz, A.; Alfredsen, G.; Eikenes, M.; Terziev, N. *Bioresour. Technol.* **2008**, *99*, 2102–2106.
51. Amburgey, T. L.; Sanders, M. G. *Crossties* **2009**, *90*, 20–22.
52. LeVan, S. L.; Tran, H. C. In *First International Conference on Wood Protection with Diffusible Preservatives*; Forest Products Society: Madison, WI, 1990; pp 28–30.
53. Freeman, M. H.; McIntyre, C. R. *For. Prod. J.* **2008**, *58*, 6–27.
54. Zhang, J. *Interactions of Copper-Amine Preservatives with Southern Pine*, Ph.D. Thesis, Michigan State University, East Lansing, MI, 1999.
55. Pizzi, A. *J. Polym. Sci.* **1981**, *19*, 3093–3121.
56. Pizzi, A. *J. Polym. Sci.* **1982**, *20*, 704–724.
57. Green, F., III; Clausen, C. A. *Int. Biodeterior. Biodegrad.* **2005**, *56*, 75–79.
58. Choi, S. M.; Ruddick, J. N. R.; Morris, P. I. *Int. Res. Group Wood Preserv.* **2002** Doc. No. IRG/WP/02-10422.
59. Woo, C. S.; Morris, P. I. *Int. Res. Group Wood Protect.* **2010** Doc. No. IRG/WP/10-10707.
60. Akhtari, M.; Nicholas, D. D. *Eur. J. Wood Wood Prod.* **2013**, *71*, 395–396.
61. Brooks, K. M. In *Managing Treated Wood in Aquatic Environments*; Morrell, J. J., Brooks, K. M., Davis, C. M., Eds.; Forest Products Society: Madison, WI, 2011; pp 59–152.

62. USDA, United States Department of Agriculture, Farmers Bulletin No. 1472, 1926.
63. Baileys, R. T. *Proc. Am. Wood-Preserv. Assoc.* **2002**, *98*, 63–69.
64. Laks, P. E.; Manning, M. J. *Int. Res. Group Wood Preserv.* **1995** Doc. No. IRG/WP/95-30074.
65. Morris, P. I.; Clark, J. E.; Minchin, D.; Wellwood, R. *Int. Res. Group Wood Preserv.* **1999** Doc. No. IRG/WP/99-40138.
66. Scown, D. K.; Creffield, J. W. *Int. Res. Group Wood Protect.* **2009** Doc. No. IRG/WP/09-30498.
67. Clausen, C. A.; Wang, V. W.; Arango, R.; Green, F., III *Proc. Am. Wood Protect. Assoc.* **2009**, *105*, 255–260.
68. Schultz, T. P.; Nicholas, D. D. In *Managing Treated Wood in Aquatic Environments*; Morrell, J. J., Brooks, K. M., Davis, C. M., Eds.; Forest Products Society: Madison, WI, 2011; pp 29−36.
69. Richardson, B. A. *Proc. Br. Wood Preserv. Assoc.* **1970**, 37–58.
70. Levi, M. P.; Smith, D. N. R. *Mater. Org.* **1971**, *6*, 233–240.
71. Van der Kerk, G. J. M. In *Organotin Compounds: New Chemistry and Applications*; Zuckerman, J., Ed.; American Chemical Society: Washington, DC, 1976.
72. Jermer, J.; Edlund, M.-L.; Henningsson, B.; Hintze, W. *Int. Res. Group Wood Preserv.* **1985** Doc. No. IRG/WP/2244.
73. Dorau, B.; Arango, R.; Green, F., III. *Proceedings from the Wood Frame Housing Durability and Disaster Issues Conference*; Forest Products Society: Madison, WI, 2004; pp 133−145.
74. Ellis, J. R.; Jayachandran, K.; Nicholas, D. *Int. Res. Group Wood Protect.* **2007** Doc. No. IRG/WP/07-30419.
75. European Committee for Standardisation. EN351-1,2 2007.
76. European Committee for Standardisation. EN599-1,2 2009.
77. Schultz, T. P.; Nicholas, D. D.; Preston, A. F. *Pest Manage. Sci.* **2007**, *63*, 784–788.
78. Herring, I. J.; Dickinson, D. J. *Int. Res. Group Wood Protect.* **1999** Doc. No. IRG/WP/99-20175.
79. Cook, S. R.; Dickinson, D. J. *Int. Res. Group Wood Protect.* **2004** Doc. No. IRG/WP/04-10544.
80. Schultz, T. P.; Nicholas, D. D.; McIntyre, C. R. *Recent Pat. Mater. Sci.* **2008**, *1*, 128–134.
81. Valcke, A. *Int. Res. Group. Wood Preserv.* **1989** Doc. No. IRG/WP/3529.
82. Gründlinger, R.; Exner, O. *Int. Res. Group. Wood Preserv.* **1990** Doc. No. IRG/WP/3629.
83. Wüstenhöfer, B.; Wegen, H.-W.; Metzner, W. *Int. Res. Group. Wood Preserv.* **1990** Doc. No. IRG/WP/3634.
84. Schauwecker, C. J.; Morrell, J. J. *For. Prod. J.* **2008**, *58*, 52–55.
85. Ross, A. S. In *Development of Commercial Wood Preservatives: Efficacy, Environmental, and Health Issues*; Schultz, T. P., Militz, H., Freeman, M. H., Goodell, B., Nicholas, D. D., Eds.; American Chemical Society: Washington, DC, 2008; pp 470−479.

86. Kamdem, D. P. In *Development of Commercial Wood Preservatives: Efficacy, Environmental, and Health Issues*; Schultz, T. P., Militz, H., Freeman, M. H., Goodell, B., Nicholas, D. D., Eds.; American Chemical Society: Washington, DC, 2008; pp 427−439.
87. Hughes, A. S. *Final Workshop COST Action E22 Environmental Optimisation of Wood Protection*, 2004.
88. Jiang, X.; Walker, L. *Int. Res. Group. Wood Protect.* **2007** Doc. No. IRG/WP/07-30425.
89. Bauschhaus, H.-U.; Valcke, A. R. *Int. Res. Group Wood Preserv.* **1995** Doc. No. IRG/WP/95-30092.
90. Clausen, C. A.; Yang, V. *Int. Biodeterior. Biodegrad.* **2007**, *59*, 20–24.
91. Van den Eynde, R. *Int. Res. Group Wood Preserv.* **1990** Doc. No. IRG/WP/90-3606.
92. Grace, J. K. *Proc. Entomol. Soc. Ont.* **1988**, *119*, 83–85.
93. Plackett, D. V. *Int. Res. Group Wood Preserv.* **1982** Doc. No. IRG/WP/3198.
94. Drysdale, J. A. *Int. Res. Group Wood Preserv.* **1983** Doc. No. IRG/WP/3237.
95. Hansen, J. *Int. Res. Group Wood Preserv.* **1984** Doc.No. IRG/WP/3295.
96. Freeman, M. H. In *Development of Commercial Wood Preservatives: Efficacy, Environmental, and Health Issues*; Schultz, T. P., Militz, H., Freeman, M. H., Goodell, B., Nicholas, D. D., Eds.; American Chemical Society: Washington, DC, 2008; pp 408−426.
97. Canadian Council of Ministers of the Environment. In *Canadian Environmental Quality Guidelines, 1999*; Canadian Council of Ministers of the Environment: Winnipeg, Manitoba, 1999.
98. Miyamoto, J.; Hirano, M.; Takimoto, Y.; Hatakoshi, M. *Pest Control Enhanced Environ. Saf.* **1993**, *524*, 144–168.
99. Creffield, J. W.; Woods, T. L.; Chew, N. *Int. Res. Group Wood Preserv.* **1996** Doc. No. IRG/WP/96-30101.
100. Ruddick, J. N. R. *Int. Res. Group Wood Preserv.* **1981** Doc. No. IRG/WP/2152.
101. Butcher, J. A. *Int. Res. Group Wood Preserv.* **1985** Doc. No. IRG/WP/3328.
102. Chapman, J. S.; Diehl, M. A.; Fearnside, K. B.; Leightley, L. E. *Int. Res. Group Wood Preserv.* **1998** Doc. No. IRG/WP/98-30183.
103. Rose, K.; Kozuma, J.; Sparrow, P. *Int. Res. Group Wood Preserv.* **1984** Doc. No. IRG/WP/84-3301.
104. Tsunoda, K.; Nishimoto, K. *Int. Res. Group Wood Preserv.* **1985** Doc. No. IRG/WP/85-3330.
105. Valcke, A. *Int. Res. Group Wood Preserv.* **1995** Doc. No. IRG/WP/95-30092.
106. Schultz, T. P.; Nicholas, D. D. *Phytochemistry* **2002**, *612*, 555–560.
107. Cui, F.; Preston, A. F.; Archer, K. J.; Walcheski, P. U.S. Patent 7264886 B2, 2007.
108. Langroodi, S. K.; Borazjani, H.; Nicholas, D. D.; Prewitt, L. M.; Diehl, S. V. *For. Prod. J.* **2012**, *62*, 467–473.
109. Coudert, L.; Blais, J.; Mercier, G.; Cooper, P.; Morris, P.; Gastonguay, L.; Janin, A.; Zaviska, F. *J. Environ. Eng.* **2013**, *139*, 576–587.

110. Stirling, R.; Bicho, P.; Daniels, B.; Morris, P. I. *Holzforschung* **2010**, *64*, 285–288.
111. Stirling, R.; Dale, A.; Morris, P. I. *Proc. Can. Wood. Preserv. Assoc.* **2010**.
112. Yang, V. W.; Clausen, C. A. *Int. Biodeterior. Biodegrad.* **2007**, *59*, 302–306.
113. Yang, D. Q. *For. Prod. J.* **2009**, *59*, 97–103.
114. Singh, T.; Singh, A. P. *Wood Sci. Technol.* **2012**, *46*, 851–870.
115. Kirker, G. T.; Blodgett, A. B.; Lebow, S.; Clausen, C. A. *Int. Res. Group Wood Protect.* **2013** Doc. No. IRG/WP/13-10808.
116. Schultz, T. P.; Nicholas, D. D. *Phytochemistry* **2000**, *54*, 47–52.
117. Schultz, T. P.; Nicholas, D. D.; Henry, W. P.; Pittman, C. U.; Wipf, D. O.; Goodell, B. *Wood Fiber Sci.* **2005**, *37*, 175–184.
118. Schultz, T. P.; Nicholas, D. D.; Kirker, G. T.; Prewitt, M. L.; Diehl, S. V. *Int. Biodeterior. Biodegrad.* **2006**, *57*, 45–50.
119. Little, N. S.; Schultz, T. P.; Nicholas, D. D. *Holzforschung* **2010**, *64*, 395–398.
120. Namchuk, M.; Tull, D.; Morris, P.; Withers, S. G. *Proc. Can. Wood Preserv. Assoc.* **1992**, *13*, 176–184.
121. Abraham, L.; Breuil, C.; Bradshaw, D. E.; Morris, P. I.; Byrne, A. *For. Prod. J.* **1997**, *47*, 57–62.
122. Green, F, III; Highley, T. L. *Trends in Plant Pathol.* **1997**, *1*, 1–17.
123. Green, F., III; Kustner, T. A. *Int. Res. Group Wood Preserv.* **1999** Doc. No. IRG/WP/99-10321.
124. Goodell, B.; Nicholas, D. D.; Schultz, T. P. In *Wood Deterioration and Preservation: Advances in Our Changing World*; Goodell, B., Nicholas, D. D., Schultz, T. P., Eds.; ACS Symposium Series 845; American Chemical Society: Washington DC, 2003; pp 2–9.
125. Bolin, C. A.; Smith, S. T. *J. Cleaner Prod.* **2011**, *19*, 620–629.
126. Bolin, C. A.; Smith, S. T. *J. Cleaner Prod.* **2011**, *19*, 630–639.
127. Bolin, C. A.; Smith, S. T. *Renew. Sustain. Energy Rev.* **2011**, *15*, 2475–2486.

Chapter 11

Treatment Technologies: Past and Future

Adam Taylor[1] and Jeffrey J. Morrell[*,2]

[1]Tennessee Forest Products Center, University of Tennessee,
Knoxville, Tennessee 37996
[2]Department of Wood Science & Engineering, Oregon State University,
Corvallis, Oregon 97331
[*]E-mail: Jeff.morrell@oregonstate.edu.

Impregnation with preservatives markedly extends the useful life of a variety of wood products used under adverse conditions. A variety of methods have been developed to accomplish this process. This chapter reviews the available technologies for delivering biocides into wood and outlines potential avenues for improving the treatment process.

Introduction

Compared to other biological materials, wood has substantial resistance to biodeterioration (*1*), but it can be degraded under the proper conditions (*2, 3*). Humans have long sought to prolong the useful life of wood-based materials through practices that have included charring, daubing with various oils, soaking, and ultimately impregnation. These efforts have yielded varying results, but ultimately, successful protection can help us to make more efficient use of our resources. It was once estimated that 10 % of our forest harvest was used to replace timber that had failed in service for various reasons, but mostly due to biodeterioration (*4*). While this figure seems staggering, a trip to any wood recycling center highlights the level of decayed wood removed from service. Various methods of wood protection have the potential to reduce these losses.

Wood protection entails a variety of approaches, the most effective and widely practiced being design to exclude moisture. In cases where extra protection is required, preservative treatments are commonly employed. For the purposes of

© 2014 American Chemical Society

this chapter, we will consider treatment strategies that coat or impregnate wood with biocides. Although water repellant systems can also be employed (5, 6), they will not be considered here, although many of these treatments are delivered into the wood using the same processes used for biocides. Biocidal wood protection has been the predominant industrial practice for over a century. Wood modification techniques, including heat treatment and chemical modification, are the subject of much research and increasing commercial practice, especially in Europe; however, they will not be discussed here because they involve different principles and are only use to a limited extent.

Wood treatment provides a barrier against biological agents. The extent of the barrier required depends on the biological hazard and the consequences of wood failure, and this will in turn be instrumental in determining the most appropriate process for delivering the preferred preservative system to the required depth at an effective loading. There are a variety of approaches for predicting the risk of decay (7, 8). The process of selecting a treatment involves considering the application, the environmental conditions, the treatment chemical, the solvent, and the wood species. For example, there is little value in placing large quantities of a chemical deep in the wood if it is primarily there as a surface protectant and is not expected to perform for long periods. Conversely, a shallow treatment that is easily compromised serves little purpose in an application where the product is expected to last for decades.

The Wood Substrate

We may view wood as a collection of parallel tubes running longitudinally between the roots and the foliage or perpendicular to this direction (radially) from the pith to the bark (ray tissue) (9). These tubes move fluids in the living tree and they differ among species in a number of important ways. The greatestr differences among woods arise between hardwood and softwoods (or angiosperms and gymnosperms). Softwoods have two cell types, tracheids and parenchyma, which can be oriented either longitudinally or radially. Longitudinal tracheids are most abundant and are dead, empty tubes that transport fluids from the roots to the needles. Parenchyma cells, which may remain alive in the wood for years, may contain materials such as starches, sugars, proteins and lipids for the tree. There are also epithelial cells in some species that produce resin, but these do not appreciably influence wood treatment. Wood cells are connected to one another through pits that can be simple, bordered or semi-bordered. Each pit has a semi-permeable membrane that restricts flow. The size of the openings in this membrane can be considered to be the limiting factor in movement of fluids through softwoods. Fluids with large particles or with high viscosity may plug the pits, limiting flow. In addition, the size and number of pits can affect flow. The pits in the sapwood tend to be open, allowing fluids to move between cells. As the sapwood ages and the parenchyma die, these pits can become encrusted with hydrophobic materials that block flow and/or the pit membranes can close to become aspirated. These processes result in sharply reduced permeability. The pits in the heartwood of most species are extremely resistant to fluid movement. Thus, preservative treatment of softwoods tends to occur primarily through the tracheids in the sapwood.

Hardwoods are structurally more complex than softwoods. Hardwoods have parenchyma and tracheids but liquids are moved from the roots to the leaves mostly through vessels, long series of cells that are individually called vessel elements. These short cells are connected by sieve plates that generally allow for liquid movement; however, vessels in the non-conductive regions (esp. heartwood) of the wood can be blocked by gums or tyloses. The vessels are surrounded by thick walled fibers that provide structural support for the tree but do not function in conduction; liquid movement among fibers is extremely limited. As with softwoods, the limiting factor in liquid flow will be the smallest pore size, which will be in the pits. Thus, even though vessels in a given hardwood may be open and receptive to treatment, the surrounding fibers may be highly resistant to fluid movement, resulting in inconsistent treatments that allow decay to develop in seemingly well-treated wood.

The overall ability of a liquid to move into wood can be described using the viscosity of the fluid, the length the fluid must move, the difference in pressure between the surface and the interior, and the pore size (*10, 11*). The effects of each of these components can be used to predict flow according to Poiseuille's Law where:

$$Q = (\pi r^4 \Delta P)/8\eta L$$

Where Q = Flow
R = radius of the capillary (in this case, the pits)
P = Pressure
η = viscosity
L = length of the flow path

The most influential component of this flow equation is pore size, since it is raised to the fourth power, but one can see where increasing pressure or reducing viscosity can also improve treatment and these are the two primary factors addressed in preservative treatment. In general sapwood is relatively easily treated, but heartwood of some species poses a major challenge to impregnation (Table 1). Pore size is very difficult to alter, although incising and through boring/radial drilling are employed in an attempt to create longer flow paths. Incising involves driving sharp metal teeth into the wood (Figure 1). The process increases the amount of cross section exposed to preservative flow and, because fluids flow more easily in the longitudinal direction, incising improves preservative treatment to the depth of the incisions (*12*). Incising is required for treatment of many species in the Western U.S. and is also used to help accelerate drying on hardwood railway ties (sleepers). Through-boring involves drilling slightly angled holes perpendicular to the grain through a timber or pole in areas where deterioration is most likely to occur, such as the area about the groundline. Like incising, the process exposes end grain to preservative flow, producing deeper, more uniform treatment in the drilled area.

Table 1. Relative Difficulty of Impregnating the Heartwoods of Selected Wood Species[a]

Treatability	Softwoods	Hardwoods
Less Difficult	Ponderosa pine (*Pinus ponderosa*)	Blackgum (*Nyssa sylvatica*)
	Redwood (*Sequoia sempervirens*)	Red oak (*Quercus* spp.)
		White ash (*Fraxinus americana*)
Moderately difficult	Coastal Douglas-fir (*Pseudotsuga menziesii*)	Maples (*Acer* spp.)
	Loblolly pine (*Pinus taeda*)	Cottonwood (*Populus* spp.)
	Sugar pine (*Pinus lambertiana*)	Mockernut hickory (*Carya tomentosa*)
Difficult	Grand fir (*Abies grandis*)	Sycamore (*Platanus occidentalis*)
	Sitka spruce (*Picea sitchensis*)	Hackberry (*Celtis occidentalis*)
	White spruce (*Picea glauca*)	Yellow poplar (*Liriodendron tulipifera*)
Very Difficult	Inter-Mountain Douglas-fir (*P. menziesii*)	Sweetgum (*Liquidambar styraciflua*)
	Tamarack (*Larix laricina*)	Black locust (*Robinia pseudoacacia*)
	Western redcedar (*Thuja plicata*)	White oaks (*Quercus* spp.)

[a] Information taken from source (*13*).

The other primary factor affecting treatability is the wood moisture content. Freshly cut wood has a moisture content ranging from 40 to over 100% depending on wood species (Table 2). Some of this moisture must be removed prior to treatment in order to create space for the treatment chemical. Moisture can be removed from wood by air-seasoning, kiln drying, steam conditioning, or Boulton seasoning (heating under vacuum). The amount of moisture that must be removed depends on the wood as well as the treatment chemical. In theory, an individual wood cell that is at or below the fiber saturation point (~25-30% moisture content) will be sufficiently free of liquid water and therefore accepting of treatment liquids. In reality, there is almost always a gradient of moisture between the wetter core and the drier surface of the wood. Generally, wood is most readily treated at bulk moisture contents between 20 and 40 %. Wood can become more difficult to treat as moisture contents decline below 20 % MC because the drying can be accompanied by pit aspiration.

Figure 1. Incisor teeth penetrate into the wood surface to increase the amount of end-grain exposed to potential preservative flow.

Some preservatives may bond chemically to the wood substrate ("fixed" biocides, eg. Copper, chromium or arsenic) and are relatively immobile in the wood in service. Others are non-reactive but also insoluble in water and thus, once deposited in the wood, mostly remain in place. Examples include the oil-borne preservatives such as pentachlorophenol or copper naphthenate. Creosote is similar in this regard, but the preservative is also the solvent. Still others, notably boron-based preservatives, are water-soluble and will diffuse into (and out of) wet wood. For diffusible preservatives, movement within the wood can be described according to Fick's law (*11*):

$$J = -D\frac{\partial \phi}{\partial x}$$

where

- J is the "diffusion flux" [(amount of substance) per unit area per unit time], example $\left(\frac{mol}{m^2 \cdot s}\right)$. J measures the amount of substance that will flow through a small area during a small time interval.
- D is the **diffusion coefficient** or **diffusivity** in dimensions of [length^2 time^{-1}], example $\left(\frac{m^2}{s}\right)$
- ϕ (for ideal mixtures) is the concentration in dimensions of [amount of substance per unit volume], example $\left(\frac{mol}{m^3}\right)$
- x is the position [length], example **m**

In practice, the diffusivity of the treated wood is affected by temperature and, especially, by moisture content. Diffusible preservatives will not move appreciably in dry wood, but diffusion will continue as long as the wood is wet. Wood that is dip or spray treated with diffusible preservatives must be kept wet to achieve deep penetration in the wood, but moisture contact with treated wood in service will leach diffusible preservatives. The concentration gradient is also an important practical consideration in diffusible treatments. For example, borates can generally only be dissolved in water to concentrations of 15 to 20 %. Thus, dipping wood in this chemical can only deliver a limited amount to the surface and the resulting concentration in the wood (assuming complete diffusion) will be dependent on the wood dimensions. Surface application of high-concentration colloids and the insertion of borate solids (e.g. fused rods) into the wood are two approaches for achieving higher gradients.

Choice of Treatment Process

Treatment processes can be divided into non-pressure and pressure. Non-pressure processes include spraying, dipping and soaking, while pressure processes use combinations of vacuum and pressure to force the chemical more deeply into the wood.

Dip or Spray on Systems

There is a long history of protecting wood by dipping or spraying with various concoctions and these techniques are still commonly used for providing temporary protection, e.g. sapstain prevention prior to lumber drying (*12*). This approach delivers a relatively small amount of chemical to the wood surface. The chemical can move into the wood via capillary flow; however, the depth to which fluids move into wood in this process is typically limited to a few mm, depending on the wood species, grain orientation (greater penetration on end-grain) and whether the wood is sapwood or heartwood. Dipping or soaking typically are used where minimal preservative penetration is required. These applications include temporary (<6 months) surface protection of freshly sawn lumber against stain and mold fungi. Dipping is also used for treatment of window and door frames where a shallow treatment across the grain plus deeper uptake along the grain is sufficient to protect wood used in above-ground applications in combination with a protective paint film.

Table 2. Moisture Contents of Sapwood and Heartwood of Various Species at the Time of Cutting[a]

Wood Species	Moisture Content(%)	
	Heartwood	Sapwood
White ash (*Fraxinus americana*)	46	44
American Beech (*Fagus grandifolia*)	55	72
Cottonwood (*Populus* spp.)	162	146
Sugar Maple (*Acer saccharum*)	65	72
Southern Red Oak (*Quercus* spp.)	83	75
White oak (*Quercus alba*)	64	78
Yellow poplar (*Leriodendron tulipifera*)	83	106
Western redcedar (*Thuja plicata*)	58	249
Coastal Douglas-fir (*Pseudotsuga menziesii*)	37	115
Loblolly pine (*Pinus taeda*)	33	110
Black Spruce (*Picea mariana*)	52	113

[a] Information taken from source (*13*).

Soaking is sometimes used where users have time to allow the materials to soak for long periods. Fence posts represent an excellent example of this type of application and oilborne preservatives are usually used in these applications. Preservative uptake can be improved slightly by heating the oil thereby reducing viscosity and improving flow into the wood. As the wood is heated in the treating solution, the air inside expands. The air contracts as the wood cools, drawing additional chemical into the wood. These processes may increase wood service life 3 to 5 fold over non-treated wood (*14, 15*).

As introduced above, dip treatment can also be used for water-diffusible chemicals such as boron. The potential advantages of this approach are greater penetration of preservative than is possible with dip treatments with non-diffusibles and the ability to treat wet wood. The wood is dipped for a short time in a concentrated aqueous solution of borate and, if the wood is wet initially or becomes wet after treatment, the boron diffuses inward with moisture (*16*). This approach is most commonly used in applications where the wood will be protected from wetting and not in soil contact since the boron can diffuse out of the wood under these conditions. However, another use for diffusibles is for dual treatment of railway crossties where borates are dip/diffusion-applied and this is followed by pressure application of an industrial preservative such as creosote. Ties treated in this manner have performed well in field trials and the process is becoming widely used in North America (*17*).

Pressure Treatments

In most cases, non-pressure processes cannot deliver a sufficient amount (retention) of chemical deeply enough into the wood (penetration) to provide the protection required for longterm performance in industrial or residential applications. In these cases, preservative treatment can be improved using combinations of vacuum and pressure, sometimes with heating. These processes are performed in a vessel or cylinder, often called a retort, capable of withstanding pressure/vacuum conditions (Figure 2). Attached to the retort are vacuum and pressure pumps along with tanks for containing the treatment chemical, and heaters for maintaining treatment temperatures.

Vacuum treatments use only atmospheric pressure to treat poles of thin sapwood species such as western redcedar or to impregnate windows, doors, and other products that will be used out of direct soil contact in somewhat protected applications. The vacuum removes air from the wood and the preservative solution is introduced as the vacuum is relieved. The preservative is then drawn into the wood by atmospheric pressure.

Thermal oscillations can also be used to induce vacuum conditions that pull preservative into the wood. This can be achieved by alternately submerging wood in heated and ambient temperature preservative solutions. This avoids the need for a treatment cylinder and makes it possible to treat only part of the wood piece. An example of this is the "butt-treatment" of cedar utility poles. The bottom ends of the poles are submerged in a tank and hot and cold creosote solutions are pumped through the tank. The vacuum created when the air inside the wood expands and contracts with temperature changes pulls the preservative oil into the relatively thin sapwood shell. Because the heartwood of the pole is naturally durable, the above-ground section of the pole doesn't require preservative treatment. This process is now performed in closed vessels because of concerns about losses of volatile organic compounds into the surrounding environment.

Figure 2. Example of a wood treating vessel or retort.

Pressure treatment was first patented in the 1830's by John Bethel (*18*). The first process used an initial vacuum to remove air from the wood. Preservative was then introduced and then pressure was raised. The amount of preservative injected into the wood is monitored using various gauges and, once the desired amount has been delivered, the pressure is released. The pressure release results in a certain amount of chemical being expelled from the wood. This material, called kickback, is recovered and reused. The full cell or Bethel process results in maximum uptakes of chemical to the maximum depth of treatment. The full cell process is typically used to treat wood to high retentions for marine applications or for use with treatment of water borne materials where the concentration of preservative can be adjusted for the intended application. There is also a modified full cell process that uses smaller vacuums prior to introduction of the treatment solution and longer vacuums at the end of the process. These modifications are designed to produce lower solution uptakes that reduce weight and decrease transportation costs.

Full cell treatments are useful, but they can deliver too much chemical or organic solvent for many applications. Two other processes, called empty cell processes, were developed after the full cell treatment to reduce the levels of preservative retention. In the Lowry process, the preservative is added to the treatment cylinder without any initial vacuum. As the preservative is forced into the wood, a small amount of air is compressed at the center of the wood. Once the desired preservative uptake is achieved, the pressure is released and the compressed air expands, forcing out some of the preservative solution originally impregnated into the wood. This added kickback results in much lower retentions of preservative. The Rueping process introduces air pressure into the treatment cylinder before the treatment chemical is added. The pressure is then raised and held as in the other processes. This initial air pressure results in more air trapped at the center of the wood, which results in more kickback and an even lower preservative retention.

The empty cell processes are typically used to impregnate wood with oilborne preservatives for terrestrial applications. These include utility poles, railroad ties and timber bridges. Wood in these applications is exposed to moderate to high risk of decay from terrestrial organisms but the levels of preservative required are much lower than those that might be required in a more severe exposure such as a marine environment. The empty cell processes tend to be very adjustable, although most plants use similar initial air pressures (30 to 50 psi; 207 to 345 kPa). Maximum pressure levels are usually set by the American Wood Protection Association based upon the ability of the wood species to withstand pressure. For example, the maximum pressure for species susceptible to collapse such as western redcedar are low (100 psi or 700 kPa) while pressures up to 250 psi (1750 kPa) can be used for oaks.

Treatment Results

Wood treatments are usually evaluated in terms of the depth of preservative penetration and the amount of chemical in the wood. Penetration is important because the preservative treatment produces a barrier or envelope of protection against attack by wood attacking agents. Poor penetration will result in a shallower barrier that is more easily compromised in service. The amount of preservative or retention can be expressed as a % wt basis or on a wt of chemical per unit volume (kg/m^3 or lb/ft^3).

Even with pressure treatment processes, the depth of treatment is usually limited by the depth of sapwood. In some countries, retentions are based upon the entire cross section of the wood, ignoring whether the wood is sapwood or heartwood. In other cases, the retentions are based upon a analysis of wood at a specific depth from the surface or assay zone. The assay zones differ with wood species and intended end-use, with deeper assays zones for larger material used under more critical applications.

Treatment quality can be assessed in a number of ways. The simplest is to use the gauges on a treating plant to measure the total amount of preservative delivered into the wood (representing the gross amount of preservative injected minus the kick-back). This can be divided by the total volume of wood in the cylinder to provide "gauge retention." The precision of this measurement can be improved if the relative amount of wood that was actually treated is known, since most of the treatment will be limited to the sapwood or, in the case of more difficult to treat materials, to the incised zone. This approach is useful if the preservative is difficult to detect or measure, however, it provides little knowledge about the variation in treatment among individual pieces. Treatment results can vary widely among different pieces of wood and ensuring some uniformity in treatment is important for producing materials that will perform well in service. For difficult-to-treat species, some treatment standards specify a gauge retention value but provide the option of accepting treatment 'to refusal', i.e. the wood will accept no more treatment chemical, as judged by gauge readings.

As an alternative to gauge measurements, the treatment can be assessed on samples removed from the treated pieces for preservative penetration and retention. These samples can be obtained in a number of ways. For shallow treatments such as those applied by dipping or spraying, chips can be removed from the surface for analysis. This approach is most often used for chemicals employed for mold and sapstain prevention. Samples for deeper treatments can be obtained using increment borers, which are hollow tubed cutters that remove a solid core of wood. This core can be examined for preservative penetration and the appropriate segment of the core (i.e. the assay zone) can be ground and, in combination with cores removed from other pieces of wood, analyzed for preservative retention. This approach is widely used in North America for assessing treatment quality of pressure treated wood. Alternatively whole production samples can be analyzed, but many producers object to this approach because of the added cost.

The processes, preservatives and treatment levels used in North American have traditionally been set by the American Wood Protection Association (AWPA) in the U.S. and Canadian Standards Association (CSA) in Canada. Founded in 1904, the AWPA is a non-governmental consensus standards writing body consisting of wood treaters, chemical suppliers, users of the treated products and general interest members who meet to review technical data supporting various standards. The AWPA Standards are results based, meaning that, within broad parameters, the treater can use varying levels of vacuum, pressure, heat or other process parameter to meet a target preservative penetration and retention set for a given degree of risk of decay (7). The risk of deterioration or "Use Category" ranges from UC1 (interior of a house) to UC 5 (marine environment) with increasing amounts of preservative specified for each increased level of risk. CSA plays a similar role in Canada. Globally, virtually all regions have either country-specific standards or in the case of Europe, European Union Standards and in Austral-Asia- the Australia/New Zealand Standards. Each set of standards differs in approach, but the goals are to produce reasonably uniformly treated material that will perform reliably in the intended application under the climatic conditions present.

Future Wood Treatments

The vast majority of preservative treated wood is produced using processes that date to the 1830's or early part of the 20th century. These processes remain viable because they meet the needs of the user, but there remains a desire to develop improved methods for delivering chemicals into wood. Specifically there are three fundamental challenges that need to be addressed:

1) Refractory woods that resist impregnation. Many species are exceedingly difficult to impregnate with conventional preservative solutions. Expanded use of these species under adverse environmental conditions will require effective treatment. Even in species that are relatively amenable to treatment, variability in treatment is high. Uniform preservative retention and penetration is critical for reliably producing durable wood products that can compete with alternative materials.

2) Treatment of composites. Although structural plywood is often pressure treated with water borne preservatives and glued laminated timbers are treated with oil borne systems using processes similar to those used for solid lumber, many composite materials would experience unacceptable dimensional changes with such treatments. The most common methods for protecting other composites involve glue-line additives, i.e. the preservative is added to the furnish along with the adhesives and other additives during the blending step, prior to the panel layup and consolidation. Zinc borate is the most common biocide used for protection of oriented strandboard against termite attack in the U.S., while a variety of organic insecticides are added to resin for plywood and laminated veneer lumber. All of these products, however, are intended for interior use where they are protected from wetting. Expansion of composite wood applications into adverse environments will require the development of processes capable of fully impregnating these products without adversely affecting other composite properties. This is a daunting challenge.

3) Reduce the potential for preservative migration into the surrounding environment. Almost all preservatives have some degree of water solubility and this mobility has the potential to affect non-target organisms. A variety of methods, termed Best Management Practices, have been developed to reduce the risk of migration, but further research will be needed to continue to improve the environmental footprint of these systems (*19, 20*).

A number of processes have been evaluated for improving treatment. Most involve either altering the pressure cycle to overcome the inherent resistance of wood to fluid flow or altering the fluid characteristics.

Process developments such as the oscillating pressure and alternating pressure processes have been available for decades. Both use varying pressures to either disrupt aspirated pits or remove debris and clear flow paths for the preservative solution. These processes produce marginal gains in treatment that may allow the use of shorter treatment cycles, but are not capable of overcoming the refractory nature of some wood species. Attempts have also been made to increase pore size using fungi and bacteria. Bacteria are known to degrade pits in logs stored for long periods in water and many fungi can produce similar effects (*21*). A number of studies have explored methods for encouraging fungi to colonize wood to remove pits. Most studies have used *Trichoderma* spp or other non-wood degrading molds, but less aggressive white rot fungi have also been explored for this purpose. At present, none of these processes are commercially used.

Vapour phase treatments were simultaneously developed in the 1980's in New Zealand and the United Kingdom (*22*). These processes introduce trimethyl borate under a vacuum. The trimethyl borate reacts with water in the wood, leaving boric acid and methanol. The latter is recovered using vacuum and the boron is uniformly distributed throughout the wood. This process is only suitable for very dry wood such as panel products and these products could only be used in interior applications because of the susceptibility of boron to leaching. An alternative process using vapour copper also showed some promise on the laboratory scale, but neither process proved to be commercially feasible. The vapour approach, however, does suggest that gas-phase treatments can readily penetrate seemingly refractory wood.

Supercritical fluid (SCF) processes represent another approach for altering the treatment fluid (*23*). SCF's are materials with properties somewhat in-between those of a conventional liquid and a gas. Some SCF's have solvating properties that approach those of a liquid but gas-like abilities to penetrate materials. SCF processes are infinitely adjustable although the pressures employed are 10 to 15 times higher than those used in conventional treatment processes and care must be taken to limit the risk of developing steep pressure gradients that can damage the wood. Extensive laboratory trials have shown that supercritical carbon dioxide can be used to deliver a range of biocides into solid wood and composites at levels that will provide protection against fungal and insect attack. Although the process shows great promise, the initial costs for a treating facility have limited application and there is currently only one commercial treatment facility in the world.

While the development of new treatment processes for impregnating wood-based materials has proven daunting, there remains a perhaps more important challenge. Traditionally, wood protection has involved delivery of chemicals into wood to protect against biological attack. While biological attack is important, wood is also susceptible to physical damage, most notably from the wet/dry cycles that lead to warping and ultraviolet light that leads to weathering. There is compelling evidence that poor appearance caused by physical degradation leads to substantial premature replacement of treated wood. Some commercial water-borne wood preservative systems include water repellents, presumably to reduce rates of wetting, but it is unclear if this has much real impact on the performance of the treated wood, or on the customers' perception of the product's quality. If wood is to remain a viable construction material in exterior applications while taking advantage of its renewability, then treatments must protect against both biological and physical degradation. The development of the plastic and wood/plastic decking markets clearly attest to the role of appearance in performance and the willingness of consumers to pay for premium products with better performance attributes. Effectively treated wood can also fill this niche, but we need to develop systems that provide water repellency, UV protection, and resistance to biological attack.

References

1. Scheffer, T. C.; Cowling, E. B. *Annu. Rev. Phytopath.* **1966**, *4*, 147–170.
2. Eaton, R. A.; Hale, M. D. C. *Wood: decay, pests and protection*; Chapman & Hall: New York, 1993.
3. Zabel, R. A.; Morrell, J. J. *Wood Microbiology: Decay and its Prevention*; Academic Press: San Diego, CA, 1992; 474 pages.
4. Boyce, J. S. *Forest Pathology*; McGraw-Hill: New York, 1961.
5. Banks, W. B.; Voulgaridis, E. V. In *Record Annual Convention British Wood Preserving Association*; 1988; pp 43–52.
6. Hill, R. R. *J. Oil Colour Chem. Assoc.* **1973**, *56*, 251–258.
7. American Wood Protection Association (AWPA). *Annual book of Standards*; AWPA: Birmingham, AL, 2010.
8. Scheffer, T. C. *For. Prod. J.* **1971**, *21* (10), 25–31.
9. Panshin, A. F.; deZeeuw, C. *Textbook of Wood Technology*; McGraw-Hill: New York, 1980; 703 pages.
10. MacLean, J. D. *USDA Agricultural Handbook 40*; USDA: Washington, DC, 1952; 160 pages.
11. Siau, J. F. *Wood: Influence of moisture on physical properties*; Virginia Tech Inst and State Univ.: Blacksburg, VA, 1995; 127 pages.
12. Hunt, G. M.; G.A. Garratt, G. A. *Wood Preservation*; McGraw-Hill: New York, 1967; 457 pages.
13. U.S. Department of Agriculture. *Wood Handbook: Wood as an Engineering Material*; General Technical Report FPL-GTR-190; U.S. Forest Products Laboratory: Madison, WI, 2010; 508 pages.
14. Crawford, D. M; Woodward, B. M.; Hatfield, C. A. *Res. Note FPL−RN−02*; U.S. Department of Agriculture, Forest Service, Forest Products Laboratory: Madison, WI, 2002.
15. Morrell, J. J.; Miller, D. J.; Schneider, P. F. *Research Contribution 26*; Forest Research Laboratory, Oregon State University: Corvallis, OR, 1999; 24 pages.
16. Lloyd, J. D. Second International Conference on Wood Protection by Diffusible Preservatives and Pesticides, Mobile, AL, November 6–8, 1996; For. Prod. Soc.: Madison, WI, 1997; pp 45–54.
17. Amburgey, T. L.; Watt, J. L; Sanders, M. G. *Crossties* **2003** May/June, 1–5.
18. Graham, R. D. In *Wood Deterioration and its Prevention by Preservative Treatments*; Nicholas, D. D., Ed.; Syracuse University Press: Syracuse, NY, 1973; Volume I, pp 1–25.
19. Western Wood Preservers Institute. *Best management practices for the use of treated wood in aquatic environments*; Western Wood Preservers Institute: Vancouver, WA, 2006; 34 pages.
20. Western Wood Preservers Institute. *Treated wood in aquatic environments*; Vancouver, WA: Western Wood Preservers Institute, 2006; 32 pages.
21. Elwood, E. L.; Ecklund, C. A. *For. Prod. J.* **1959**, *9* (9), 283–292.
22. Turner, P.; Murphy, R. J.; Dickinson, D. J. *International Research Group on Wood Preservation*; 1990; Document No IRG/WP/3616.
23. Morrell, J. J.; Levien, K. L. In *Supercritical Fluid Methods and Protocols*; Williams, J. R., Clifford, A. A., Eds.; Humana Press: Totowa, NJ, 2000; pp 221–226.

Chapter 12

Copper-Based Wood Preservative Systems Used for Residential Applications in North America and Europe

Stefan Schmitt,[1] Jun Zhang,[2] Stephen Shields,[3] and Tor Schultz[*,4]

[1]Rütgers Organics, GMBH, Oppauer Strasse 43,
Mannheim D-68305, Germany
[2]Osmose, Inc., 1016 Everee Inn Road, Griffin, Georgia 30224, U.S.A.
[3]Lonza Wood Protection, 5660 New Northside Drive NW,
Atlanta, Georgia 30328, U.S.A.
[4]Silvaware, Inc., 303 Mangrove Palm, Starkville, Mississippi 39759, U.S.A.
[*]E-mail: tschultz.silvaware@bellsouth.net.

For many years chromated copper arsenate (CCA) provided excellent and economical protection for treated wood products used in residential and industrial applications. However, CCA was limited to non-residential applications in Europe and Japan in the 1990s and in North America in 2004, and replaced with other copper-based systems. This chapter reviews the major copper-based wood preservatives used commercially to pressure treat lumber and other wood products for the large residential markets in North America and Europe. Residential systems which currently are standardized but have only minor use in North America, and the likely short-term future trends in copper-based residential systems, are also discussed.

Introduction

Waterborne chromated copper arsenate (CCA) was developed in the 1930s to treat utility poles. CCA treated wood proved very effective in above-ground and ground-contact applications and, being waterborne, had no residual petroleum odor or oily surface. Thus, although originally developed for industrial

© 2014 American Chemical Society

applications, CCA also proved ideal for residential applications. Starting in the 1960s the use of CCA expanded rapidly with the growth of the residential treated wood market and over a 40-year period grew from 30% to 95% share in that market.

Numerous studies found that CCA treated wood products posed negligible risk in residential applications except when improperly burned. However, public perceptions alleging possible arsenic exposure and concerns regarding disposal of CCA treated wood led to the replacement of CCA preservative for use in residential products with other copper-based, waterborne systems in the 1990s in Japan and Europe and in 2004 in North America. These new systems have all undergone many years of testing, including extensive field trials, prior to approval by a standard-setting or product evaluation organization and subsequent commercialization. Further, third-party inspection protocols for treated wood help to ensure that the products meet the established requirements. These treated wood products provide home owners with effective, economical and renewable building materials that will give many years of satisfactory service in applications such as outdoor decks, privacy fences, benches, and fresh water docks.

This chapter discusses the major copper-based systems used for residential applications in North America in chronological order and briefly discusses other lesser-used copper systems which have been recently standardized, copper-based systems in Europe, and the likely near-term trends in preservative systems for residential uses. Many of the waterborne systems discussed below are also standardized for freshwater, industrial and agricultural applications.

Systems Employed in North America

Dissolved Copper Systems

Copper in the cupric or copper(II) oxidation state is effective against most wood destroying fungi and insects. As with all preservatives intended for pressure treatment, the system must be formulated in a liquid carrier as solubilized, emulsified, or as extremely small dispersed particles so that the biocide(s) deeply impregnate the porous wood matrix during the treatment process. All residential wood preservatives for pressure treatment are formulated with water as the carrier and are generally referred to as waterborne. Systems where the copper is in a soluble form are formulated using ammonia and/or monoethanolamine (MEA), an organic amine, which reacts with copper(II) to form water soluble copper-ammonia or copper-MEA complexes. MEA has become the dominant amine complexing agent as it eliminates or minimizes ammonia odor and reduces metal corrosion in treating plants from the treating solutions. Once impregnated into wood, the copper solution interacts with the wood lignin and cellulosic groups to form water insoluble copper compounds/complexes or precipitates, which render the copper leach resistant. While copper is effective against most wood-destroying fungi and insects, it is weak against some copper-tolerant fungi. To prevent infestation and subsequent deterioration by copper-tolerant fungi, a carbon-based co-biocide is added to all residential copper-based systems.

Alkaline Copper Quat (ACQ)

The first major waterborne system to replace CCA was ACQ. The co-biocide in ACQ is a quaternary ammonium compound (quat). Quats are commonly used as disinfectants in a variety of household cleaning products due to their low mammalian toxicity. In addition, quats are water soluble and can be readily formulated with waterborne copper solutions to make a wood preservative treating solution. When impregnated into wood, quats fix onto the acidic wood groups by ion-exchange mechanisms to become more leach resistant. Quats are effective against a broad range of wood-destroying fungi and insects and also control wood-inhabiting molds and stains.

Four different types of ACQ formulations are currently listed in the American Wood Protection Association (AWPA) 2013 Book of Standards (*1*). The classification of ACQ formulations is determined by the type of copper complexing agent, the type of quat and the ratio of copper to quat. All ACQ formulations, with the exception of ACQ-B, employ either monoethanolamine (MEA) or MEA/ammonia combination to solubilize copper. ACQ-A [AWPA P26] (*1*) consists of 50% copper as CuO and 50% quat as didecyldimethyl ammonium compound with chloride or carbonate/bicarbonate anions (DDAC). ACQ-B [AWPA P27] (*1*) is formulated with only an ammonia complexing agent and has a CuO:DDAC ratio of 2:1. Ammonia is used as the solvent as it facilitates the penetration of copper into refractory western species, especially Douglas fir. ACQ-C [AWPA P28] (*1*) is formulated with 66.7% CuO and 33.3% of alkylbenzyldimethyl ammonium compounds (BAC). ACQ-D [AWPA P29] (*1*) is formulated with 66.7% CuO and 33.3% DDAC.

The Use Class 4A (UC4A) ground contact retention used for general residential applications in North America for all ACQ formulations is 6.4 kg/m^3 (0.40 pcf). Lower retentions are specified for above-ground UC1-UC3B applications, due to the lower deterioration hazard for treated wood in above ground use.

Copper Azole (CA)

The second waterborne copper system introduced to replace CCA was copper azole. Copper azole (CA) is also based on copper solubilized in ethanolamine, but employs azoles as the co-biocide(s). Azoles, either tebuconazole or a mixture of tebuconazole and propiconazole, are extremely effective against basidiomycete brown and white-rot fungi, so relatively low levels are necessary to protect wood against fungi in comparison to quats. Azole fungicides are commonly used in agricultural applications and many home gardening sprays.

Currently, two CA systems are standardized by the AWPA. CA-B [AWPA P32] (*1*) is formulated with 96.1% copper as metal and 3.9% tebuconazole. CA-C [AWPA P48] (*1*) has the same copper to total azole ratio, but has a 1:1 mixture of propiconazole:tebuconazole.

The Use Class 4A (UC4A) ground contact retention used for general residential applications in North America for CA formulations ranges from 2.4 kg/m³ (0.15 pcf) to 3.3 kg/m³ (0.21 pcf).

Just as for ACQ, lower retentions are specified for above-ground UC1-UC3B applications due to the lower deterioration hazard for treated wood in above ground use.

Dispersed Particulate Copper Systems

In 2006 an entirely new type of copper based wood preservative was introduced. Instead of the copper being solubilized and stabilized by ammonia or MEA, basic copper carbonate is milled into fine particles. A dispersant is added during the milling process to keep the copper particles from clumping together and to maintain the dispersion when diluted in water for pressure treatment. The size of the basic copper carbonate particles for the commercial dispersed copper systems used in the U.S. is greater than 100 nanometers, with a mean particle size of about 300 – 500 nanometers. These systems are referred to as dispersed, particulate or micronized copper preservatives. They offer several advantages compared to soluble copper systems which include reducing the copper leaching and eliminating the nitrogen containing amine or ammonia which reduces the formulation cost and may also reduce undesirable surface mold. In addition, the concentrates can be formulated at much higher copper concentrations compared to soluble copper formulations which reduces shipping costs, and wood treated with dispersed copper wood preservatives have reduced corrosion properties to metal fasteners (*2–5*).

Dispersed or Micronized Copper Quat (MCQ)

The first commercial micronized copper system was MCQ. Much like ACQ-D and ACQ-A, this system is formulated with two CuO:quat ratios, 66.7% to 33.3% and 50% to 50%. Within a few years, this system was superseded by copper azole systems and MCQ is no longer commercially available in the U.S. market but may still be available in Canada. MCQ is listed in ICC Evaluation Service report (*6*) ESR-1980 with a ground contact retention of 5.44 kg/m³ (0.34 pcf).

Dispersed Copper Azole (DCA)

There are two commercial dispersed, or micronized or particulate, copper azole systems available in the US market, known as MCA (*6*) [ICC ESR-2240] or µCA-C (*6*) [ICC ESR-1721]. The latter system is sometimes formulated with both micronized and soluble copper; these systems will be generally referred to as DCA in this chapter. DCA has a formulation similar to soluble copper azoles, with 96.1% copper and either 3.9% tebuconazole or with 3.9% of a 1:1 ratio of tebuconazole:propiconazole. There is one commercial DCA preservative available in the Canadian market, using tebuconazole as the co-biocide (7). The azole co-biocide(s) can be formulated as a particulate dispersion or an emulsion as commonly used in the solubilized formulations. At the present time DCA systems reportedly account for about 70% of the total volume of copper-based waterborne residential systems employed in North America (5), although this value may relate only to the relatively large eastern US production. The ground contact retention specified by the ICC evaluation reports (6) for MCA and μCA-C in the US market is 2.4 kg/m³ (0.15 pcf) and 2.2 kg/m³ (0.14 pcf), respectively, with lower retentions employed for above-ground residential applications.

Other Standardized Copper-Based Waterborne Systems

Several other copper-based systems have been standardized by the AWPA (1) or listed in ICC Evaluation Service reports (6). Some of these systems, described below, have been recently or may be currently employed at a few treating facilities and their use may expand further.

Copper HDO Type A (CX-A), AWPA P33 (1) contains 61.5% copper as CuO, 24.5% boron as H_3BO_3, and 14% N-cyclohexyldizeniumdioxide (HDO) which complexes with some of the copper. This system is formulated as a soluble copper system employing ethanolamine. It is standardized by AWPA only for above-ground applications, UC1-UC3B.

Waterborne copper naphthenate (CuN-W), AWPA P34 (1), contains 5% copper as metal and 48% copper naphthenate formulated in ethanolamine. An oilborne formulation of copper naphthenate has a lengthy record of safe and effective use. Copper naphthenate has long been available at local hardware stores for homeowners to brush on the end-cuts of treated wood to give added protection to any untreated heartwood in the middle of the lumber. The waterborne system was recently standardized by AWPA as a pressure-treatment system. The UC4A ground contact residential retention is 1.76 kg/m³ (0.11 pcf).

Another ethanolamine based, soluble copper system (ACD) contains the effective isothiazolone biocide 4,5-dichloro-2-n-octyl-4-isothiazolin-3-one (DCOI) and has been standardized as ACD, AWPA P54 (1), but retentions have not yet been standardized for either above ground or ground contact use categories. ACD is formulated with 95.8% copper as CuO and 4.2% DCOI.

Two formulations of alkaline copper betaine (KDS, KDS-B) have been standardized in AWPA P55 and P56 (1). KDS is formulated with 47.2% copper as CuO, 30.2% borate as boric acid, and 22.6% betaine (polymeric betaine, dodecyl-bis, 2-hydroxyethyl ammonium borate, didecylpolyoxethylammonium borate, DPAB). Betaine is an oligomer based on alternating quat and borate ether units. Both the quat and borate groups in this oligomer can fix to wood. KDS-B is formulated with 67.7% copper as CuO and 32.3% DPAB. The standardized AWPA UC4A ground contact retention for KDS is 7.5 kg/m³ (0.47 pcf). The ground contact retention specified by the ICC Evaluation Services report ICC ESR-2500 (6) for KDS-B is 4.3 kg/m³ (0.27 pcf).

Non-Biocidal Additives

As with all products, manufacturers of wood preservatives seek value-added improvements to their products which will result in greater commercial appeal. Most of these improvements involve non-biocidal additives. One type of commonly used additive is waterborne wax emulsion water repellents, which are formulated as part of the preservative system for above-ground applications. Water repellents reduce undesirable dimensional change and checking of lumber used for decking and other above-ground applications by reducing water absorption during rainstorms. Another common additive is colorants or pigments to give enhanced visual appeal and slow surface graying of treated wood. Researchers are also examining additives to reduce photodegradation of the lumber surface, such as nano zinc oxide which is colorless to visible light but blocks UV radiation (8).

Trends in North America

In very general terms, treated wood in residential use is employed in either above-ground or ground-contact applications. We believe that for ground-contact applications, waterborne copper-based systems will continue to be the dominant preservatives employed in North America for the foreseeable future, for several reasons. First, several carbon-based waterborne systems are standardized and commercially available for above-ground use, such as EL2 and PTI (1). However, at this time no carbon-based or non-metallic system is effective for ground contact applications unless formulated in a heavy oil carrier but such formulations are not suitable for residential applications. Some nonbiocidal additives are available which enhance the efficacy of organic biocides, such as the antioxidant BHT or water repellents, but the high levels of these additives required to be effective for ground-contact applications apparently makes them uneconomical for waterborne systems. Wood-plastic composites (WPCs) are mainly used for above ground applications and this is likely to remain the case for the foreseeable future. Timbers cut from the heartwood of naturally-durable species, such as western red cedar, are available, but unless treated with a preservative these timbers do not perform satisfactorily in long-term ground-contact applications.

Copper-based systems are also the dominant product used in above-ground residential applications; however, there is a wider range of available products. In recent years, two carbon-based preservatives have been introduced in the U.S. market. Systems currently standardized by the AWPA include propiconazole / tebuconazole / imidacloprid (PTI) [AWPA P45] and DCOI / imidacloprid (EL2) [AWPA P47] (1). Both systems use a water repellent to give treated lumber improved dimensional stability and reduced decay hazard in above-ground uses. Further, while WPCs have found only a very limited market for ground-contact, in the past decade they have enjoyed a steadily increasing share of the above-ground decking market although this trend has recently ended. Finally, several plants, one in the US and two in Europe, have recently started producing chemically-modified wood as a premium decking product.

It might also be noted that a modified copper azole formulation was recently released which includes a third biocide, a quat, to provide greater protection against certain copper-tolerant fungi (*9, 10*). In this system the quat is used as an additive, that is, in addition to the required components in the standardized system. Another laboratory independently found that the combination of copper/azole and the quat DDAC was synergistic against all three copper-tolerant fungi tested (*11*). It is expected that evolution of copper-based products will continue with additional use of biocidal and non-biocidal additives to enhance the treatment and protection of wood.

In 2013, soluble azole-based systems gained some market share at the expense of the quat-based preservatives, especially in the eastern U.S., and this trend may continue. Also, dispersed copper systems may gain some market share in ground-contact applications in Canada with the recent CSA listing (*7*) for dispersed copper tebuconazole, and if one or more DCA systems are standardized by the AWPA they may gain further market share in the eastern U.S. market. In the past decade WPC products have gained some market share from copper-based systems for decking applications, but recently WPCs apparently have lost some decking market share, which suggests that the use of WPCs may be relatively level or slightly decrease over the next few years. For treating the refractory western US and Canadian softwoods, it is likely that soluble copper ammonia/amine systems will continue to be dominant. For the foreseeable future, unless unexpected governmental policies or regulations restrict their use, waterborne copper-based systems will likely continue to be the dominant preservatives for ground-contact residential applications in North America. Copper-based systems currently are also the major preservative for above-ground applications but carbon-based systems have had a recent increase in market share.

Waterborne Copper-Based Systems in Europe

The North American market is estimated to be about 60% of the world-wide residential market, while Europe is estimated to have about a 20% share (*12*). As discussed below, the environmental concerns which North America is just now facing concerning leaching and disposal of metallic-treated lumber has been an issue in Europe for a longer period.

In 1998 the Biocidal Product Directive (BPD) (*13*) went into force and was superseded by the Biocidal Product Regulation (BPR) in 2013. This legislation focused on environmental and toxicological evaluations of biocidal products in a two-step process: first the active ingredients and, second, the products. Low critical values in the assessments has resulted in a limited use of some active ingredients such as boric acid in use class 3 or 4 due to its high leaching potential. The high cost associated with the development of a new biocide followed by the lengthy registration process has generally inhibited the development of new active ingredients for the relatively small wood protection market (*14*). As a result, during the past decade few new wood preservatives have become available while many of the traditional biocides have been restricted. The leaching potential of borates and their possible health effects (*15*) has also resulted in suppliers removing this biocide from their systems.

Typical copper-based wood preservatives for use class 3, above ground, are products which contain azoles, quats, betaine or HDO as the co-biocide, much as in North America. However, as Europe generally has a lower decay hazard than many parts of North America, wood for above ground use can sometimes be effectively protected with copper-free wood preservatives, although weather degradation for these metal–free systems remains a drawback. Consequently, while copper-based systems remain the dominant above ground systems employed throughout Europe, effective waterborne carbon-based systems for above-ground applications are gaining market share, with Germany, France, Scandinavia and Switzerland already having registered products on the market. While this remains a limited trend in Europe the practice is somewhat ahead of what is currently occurring in North America where carbon-based systems have only been available in the mass market for the past one or two years.

In use class 4, ground-contact, copper used with carbon-based co-biocides (azoles, quats, betaine or HDO) is still employed for ground-contact residential applications. Due to the milder climatic conditions in most of the European areas, the copper level is typically lower than in North America. The copper concentrations approved in Europe do however differ from country to country, depending on the different evaluation criteria, the effectiveness of the carbon co-biocide, severity of use and actual performance in use. While regions like Scandinavia have set retentions based on field test data for many years, there has been a clear recent trend in other countries to also set retentions based on long-term field performance as opposed to laboratory tests. France and the UK now require five or even ten years field data to support certain approvals, and both countries have more than one use class 4 retention to take into account severity of use and desired service life.

In the last few years, micronized copper systems have been introduced to European market, and there are currently several UK and Scandinavian treating companies using a micronized copper system. Micronized pigment additives are often used in the preservative treating solutions to improve the esthetic appearance as well as enhance the UV stability of the treated wood.

Even greater focus is now being given to preservative penetration. In addition to the ability of a preservative to fully penetrate permeable sapwood, high intensity incising is emerging as a technique that delivers improved penetration of heartwood and European refractory species. Besides the common vacuum-pressure treatment process, an oscillating pressure/vacuum process is employed in the Germanic regions of Europe. A continuous change between pressure and vacuum allows the treatment of wet wood with a moisture-content above 60 % without any negative impact on the preservative penetration.

Even though unpublished studies have shown a better environmental footprint for preservative treated wood durable wood species are an alternative in Europe, mainly in use class 3. Further, to avoid the in-ground contact of treated or naturally-durable wood materials other than wood such as metal, plastics or stone are often used to build fences.

References

1. *AWPA 2013 Book of Standards*; AWPA, P.O. Box 361784, Birmingham, AL 35236.
2. Freeman, M. H.; McIntyre, C. R. *For. Prod. J.* **2008**, *58* (11), 6–27.
3. McIntyre, C. R.; Freeman, M. H.; Shupe, T. F.; Wu, Q.; Kamdem, D. P. *Int. Res. Group Wood Preserv.* 2009, IRG/WP 09-30513.
4. McIntyre, C. R.; Freeman, M. H. *Int. Res. Group Wood Preserv.* 2011, IRG/WP 11-30564.
5. Freeman, M. H.; McIntyre, C. R. *Int. Res. Group Wood Preserv.* 2013, IRG/WP 13-30609.
6. International Code Council (ICC) Evaluation Services, Inc., reports available at www.icc.es.org (accessed January 17, 2014).
7. Canadian Standards Association (CSA) CSA-080 Series 08 (R2012), available at www.shopcsa.ca (accessed January 17, 2014).
8. Evans, P. D. In *Development of Commercial Wood Preservatives: Efficacy, Environmental, and Health Issue*; Schultz, T. P., Militz, H., Freeman, M. H., Goodell, B., Nicholas, D. D., Eds.; ACS Symposium Series 982; American Chemical Society: Washington, DC., 2008; pp 69–117.
9. Arch/Lonza. *Wood Treated Right*; http://www.wolmanizedwood.com/Products/Preservative/genuine/default.htm (accessed January 17, 2014)
10. International Patent Application, WO 2100/161404 A1.
11. Nicholas, D. D.; Schultz, T. P. *Wood preservative composition which is synergistic against copper tolerant fungi*; Miss. State Univ. Invention Disclosure 2012.
12. Preston, A. F. Personal Communication 2013, Apterus Consulting.
13. Leithoff, H.; Blancquaert, P.; van der Flass, M.; Valcke, A. In *Development of Commercial Wood Preservatives: Efficacy, Environmental, and Health Issue*; Schultz, T. P., Militz, H., Freeman, M. H., Goodell, B., Nicholas, D. D., Eds.; ACS Symp. Series 982, ACS, Washington, D.C. 2008, pp 564 – 581.
14. Preston, A. F. In *Wood Deterioration and Preservation: Advances in Our Changing World* Goodell, B., Nicholas, D. D., Schultz, T. P., Eds.; ACS Symposium Series 845; American Chemical Society: Washington, DC., 2008; pp 372 – 377.
15. *Reproductive and General Toxicology of some Inorganic Borates and Risks Assessments for Human Beings*; European Centre for Ecotoxicology and Toxicology of Chemicals Tech. Report No. 63, Brussels, Belgium, 1995.

Chapter 13

Microdistribution of Copper in Southern Pine Treated with Particulate Wood Preservatives

Philip D. Evans,*,[1,2] Hiroshi Matsunaga,[3] Holger Averdunk,[2] Michael Turner,[2] Ajay Limaye,[4] Yutaka Kataoka,[3] Makoto Kiguchi,[3] and Tim J. Senden[2]

[1]Department of Wood Science, University of British Columbia, Vancouver, V6T1Z4, Canada
[2]Department of Applied Mathematics, The Australian National University, Canberra, ACT 0200, Australia
[3]Forestry and Forest Products Research Institute, Tsukuba 305-8687, Japan
[4]VizLab, Supercomputing Facility, The Australian National University, Canberra, ACT 0200, Australia
*E-mail: phil.evans@ubc.ca.

The micro-distribution of preservatives in wood has fascinated many scientists. Their fascination with this subject stems from the challenges of mapping and quantifying preservative elements in the complex porous micro-structure of wood and the even greater challenge of linking the micro-distribution of preservatives to the performance of the treated wood in the field. Our studies attempt the former using field emission scanning electron microscopy and energy dispersive analysis of X-rays in combination with X-ray micro-computed tomography. We have focused on the micro-distribution of copper in southern pine treated with commercial aqueous wood preservatives containing micron and nanometer-sized particles of basic copper carbonate. These preservatives were commercialized in North America in 2006 and we were the first to study their micro-distribution in treated wood. This chapter describes the results of our studies and the analytical techniques that have been used to study the micro-distribution of preservatives in treated wood.

© 2014 American Chemical Society

Particulate Wood Preservatives

Particulate copper preservatives are well described in both the scientific (*1*, *2*) and patent literature (*3–6*). The patent literature describes a variety of aqueous preservative formulations, typically containing small 'micronized' particles ranging in size from 0.005 microns (5 nm) to 25 microns (Figure 1). The formulations may contain micronized metal or metal compounds on their own or in combination with soluble or water-dispersible or micronized organic biocides. Formulations are also mentioned that contain both micronized organic biocides and soluble metal compounds. Commercial formulations in use today typically consist of micronized basic copper carbonate, an organic co-biocide and unspecified dispersants (*2*). Micronized particles are produced by grinding the metal compound together with a dispersant. Particles with the desired size distribution are obtained by adjusting the ratio of dispersant to metal compound, and/or altering grinding times or the size of grinding media. Grinding should eliminate large particles that can block flow paths in wood and prevent micronized particles from penetrating wood. The importance of adjusting the sizes of micronized particles to increase degree of penetration and uniformity of distribution in wood is mentioned in the patent literature. For example, the patent by Zhang and Zhang (*3*) mentions the need to adjust particle size distribution toward smaller particles when treating species that are more refractory than southern pine. The penetration and uniformity of distribution of micronized copper preservatives referred to in the aforementioned patents has been assessed macroscopically by spraying the treated wood with a reagent that changes color in the presence of copper (*7*). The micro-distribution of copper and other metal elements in treated wood, however, is assessed using quite different techniques. We briefly describe these techniques below, but first we explain why there is ongoing commercial interest in the micro-distribution of preservatives in wood.

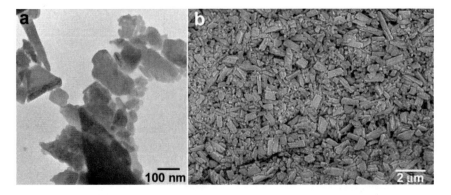

Figure 1. Appearance of micronized particles in: (a) preservative solution and; (b) accumulating in a bordered pit chamber in treated wood.

Microdistribution of Preservatives in Treated Wood

The effectiveness of preservatives at preventing the biological deterioration of wood depends, as pointed out by Arsenault, on the preservative system and also the penetration, retention and distribution of the preservative in wood (*8*). The distribution of preservatives in wood can be viewed at various scales-macroscopic, microscopic and at the cell wall level. For example, at the macroscopic level, Arsenault points out the need for higher concentrations of preservative in outer treated zones for wood in ground contact where soft-rot occurs at the interface between soil and treated wood (*8*). Clearly, in such situations a uniform distribution of preservative in treated wood is not desirable. Conversely, at the microscopic level more uniform distribution of preservative is desirable because decay has been observed in wood where levels of preservative vary in different tissue types. For example, uneven distribution of creosote in treated southern pine was found to be associated with decay of poorly treated earlywood bands (*8*). Similarly, decay of fibers in some hardwoods treated with chromated-copper-arsenate was associated with lower levels of fungitoxic metals in the cell walls of this tissue type (*9*). As a result of these and other observations it has been argued that wood preservatives need to be evenly distributed in woods' tissues and cell types and, in the case of inorganic preservatives, have the capacity to penetrate wood cell walls (*9–12*).

The micro-distribution of preservatives in wood has been examined using a variety of analytical techniques, which we describe below. Most of these techniques are only capable of analyzing small areas of wood. Since a cubic inch-sized (16.4 cm^3) sample of coniferous wood contains approximately 4 million cells (*13*) it can be readily appreciated that none of the techniques provide a complete picture of the micro-distribution of preservatives in wood.

Electron Microscopy and Energy Dispersive Analysis of X-rays

The micro-distribution of preservatives in treated wood has mainly been studied using electron microscopy in combination with energy dispersive analysis of X-rays (EDX). This technique involves placing treated wood under vacuum in a scanning or less commonly a transmission electron microscope and bombarding the sample with electrons. As a result, X-rays are emitted by the sample and 'from the wavelength (or photon energy) and intensity of the lines in the X-ray spectrum, the elements present in wood may be identified and their concentrations estimated' (*14*). Both scanning and transmission electron microscopy in combination with EDX has been used to assess whether arsenic, chlorine, chromium, copper, potassium, silicon, sulfur, titanium and zinc found in various wood preservatives can penetrate woods' micro-structure (*9–12, 15–32*). Some of these studies have used scanning electron microscopy (SEM) and EDX to look at differences in the distribution of preservative elements in the earlywood and latewood of softwoods, and also the vessels, fibers and rays of hardwoods (*11, 12, 23, 26, 28, 29*). Prior to our studies SEM-EDX had not been used to map particles in treated wood even though some older preservatives, for example copper dimethyldithiocarbamate, deposit crystalline particles in wood (*27*).

EDX can be used to detect elements at a particular location (point analysis) or map the distribution of elements in a selected area. Point analysis has mainly been used to detect variations in the concentrations of metal elements in different tissue types and within wood cell walls. Some studies have also included dot maps of treated wood, which show higher concentrations of metals in different tissues types or within wood cell walls compared to lumens (*11*, *24*, *29*, *30*, *32*). The resolution of these maps, however, is much lower than secondary electron images, and they only reveal large variations in the concentrations of metal elements. The resolution of X-ray maps can be improved by increasing the scanning time, but this can lead to specimen drift and blurred images (*14*, *33*). Alternatively, beam current and energy can be raised, which increases the number and energy of electrons hitting the target and the possibility of X-ray photons being emitted. However, this can lead to specimen damage (*14*, *33*). Furthermore, as the accelerating voltage is increased the size of the volume of sample excited by the electron beam increases which makes it difficult to map small surface features. These limitations explain why it has always been difficult to obtain good X-ray maps at high magnifications (*33*). Accordingly, mapping of metal elements in wood treated with copper particles is challenging because some of the nanoparticles lie near the resolving power of SEM. Recently, however, electron microscopes and EDX systems have been developed that make it easier to obtain X-ray maps at high magnifications, and these have found application in the mapping of low concentrations of metals in doped nano-scale semi-conductor devices (*34*). We used a similar system to examine the micro-distribution of metal elements in wood treated with particulate wood preservatives and describe our findings in the section below on micro-distribution of particulate copper in treated wood.

X-ray Fluorescence Microscopy

X-ray fluorescence (XRF) microscopy is another tool that can be used to identify and estimate the concentrations of inorganic elements in materials. XRF is related to EDX (*14*), but the X-ray spectrum emitted by the sample is produced by bombarding it with an intense beam of X-rays, rather than electrons. Recently, the development of methods of producing narrow, high intensity, X-ray beams and the availability of high resolution optical cameras and computational techniques for image processing has allowed bench-top X-ray fluorescent microscopes to be produced. These XRF microscopes are being used to map the concentrations of metals in a range of biological materials including treated wood (*35–37*).

XRF was used by Zahora (*37*) to examine the distribution of copper in southern pine treated with two different commercial particulate wood preservatives (*37*). He found high concentrations of 'copper in the resin canals and rays of treated wood' Figure 2. He also observed that the 'concentration of copper was lower in latewood than in earlywood, and high near the surface of the treated wood' Figure 2.

Figure 2. Cross-section of a sample from a southern pine board treated with a particulate preservative. The slice at the top of the image is stained for copper with chrome azurol and shows bands of lighter-colored latewood and darker-stained earlywood. The remainder of the figure is an X-ray fluorescence image indicating higher concentrations of copper (brighter color) in earlywood, vertical resin canals, rays (especially in earlywood), and at the outer surface of treated wood. (Courtesy of Dr. A. Zahora)

X-ray Microcomputed Tomography

X-ray micro-computed tomography (CT) shares many similarities with medical CT, but operates at the micron scale. X-ray micro-CT is being used to describe the structure of many different materials including rocks, bone, coal and wood to name just a few (*38*). These studies of the micro-structure of materials place a small sample on a stage where it is irradiated with a focused beam of X-rays (*39*). The sample will attenuate the X-rays depending on its thickness, chemical composition (atomic number) and physical density (*40*). Greater attenuation of X-rays results in fewer X-ray photons passing through the sample producing a whiter image on the capture device (film or X-ray camera). Conversely, voids (air) within the material appear black because there is relatively little attenuation of X-rays. The resulting grey scale pattern recorded by the X-ray camera is called a radiograph. A series of radiographs are collected by rotating the specimen on a sample stage. These radiographs are processed using software to generate a tomogram of the specimen. The tomogram is in effect a three-dimensional representation of the porous structure and variation of composition within a specimen (*39*). X-ray micro-CT has been used to characterize wood treated with water-repellent 'silicones', but first the chemicals had to be doped with 3-bromopropyltrimethoxysilane to increase their X-ray opacity (*32*). Once this was done it was possible to visualize the distribution of the 'silicones' in treated wood. We also found it necessary to dope a urea formaldehyde adhesive to visualize its distribution in particleboard (*41*). We selected copper sulfate as a doping agent because it bonds covalently to UF resin and has a higher atomic number and X-ray opacity than wood flakes and the voids in particleboard. Accordingly, it is possible to visualize copper in treated wood using X-ray micro-CT without any preliminary doping.

Microdistribution of Particulate Copper in Treated Wood

We carried out the first studies of the micro-distribution of copper in wood treated with particulate wood preservatives, as mentioned above (*42, 43*). These studies used scanning electron microscopy (SEM) and energy dispersive analysis of X-rays (EDX) to examine the micro-distribution of copper in southern pine wood treated commercially with a particulate copper preservative. We found that some of the rays in treated wood contained an amorphous white material (Figure 3a). This material contrasted very strongly with wood cell walls when treated samples were viewed using back-scattered electron imaging, suggesting that the material's atomic number was greater than those of elements in cell walls (carbon, hydrogen and oxygen). This suggestion proved to be correct when samples were analyzed using EDX. This technique confirmed that the white material contained copper and also iron, which is used as a pigment in some particulate preservative formulations (Figure 3b,c).

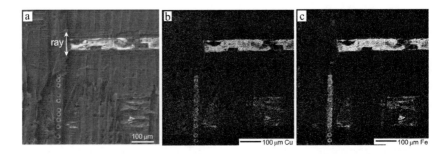

Figure 3. Radial longitudinal surface of southern pine wood treated with a particulate preservative: (a) Back-scattered electron image showing an amorphous white deposit within a ray; (b) Map of the Cu-Kα X-ray peak of the same region shown in Figure 3a; (c) Map of the Fe-Kα X-ray peak of the same region shown in Figure 3a.

The same white material seen in rays also accumulated in the chambers of bordered pits (Figure 4a), and it was possible at higher magnification using field emission scanning electron microscopy (FESEM) to see individual particles (Figure 4b). These particles were rectangular in shape and varied in size from ~50 to 700 nm. FESEM in combination with EDX confirmed that these particles contained either copper or iron (Figure 4c,d).

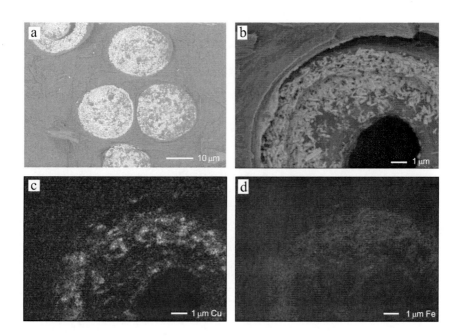

Figure 4. Radial longitudinal surface of southern pine wood treated with a particulate preservative: (a) Back-scattered electron image showing inorganic material in bordered pit chambers; (b) Back-scattered electron image showing inorganic particles in a bordered pit chamber; (c) EDX color mapping of copper in the bordered pit chamber shown in Figure 4b; (d) EDX color mapping of iron in the bordered pit chamber shown in Figure 4b.

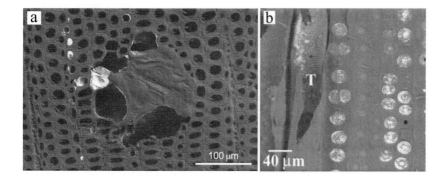

Figure 5. Transverse and radial longitudinal surfaces of southern pine wood treated with a particulate preservative: (a) Transverse surface showing white deposits in tracheid lumens and in a vertical resin canal; (b) Radial longitudinal surface showing white deposits in bordered pit chambers and in a tracheid (T) lumen.

White inorganic deposits in the rays and bordered pits were very prominent, but we also observed them in vertical resin canals and on, but not in, tracheid cell walls (Figure 5).

We used EDX to map the distribution of copper in rays and in bordered pit chambers, but point analysis was used to examine whether copper was present in the walls of latewood tracheids. Figure 6 shows representative spectra obtained from the secondary cell wall layer and the interfacial zone between cell walls (middle lamella). The level of copper was greater in the middle lamella than in the secondary cell wall layer in accord with results of previous studies that have examined the concentrations of copper in cell walls of wood treated with aqueous preservatives containing dissolved copper (28, 29).

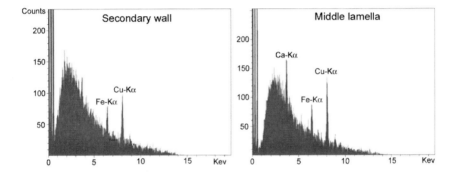

Figure 6. EDX spectra (point analyses) of the secondary cell wall and middle lamella in latewood tracheids in southern pine treated with a particulate preservative containing copper and iron.

To complement our SEM and EDX studies we used X-ray micro-CT to examine the variation in distribution of copper with depth of treatment and also in different tissues (rays, resin canals, and earlywood and latewood). The presence of copper in rays and resin canals was very apparent when a treated sample was imaged using X-ray micro-CT. Figure 7a shows a transverse section of part of a treated sample. A lighter yellow color is used to highlight copper. There is part of a band of low density earlywood at the top of the sample (Figure 7a). Then there is a complete growth ring consisting of the first band of latewood at the top of the image and a band of earlywood. Embedded in this band of earlywood is a thin band of latewood, possibly a false growth ring. Thereafter there are two complete growth rings each containing latewood and earlywood bands, and a partial growth ring at the bottom of the image (Figure 7a). The pattern of growth is from the bottom of the image to the top. The top of the block contains a diagonally angled hole at the end of which is a small mound of wood that sits on the second band of latewood (Figure 7b). These structures were created by a staple prong used to fix a label to the treated wood.

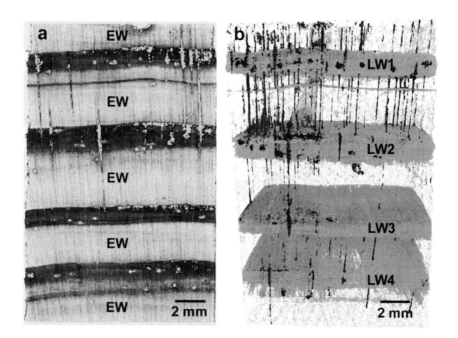

Figure 7. Two and three-D X-ray micro-CT images of southern pine wood treated with a particulate preservative: (a) Transverse section showing lighter colored copper located in rays running from top-to-bottom and within resin canals in the latewood; (b) Three-D image of the wood block in Figure 7a showing darker colored copper and bands of light colored latewood.

The two-D image in Figure 7 shows that copper is concentrated in rays running perpendicular to growth rings and within resin canals in latewood (Figure 7a). The adjacent 3-D image shows the same sample, but the bands of earlywood have been made transparent, allowing higher density copper to be more easily visualized in this tissue type (Figure 7b). Higher density regions of copper are colored a darker blue. Numerous 'pillars' of higher density material created by the accumulation of copper particles in rays run radially from the top of the specimen to the bottom, in-between the lighter horizontally aligned latewood bands (Figure 7b). These pillars are occasionally contiguous running across a number of bands of latewood. Copper also accumulates at the interface between earlywood and latewood and occasionally as 'tubes' within resin canals that run at right angles to the rays (see 3rd band of earlywood in Figure 7b).

Figure 8 shows 3-D images of the same block of treated wood shown above in Figure 7. In Figure 8a the latewood rather than the earlywood has been made optically transparent allowing copper to be seen within latewood.

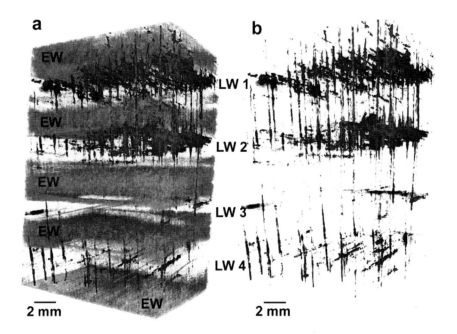

Figure 8. X-ray micro-CT images showing the location of copper in a block of southern pine wood treated with a particulate preservative: (a) Image of the whole wood block showing copper in latewood and five bands of lighter colored earlywood; (b) Image of the copper skeleton in the treated block (earlywood and latewood removed).

The first two bands of latewood in Figure 8a contain a large amount of copper resulting from the accumulation of copper in rays and resin canals. This is also apparent in Figure 8b. In this figure all of the wood tissues have been made transparent allowing the copper skeleton in the treated block to be seen. Less copper is present in the two lower bands of latewood because some of the resin canals do not contain copper, and fewer rays contain copper (Figure 8b).

Summary and Concluding Remarks

We used scanning electron microscopy in combination with energy dispersive analysis of X-rays (EDX) and also micro-computed tomography (CT) to examine the micro-distribution of copper in southern pine treated with particulate preservatives. We found inorganic material containing copper in the rays and resin canals of treated wood. Rays and resin canals containing copper particles formed a grid-like network, which was more complete in growth rings near the surface of treated wood. Copper particles accumulated in pit chambers and on the cell walls of some tracheids. We were unable to observe nanoparticles in tracheid cell walls using a field emission scanning electron microscope with a resolution of ten to twenty nanometers. Nevertheless, we detected elemental copper in the cell walls of latewood tracheids using EDX-point analysis. The concentration of copper was greater in the middle lamella than in the secondary cell wall layer, in accord with results of previous studies that have examined the concentrations of copper in cell walls of wood treated with aqueous preservatives containing dissolved copper.

We conclude that scanning electron microscopy and EDX in combination with X-ray micro-CT are powerful and complementary techniques for examining the micro-distribution of copper in wood treated with particulate wood preservatives. X-ray micro-CT has the potential to quantify the spatial distribution of copper in treated wood, and because it is a non-destructive technique it could be used to evaluate if copper is redistributed when wood is exposed to moisture or micro-organisms. These features of X-ray micro-CT provide a route towards the elusive goal of linking the micro-distribution of copper in wood and the performance of treated wood in the field.

References

1. Evans, P. D.; Matsunaga, H.; Kiguchi, M. *Nat. Nanotechnol.* **2008**, *3*, 577.
2. Freeman, M. H. In *Development of Commercial Wood Preservatives. Efficacy, Environmental and Health Issues*; Schultz, T. P., Militz, H., Freeman, M. H., Goodell, B., Nicholas, D. D., Eds.; ACS Symposium Series 982; American Chemical Society: Washington, DC, 2008; pp 408−426.
3. Zhang, J.; Zhang, W. U.S. Patent 7,632,567, 2009.
4. Leach, R. M.; Zhang, J. U.S. Patent 7,674,481, 2010.
5. Zhang, J.; Zhang, W.; Leach, R. M. U.S. Patent 8,168,304, 2012.
6. Richardson, H. W.; Hodge, R. L. U.S. Patent 8,409,627, 2013.
7. American Wood Preservers Association. AWPA Standard A3-05, 2005.
8. Arsenault, R. D. In *Wood Deterioration and its Prevention by Preservative Treatments. Preservatives and Preservative Systems*; Nicholas, D. D., Ed.; Syracuse Wood Sci. Series 5; Syracuse University Press: New York, 1973; Vol. 2, 121−278.
9. Greaves, H. *Holzforschung* **1974**, *28*, 193–200.
10. Chou, C. K.; Chandler, J. A.; Preston, R. D. *Wood Sci. Technol.* **1973**, *7*, 151−160.
11. Greaves, H.; Levy, J. F. *Holzforschung* **1978**, *32*, 200–208.
12. Levy, J. F.; Greaves, H. *Holzforschung* **1978**, *32*, 209–213.
13. Hoadley, R. B. *Understanding Wood*; The Taunton Press: Newtown, CT, 2000.
14. Reed, S. J. B. *Electron Microprobe Analysis*, 2nd ed.; Cambridge University Press: Cambridge, 1993.
15. Greaves, H. *Mater. Org.* **1972**, *7*, 277–286.
16. Dickinson, D. J. *Mater. Org.* **1974**, *9*, 21–33.
17. Desai, R. L.; Côté, W. A., Jr. *J. Coat. Technol.* **1976**, *48*, 33–37.
18. Dickinson, D. J.; Sorkhoh, N. A. A. H. *Scanning Electron Microsc.* **1976**, *9*, 549–554.
19. Drysdale, J. A.; Dickinson, D. J.; Levy, J. F. *Mater. Org.* **1980**, *15*, 287–303.

20. McNamara, W. S.; Greaves, H.; Triana, J. F. *Mater. Org.* **1981**, *16*, 81–94.
21. Greaves, H.; Nilsson, T. *Holzforschung* **1982**, *36*, 207–213.
22. Yata, S.; Nishimoto, K. *Wood Res.* **1983**, *69*, 71–79.
23. DeGroot, R. C.; Kuster, T. A. *Holzforschung* **1984**, *38*, 313–318.
24. Parameswaran, N.; Wilcox, W. W.; Côté, W. A., Jr. *Holzforschung* **1985**, *39*, 259–266.
25. Miyafuji, H.; Saka, S. *Wood Sci. Technol.* **1997**, *31*, 449–455.
26. Dawson-Andoh, B. E.; Kamdem, D. P. *Holzforschung* **1998**, *52*, 603–606.
27. Kamdem, D. P.; McIntyre, C. R. *Wood Fiber Sci.* **1998**, *30*, 64–71.
28. Petrič, M.; Murphy, R. J.; Morris, I. *Holzforschung* **2000**, *54*, 23–26.
29. Matsunaga, H.; Matsumura, J.; Oda, K. *Int. Assoc. Wood Anat. J.* **2004**, *25*, 79–90.
30. Cao, J.; Kamdem, D. P. *Holzforschung* **2005**, *59*, 82–89.
31. Tingaut, P.; Weigenand, O.; Militz, H.; De Jéso, B.; Sèbe, G. *Holzforschung* **2005**, *59*, 397–404.
32. De Vetter, L.; Cnudde, V.; Masschaele, B.; Jacobs, P. J. S.; Van Acker, J. *Mater. Charact.* **2006**, *56*, 39–48.
33. Russ, J. C. *Fundamentals of Energy Dispersive X-ray Analysis*; Butterworths Monographs in Materials; Butterworths: London; 1984.
34. Huang, R. T.; Wang, M. C.; Yan, J. Y.; Wu, T. W.; Kai, J. J.; Chen, F. R. *Microsc. Microanal.* **2005**, *11*, 1924–1925.
35. Fitzgerald, S. *G.I.T. Lab. J.* **2005**, *4*, 47–48.
36. Chen, I. H.; Kiang, J. H.; Correa, V.; Lopez, M. I.; Chen, P-Y.; McKittrick, J.; Meyers, M. A. *J. Mech. Behav. Biomed. Mater.* **2011**, *4*, 713–722.
37. Zahora, A. IRG Wood Protect. Doc. 2010, IRG/WP 10-40507.
38. Sakellariou, A.; Arns, C. H.; Sheppard, A. P.; Sok, R. M.; Averdunk, H.; Limaye, A.; Jones, A. C.; Senden, T. J.; Knackstedt, M. A. *Mater. Today* **2007**, *10*, 44–51.
39. Sakellariou, A.; Sawkins, T.; Senden, T. J.; Limaye, A. *Physica A (Amersterdam Neth.)* **2004**, *339*, 152–158.
40. Golab, A.; Romeyn, R.; Averdunk, H.; Knackstedt, M. A.; Senden, T. J. *Aust. J. Earth Sci.* **2013**, *60*, 111–123.
41. Evans, P. D.; Morrison, O.; Senden, T. J.; Vollmer, S.; Roberts, R. J.; Limaye, A.; Arns, C. H.; Averdunk, H.; Lowe, A.; Knackstedt, M. A. *Int. J. Adhes. Adhes.* **2010**, *30*, 754–762.
42. Matsunaga, H.; Kiguchi, M.; Roth, B.; Evans, P. D. *Int. Assoc. Wood Anat. J.* **2008**, *29*, 387–396.
43. Matsunaga, H.; Kiguchi, M.; Evans, P. D. *J. Nanopart. Res.* **2009**, *11*, 1087–1098.

Chapter 14

Evaluating the Leaching of Biocides from Preservative-Treated Wood Products

Stan T. Lebow[*]

USDA, Forest Service, Forest Products Laboratory,
Madison, Wisconsin, 53726
[*]E-mail: slebow@fs.fed.us

Leaching of biocides is an important consideration in the long term durability and any potential for environmental impact of treated wood products. This chapter discusses factors affecting biocide leaching, as well as methods of evaluating rate and quantity of biocide released. The extent of leaching is a function of preservative formulation, treatment methods, wood properties, type of application and exposure conditions. Wood properties such as permeability, chemistry and heartwood content affect both the amount of biocide contained in the wood as well as its resistance to leaching. A range of exposure factors and site conditions can affect leaching, but the most important of these appears to be the extent of exposure to water. For wood that is immersed in water or placed in contact with the ground the characteristics of that water (pH and inorganic and organic constituents) also play a role. For wood that is used above-ground or above water, the frequency of precipitation and patterns of wetting and drying are key considerations. Current standardized methods are intended to greatly accelerate leaching but are not well-suited to estimating leaching in service. Continued research is needed to refine methods that utilize larger specimens and more closely simulate in-service moisture conditions.

Introduction

The depletion of biocides from preservative-treated wood products exposed to precipitation, placed in contact with soil, or immersed in water is generally referred to as leaching. Resistance to leaching is imparted in a variety of ways and may differ between formulations of the same biocide. Leaching of biocide from treated wood is of importance in both the long-term durability of the treated product and its potential for impacting the environment. The role of resistance to leaching in durability is clear, as wood treated with a readily leachable biocide may be only slightly more durable than untreated wood if placed in contact with soil or standing water. Even wood treated with a leach-resistant biocide may eventually fail if the concentration of biocide remaining in the wood falls below that needed to prevent biodeterioration. The significance of potential environmental impacts associated with leaching of biocides is less clear, but concerns have been expressed by governmental regulatory and advisory bodies, and use of biocide-treated wood has been limited in some situations. In essence, durability concerns are focused on the quantity of biocide remaining in the wood during long-term service, while environmental concerns are focused on the quantity of biocide lost from the wood. Although this distinction may appear trivial, there are practical consequences for the manner in which leaching is evaluated and the results are interpreted.

Obtaining useful and representative estimates of biocide leaching from treated wood can be challenging. Wood is an inherently variable material and this factor alone can make assessments of environmental impact more challenging. However, there other factors such as the treatment process, type of end-use application, and exposure environment that can also effect leaching. This chapter discusses some approaches used to evaluate biocide leaching and/or environmental accumulation and the influence of various aspects of these methods on research results. Focus is placed on evaluation of biocide release (leaching) rather than environmental impacts. For a detailed discussion of the potential environmental impacts of leached wood preservatives the reader is referred to "*Managing Treated Wood in Aquatic Environments*" (*1*) or *Environmental Impacts of Treated Wood* (*2*). In addition, leaching is distinguished from other forms of biocide depletion such as evaporative aging, UV degradation, or microbial decomposition, which have been previously reviewed by others (*3, 4*).

Wood and Treatment Factors Affecting Leaching

One of the greatest contributors to variability in biocide leaching is the complexity of wood as a material. The structure, anatomy, and chemistry of wood affect the way that preservative components and leaching medium move through and react with the wood substrate.

Wood Dimensions and Proportion of End-Grain

The volume, surface area and proportion of end-grain of wood products effect the percentage and flux of biocide leached from the wood. Thinner pieces have a larger portion of their surface area exposed for leaching and allow more rapid water penetration. Conversely, members with larger dimensions, such as timbers, not only have a lower relative surface area but also contain a larger reservoir of preservative and might be expected to release biocide over a longer period. Larger members, especially round members, have a greater tendency to develop drying checks that can increase overall surface area and facilitate water penetration. Because the rate of movement of liquids along the grain of wood is several orders of magnitude greater than that across the grain, greater leaching is likely to occur from shorter dimensions with a higher proportion of end-grain. Because of these size and grain orientation effects, care must be taken in extrapolating the results of leaching tests with small specimens to losses from the larger members used in preservative-treated structures.

Wood Anatomy and Chemistry

Species differences in the anatomy and chemistry of wood also affect the interactions of the preservative with the wood substrate and the permeability of the wood to liquid water (*5–8*). Permeability varies greatly among wood species, and those species that are more permeable tend to leach at a higher rate because of more rapid movement of water through the wood (*9, 10*). Studies also indicate that preservative components may be more leachable from hardwoods than from softwoods (*11, 12*). Wood species may also affect the distribution of preservative within the wood and the chemical reactions that occur to fix water-based preservatives within the wood (*11, 12*).

Leaching of preservatives may also be affected by the presence and amount of heartwood. In most wood species the heartwood portion of a tree is much less permeable and sometimes more hydrophobic than sapwood portion. Heartwood portions of test specimens may contain much less preservative than sapwood and may also be more resistant to penetration of the leaching medium. These effects might be expected to result in lower leaching rates from heartwood, but this generalization may be confounded by differences in preservative fixation in heartwood or by the presence of a higher concentration of preservative at the heartwood surface (*13*).

Effect of Treatment Parameters on Leaching

Retention of Biocide

The retention of preservative in biocide-treated wood is varied intentionally according to the intended end-use, as well as unintentionally as a result of variability of the wood substrate. Typically wood is treated to low retentions when intended for use above ground (such as decking) and to higher retentions for use in ground-contact or seawater (*14*). In general leaching does increase at higher retentions, but this trend does not always hold true, nor is leaching always directly proportional to retention. Several researchers have noted that the percentage of leachable arsenic from CCA- treated wood decreases with increased retention (*13*, *15–19*). Increased retention does appear to result in greater leaching for the amine copper preservatives, but it is not clear whether the leaching increase is proportional to retention. At least two studies have indicated that leaching of copper from amine-copper treated wood increases more than proportionally as retention increases (*20*, *21*). Dubai et al. (*20*) theorized that greater Cu leaching at higher retentions could result from the presence of at least two types of reactive sites in the wood. Once the limited number of strong binding sites is consumed, the remaining copper reacts with a larger number of weaker binding sites and is thus more leachable. Similarly, Humar et al. (*21*) propose that a portion of the copper is initially strongly bound to the wood, and that once those reactive sites are filled the remaining copper is simply precipitated within cell walls and lumens.

Post-Treatment Conditioning

The biocides (metals and/or organics) of waterborne wood preservatives are initially carried in water but become resistant to leaching when placed into the wood. This leaching resistance results from a range of "fixation" mechanisms that differ with preservative formulation and individual biocide. Some fixation occurs very rapidly during pressure treatment while others may take days or even weeks to reach completion, depending on post treatment storage and processing conditions. If the treated wood is placed in service before these reactions are completed, the initial release of preservative into the environment may be greater than for wood that has been adequately conditioned. Best Management Practices (BMPs) have been developed through a cooperative effort of several trade associations to ensure that commercially treated wood is produced in a manner that will minimize subsequent leaching (*22*). Research indicates that these BMPs do have practical benefit in minimizing the potential for environmental releases (*23*).

Exposure Factors Affecting Leaching

The extent of water exposure is the key to biocide depletion from preservative treated wood. Although this concept is simple, interpretation of the extent of moisture exposure for treated wood in-service is complex. Only a small fraction of the volume of treated wood in service is continually immersed in water or kept continually moist through soil contact. The greatest proportion of treated wood is used above the ground or above water where wetting is intermittent. Structures that are only intermittently exposed to precipitation will have much lower leaching rates than those continually immersed in water, especially in water or soil that contains solubilizing organic or inorganic components. In this section the role of exposure to water, and the effect of water characteristics on leaching, is discussed in more detail.

Wood Used Above-Ground or Above Water

The extent of wetting in wood used above ground or above water is not easily quantified and is dependent on construction details, precipitation characteristics, and possibly on other climatic factors such as temperature, and humidity.

Effect of Rainfall Pattern

Previous studies of treated wood exposed to simulated or natural weathering have indicated that both the pattern and rate of rainfall influence the quantity of preservative released. When expressed on the basis of mass of preservative leached per unit rainfall, greater amounts of biocide appear to be released at slower rainfall rates (*9, 13, 24, 25*), presumably because the wood is wetted for a longer period and a greater proportion of the rainfall is absorbed by the wood (Figure 1). In addition, the interval between rainfall events appears to influence leaching, with greater amounts leached after longer resting periods. This type of effect has been attributed to the allowance for a longer period for soluble preservative components to diffuse to the surface from the interior of the wood products (*3, 18, 26–30*).

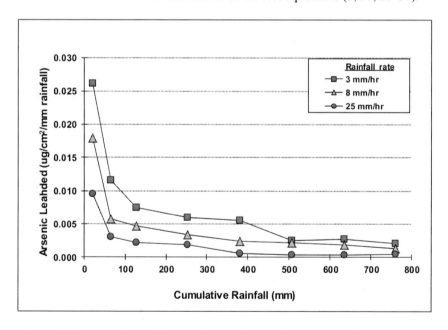

Figure 1. Effect of rainfall rate and cumulative volume on leaching of arsenic from CCA-treated wood (25).

Researchers have reported that the moisture content of pine sapwood exposed to natural weathering may range from maximums of 80% to minimums of approximately 10% (*31–38*). Average moisture contents reported for horizontal exposures ranged from 21 to 26%, whereas the averages reported for vertical exposure were 18.6 and 25.4%. Moisture contents reported for less permeable species such as spruce or Douglas-fir tended to be lower than those of pine species when exposed under similar conditions (*33*, *34*, *39*).

Other Climatic Factors

Climatic factors other than precipitation appear to play some role in leaching. For example, exposure to ultraviolet radiation in a weathering chamber strongly increased leaching from CCA-treated decking specimens exposed to artificial rainfall. Other factors, such as temperature and humidity, can affect the rate of drying after precipitation as well as the extent of cracking that may occur on the wood surface. One study did note that leaching, per unit rainfall, appeared to be greater during rain events with higher ambient temperature (*18*). While no leaching should occur from frozen wood, it is likely that the stresses developed during freeze-thaw cycles contribute to subsequent crack formation.

Construction and Site Parameters

In actual structures, wood moisture content can be a function of wood dimension and construction detailing. Larger dimension material may be slower to wet initially but is also slower to dry. Connections are likely to trap and hold more moisture because precipitation is absorbed through the end-grain but drying is slowed because of limited air movement. This effect was recently demonstrated by a study that compared the moisture contents of specimens exposed with and with-out end-grain connections (*40*).

The presence of shade has been shown to substantially increase the moisture content of specimens exposed above-ground (*39*), presumably by slowing drying. Vegetation associated with shading can also result in the deposition of leaf litter and other organic debris in connections and in spaces between deck boards. This organic debris traps moisture and can potentially contribute to higher wood moisture contents.

Application of Finishes and Wraps

In many applications some type of finish or coating is applied to preservative-treated wood, and there is evidence that these finishes can lessen biocide release. (*41–46*). A caveat with the use of finishes is the risks associated with surface preparation and application. Aggressive surface preparation techniques such as sanding or power washing might be expected to cause release of additional biocide into the environment. Although less common than finishes, wraps are sometimes applied to piles or poles to provide protection and enhance durability. Studies with marine piles indicate that these wraps can also be very effective in minimizing preservative release (*23*).

Water Characteristics

The characteristics of the leaching water can also influence leaching of preservatives. The presence of some types of inorganic ions in water has been reported to increase leaching from CCA-treated wood (*16, 47, 48*) whereas they have been reported to decrease leaching with at least one type of preservative (*49, 50*). Seawater has been reported to both increase and decrease leaching relative to purified or naturally occurring freshwater depending on the study conditions, preservative, and biocide component (*5, 20, 51, 52*). Water pH can also affect leaching of preservatives. Leaching of CCA is greatly increased when the pH of the leaching water is lowered to below 3, and the wood itself also begins to degrade (*53, 54*). Water pH ranges more typical of those found in the natural world are less likely to have a great effect on leaching (*55*), although leaching of copper from copper-azole treated wood was found to be greater at pH 5.5 than at pH 8.5 (*56*).

The presence of organic acids in surface waters may also affect leaching. Surface waters containing high levels of humic or fulvic acid can have the potential for increasing CCA leaching (*5, 11, 20, 51, 53*), while one study (*57*) reported that addition of humic acid to leaching water lowered concentrations of leached creosote components relative to deionized water.

Water temperature may also affect leaching, as some of the fixation products that immobilize biocide components in treated wood might be expected to be more soluble at higher temperatures (*51, 56*). Brooks (*58*) concluded that leaching of copper from CCA-treated wood could be substantially increased as water temperatures increased from 8 to 20° C. Subsequent research indicated that leaching of both copper and tebuconazole increased at higher temperatures, although this effect was diminished with longer leaching periods (*56*). A similar temperature effect was noted in a study of release of creosote components from treated wood (*59*).

The rate of water movement around the wood can also influence leaching, although this effect has not been well quantified. Xiao and others (*59*) reported that release of creosote was greatest at the highest flow rate tested and that turbulent flow may have greatly increased leaching and Brooks (*56*) suggests that more rapid water movement may increase leaching by promoting water exchange in checks and cracks.

Effect of Soil Properties

Studies have illustrated that soil composition may affect both leaching and subsequent mobility of preservative components (*48, 60–65*) and indicate that leaching from wood placed in soil can be greater than that of wood immersed in water (*61, 62, 64*). Increased leaching of biocides from wood in contact with the ground has been attributed to lower pH, and higher concentrations of inorganic soil constituents and organic acids. Soil pH often cannot be separated from the effect of other factors, such as the presence of organic acids that have been shown to increase leaching from wood treated with some types of biocides (*62, 66*). Cooper and Ung (*66*) compared CCA losses from jack pine blocks exposed in garden soil and organic-rich compost and found that leaching was more than doubled by compost exposure. Inorganic components in soil have also been implicated in increasing or reducing leaching. Depletion of pentachlorophenol has been reported to be greater in soils high in copper and iron (*64*), and iron has also been implicated in increased leaching from CCA treated wood (*48*). Conversely, one report suggests that iron and aluminum in soil surrounding CCA-treated wood can retard arsenic leaching because these metals may migrate into the wood and irreversibly precipitate the arsenic (*67*).

Test Methods for Assessing Leaching

Standardized Laboratory Test Methods

Conventional laboratory methods of evaluating preservative leaching were primarily developed to allow comparison between experimental formulations and provide information on leach resistance as it relates to long term durability. These methods utilize continuous immersion of small specimens with the goal of accelerating and amplifying leaching. As mentioned earlier, the rate of movement of liquids along the grain of the wood is several orders of magnitude greater than that across the grain, and so specimens with a high proportion of exposed end-grain will exhibit exaggerated rates of preservative leaching (*68, 69*).

In the United States the most commonly used standardized leaching method for biocide-treated wood is AWPA Method E11-12, *Standard Method for Accelerated Evaluation of Preservative Leaching* (*14*). This method specifies biocide treatment of small (19 mm) cubes. The Japanese (JIS K 1571) and Chinese (CNS 6717) leaching methods are weathering steps in preparing specimens for exposure to biological attack (*70, 71*). Again, the small size and grain orientation (10 by 20 by 20 mm with the 10 mm parallel to the grain) of the specimens is expected to greatly accelerate leaching. Both methods also incorporate drying events between leaching exposures. A European method (EN 84) is also intended as a conditioning step prior to biological exposure (*72*). It uses somewhat larger specimens (15 by 25 by 50 mm) with a lower proportion of end-grain than that US, Japanese or Chinese methods. Unlike the US, Japanese and Chinese methods, EN 84 does not specify agitation during leaching.

The Organization for Economic Cooperation and Development (OECD) has also developed guidelines for evaluating biocide release from preservative-treated wood, and these methods are intended for use in estimating release from in-service products. Separate methods are recommended for wood that is intended for use immersed in water versus wood that is to be used above-ground or above water. For wood to be immersed in water, the method is similar to EN -84 (*73*). For wood used above-ground, OECD guidelines describe an approach involving a brief dip immersions also utilizing small (15 by 25 by 50 mm) specimens (*74*). Although intended to simulate in-service leaching, there is some concern that this approach may not represent commercially produced lumber (*75*) or produce the moisture conditions reported for wood products exposed to natural weathering (*69*). One study which compared outdoor leaching to the OECD method concluded that the laboratory method risked underestimating in-service leaching (*76*). Use of simulated rainfall is also mentioned in the guideline, but only general guidance is provided for this approach.

In the US, the AWPA has also standardized a laboratory method (E20) to evaluate preservative depletion from wood placed in ground contact (*14*). This method was developed in recognition of research indicating that soil properties can affect biocide leaching (*62, 63*). It involves burying small (14 by 14 by 250 mm) stakes in moist soil for 12 weeks. The smaller stake dimensions and the maintenance of saturated soil conditions are intended to accelerate loss of preservative. Unlike other laboratory methods, where leaching is quantified by analyzing leaching water, extent of leaching with the AWPA E20 method is determined by assaying end-matched portions of the stakes before and after exposure.

Non-Standard Test Methods

Numerous non-standard methods have been used to evaluate preservative leaching, in part because it is recognized that standardized methods are not well suited to for providing estimates of leaching from treated wood in service. Because it is often not practical to conduct leaching studies using full-length lumber, poles, or piles, shorter specimens are typically cut from commodity-sized material. To avoid the problem of increased leaching from end-grain, specimens may be end-sealed with a waterproof sealer prior to leaching.

Many of these approaches involve immersion of specimens of varying dimensions in water for varying periods. Movement of the leaching water may be achieved by agitation (*52*) or pump circulation (*56, 77, 78*). Leaching water is either periodically replaced (*52, 76*) or continually replaced using flow-through systems (*56, 78*). Brooks (*56*) has conducted several studies of leaching from pile and lumber sections using large (40 L) tanks with a pump providing constant circulation of the leaching water. Fresh leaching water is steadily added to the tanks and samples for analysis are collected from the overflow. Brooks notes that methods without continuous water replacement risk underestimating leaching (*56*).

A variety of non-standard methods have also been used in an attempt to evaluate leaching of biocides from treated wood exposed to precipitation. The most common approach has been to expose specimens cut from product-size material to natural weathering and collect the leachate for analysis. Numerous studies have measured biocide concentrations in rainwater run-off from treated products including deck boards (*18, 46, 79–82*) fence boards (*83*), deck sections (*26, 84–88*) and shingles (*24*). An advantage of this approach is that it incorporates all of the weathering and exposure factors that may affect leaching, and provides "real world" leaching data under the test conditions. These tests are also relatively simple and inexpensive to setup. A disadvantage of this approach is that the exposure conditions are uncontrolled and unpredictable, thus making it difficult to replicate an evaluation or apply the findings to other conditions. It is also difficult to accelerate testing with this approach, and depending on the weather pattern, it may take substantial time to obtain results.

Another approach to evaluating leaching from wood exposed to precipitation is through some form of simulated rainfall (*25, 43, 83, 89*). This approach allows control over rainfall rates and schedules, but the methodology and equipment are more complex than that needed for natural exposures. It is also difficult to simulate the lower rainfall intensities while maintaining realistic droplet sizes and uniform coverage of replicate specimens. Simulated rainfall also may not realistically incorporate other exposure factors, such as check formation, that potentially contribute to leaching (*3, 18, 43, 84*).

Models of Biocide Leaching

The leaching methods discussed above provide information on quantities of biocide leached under certain experimental conditions, but do not necessarily allow ready estimation of leaching from treated products in service. One proposed approach allows leaching estimation based on laboratory determination of the amount of biocide component available for leaching, the equilibrium dissociation of the biocide component into free water in the wood, and diffusion coefficients for movement in the radial, tangential, and longitudinal directions (*30, 90, 91*). Once these parameters are determined for a particular preservative, leaching can be estimated as a function of product dimensions and the length of time that the wood is sufficiently wet to allow diffusion. More recently a series of studies has been conducted to model leaching of copper azole biocide based on the chemical interactions of the biocide with reactive sites in the wood as well as with constituents of the leaching water (*92–94*). It uses commercially available chemistry modeling software to allow prediction of solubility, complexation and transport of biocide components under a range of conditions. A major limitation for all these modeling efforts is lack of information on the extent of time that wood products in service have sufficient moisture to allow diffusion to occur.

Evaluations of Leaching in Service

Evaluation of leaching from in-service structures offers the promise of long-term leaching data under real world conditions. A disadvantage of these types of studies is that they are specific to the conditions at that site and are difficult to relate to other exposures. It is also difficult to quantify preservative leaching from in-place structures. For in-service evaluations, leaching is generally evaluated by either assaying the treated wood to determine the quantity of preservative remaining, by collecting and analyzing environmental samples adjacent to the treated wood, or by collecting precipitation run-off from the structure.

Determining preservative loss by assaying wood after exposure requires knowledge of original preservative retention in the wood. Often original retention is assumed based on the specified target or standard retention for treated wood used in that application. This assumption can be problematic, as the initial preservative retention in a treated product can be substantially higher or lower than the target retention. Retention can vary within a single piece, more greatly between material in a single charge and even more greatly between charges and treating plants (*95*). Variability in retention also makes it difficult to accurately assess the quantity of preservative remaining in a structure after exposure. This type of sampling is destructive, and efforts to evaluate changes in retention over time require analysis of different samples. Because of the variability in retention within wood products, it is often difficult to draw strong conclusions about leaching based on analysis of the amount of preservative remaining in a structure.

Researchers may also attempt to evaluate leaching by collecting environmental samples adjacent to a treated structure. Because metallic preservative components such as copper, chromium, and arsenic are reactive with soil constituents and accumulate near the structure (*55, 60, 96*), soil concentrations can potentially provide an indication of the quantity of these components leached. However, most metallic preservative components have some mobility in soil, and thus levels of accumulation are a function of both the leaching rate of preservative components and their subsequent mobility in the soil. For organic biocides, decomposition also plays a role in soil concentrations. Environmental sampling also introduces a range of sources of variability into a leaching study. In addition to leaching rate, environmental concentrations of preservative components will be a function of background concentrations, sampling location, and soil or water characteristics.

A third, and less common, approach to evaluating in service leaching is the collection of precipitation run-off from sections of a structure. This approach was used to quantify leaching from utility poles (*97*) and roofing materials (*24, 98*). Key factors in this approach are determining and limiting the surface area of treated wood, and quantifying the volume of run-off contacting the wood surface area.

Summary

Resistance to leaching is a key attribute for treated wood products intended for use outdoors. The rate and quantity of leaching is dependent on a range of factors including preservative characteristics, wood properties, treatment methods, type of structure, and exposure conditions. The volume, surface area and proportion of end-grain of wood products effect the percentage and flux of biocide leached from the wood. Differences in the anatomy and chemistry of wood also affect the interactions of the preservative with the wood substrate and the permeability of the wood to liquid water. Treatment methods and post-treatment conditioning steps can also affect leaching, especially for oil-type preservatives or those water-based formulations that rely on drying or lengthy chemical reactions to minimize solubility.

A range of exposure factors and site conditions can affect leaching, but the most important of these appear to be the extent of exposure to water. For wood that is immersed in water or placed in contact with the ground the characteristics of that water (pH and types of inorganic and organic constituents) may also play a role. For wood that is used above-ground or above water the frequency of precipitation and pattern of wetting and drying is a key consideration.

The number of factors that can affect biocide leaching has made it challenging to develop accelerated test methods that provide realistic estimates of leaching that may occur in service. Current standardized test methods use small specimens that have an unrealistic surface area to volume ratio and tend to exaggerate short-term leaching. Although small specimens produce the greatest percentage loss of biocide, for more leachable biocides the small reservoir of available preservative may result in lower releases when expressed on the basis of mass-per-unit surface area. Because the extent and pattern of preservative release is dependent on both test method and type of preservative, it is difficult to anticipate how well these test methods will estimate long-term release from a new type of preservative. For wood that is intended for use in water, there is potential for utilizing immersion tests using larger specimens that are end-sealed to prevent loss of preservative through the end-grain. Water circulation and frequent water changes are needed to simulate exposure conditions and ensure that biocide accumulation in the water does not inhibit further leaching.

Developing test methods for wood exposed to leaching due to rainfall is more complex. Artificial rainfall exposures have the potential for relatively close simulation of natural rainfall events and have the additional advantage of allowing extrapolation based on volume of rainfall. However, they do not necessarily incorporate the wetting/drying cycles experienced by treated wood in service. The dip-immersion methods are simple to conduct and have the potential for simulating natural wetting and drying conditions with adjustment of immersion scenarios. However, the current use of small specimens and limited water uptake makes extrapolation to in-service leaching rates difficult. Methods that more closely simulate natural wetting and drying conditions will help to minimize the under or over-estimation that is likely to occur when extrapolating results to long-term natural exposures. Ideally, test methods would use large enough specimens and sufficient moisture changes to induce a degree of checking similar to that exhibited by treated products exposed in service. However, these conditions may be difficult to achieve in accelerated testing because large specimens are slow to gain and lose moisture. In contrast, field exposures provide realistic leaching results but are time-consuming and dependent on the weather conditions during the test. However, field exposures remain an important tool for evaluating new test methods.

References

1. Morrell, J. J.; Brooks, K. M.; Davis, C. M. *Managing Treated Wood in Aquatic Environments*; Forest Products Society: Madison, WI, 2011; pp 1–503.
2. Townsend, T.; Solo-Gabriele, H. *Environmental Impacts of Treated Wood*; CRC Press, Taylor and Francis Group: Boca Raton, FL, 2006; pp 1–501.
3. Ruddick, J. N. R. In *Development of Commercial Wood Preservatives: Efficacy, Environmental and Health Issues*; Schultz, T. P., Militz, H., Freeman, M. H., Goodell, B., Nicholas, D. D., Eds.; ACS Symposium Series 982; American Chemical Society: Washington, DC, 2008; pp 285–311.

4. Wallace, D. F.; Cook, S. R.; Dickinson, D. J. In *Development of Commercial Wood Preservatives: Efficacy, Environmental and Health Issues*, Schultz, T. P., Militz, H., Freeman, M. H., Goodell, B., Nicholas, D. D., Eds.; ACS Symposium Series 982; American Chemical Society: Washington, DC, 2008; pp 312–323.
5. Humar, M.; Pohleven, F.; Zlindra, D. *Wood Res.* **2006**, *51* (3), 69–76.
6. Kartal, S. N.; Lebow, S. T. *Wood Fiber Sci.* **2001**, *33* (2), 182–192.
7. Radivojevic, S.; Cooper, P. A. *Wood Fiber Sci.* **2007**, *39* (4), 591–602.
8. Stevanovic-Janezic, T.; Cooper, P. A.; Ung, Y. T. *Holzforschung* **2001**, *55* (1), 7–12.
9. Cockcroft, R.; Laidlaw, R. A. *International Res. Group on Wood Preservation*; Doc No. IRG/WP/3113, 1978.
10. Wilson, A. *J. Inst. Wood Sci.* **1971**, *5* (6), 36–40.
11. Cooper, P. A. *Proc. Can. Wood Preserv. Assoc.* **1990**, *11*, 144–169.
12. Yamamoto, K.; Rokova, M. *International Res. Group on Wood Preservation*; Doc No. IRG/WP/3656, 1991.
13. Cooper, P. A. *Proc. Am. Wood-Preserv. Assoc.* **2003**, *99*, 73–992003.
14. AWPA. *Book of Standards*; American Wood Protection Association: Birmingham, AL, 2013; pp 1–658.
15. Arsenault, R. D. *Proc. Am. Wood-Preserv. Assoc.* **1975**, *71*, 126–146.
16. Irvine, J.; Eaton, R. A.; Jones, E. B. G. *Mater. Org.* **1972**, *7*, 45–71.
17. Lee, A. W. C.; Grafton, J. C.; Tainter, F. H. In *Chromium-containing waterborne wood preservatives: Fixation and environmental issues*; Winandy, J., Barnes, M., Eds.; Forest Products Society: Madison, WI, 1993; pp 52–55.
18. Taylor, J. L.; Cooper, P. A. *Holzforschung.* **2005**, *59*, 467–472.
19. Wood, M. W.; Kelso, W. C.; Barnes, H. M.; Parikh, S. *Proc. Am. Wood-Preserv. Assoc.* **1980**, *76*, 22–37.
20. Dubey, B.; Townsend, T.; Solo-Gabriele, H.; Bitton, G. *Environ. Sci. Technol.* **2007**, *41* (10), 3781–3786.
21. Humar, M.; Žlindra, D.; Pohleven, F. *Build. Environ.* **2007**, *42* (2), 578–583.
22. Anon. *Best Management Practices for the Use of Wood in Aquatic and Wetland Environments*; Western Wood Preservers Institute: Vancouver, WA, 2011; pp 1–36.
23. Hayward, D.; Lebow, S. T.; Brooks, K. M. In *Managing Treated Wood in Aquatic Environments*; Morrell, J. J., Brooks, K. M., Davis, C. M., Eds.; Forest Products Society: Madison, WI, 2011; pp 407–433.
24. Evans, F. G. *International Res. Group on Wood Preservation*; Doc No. IRG/WP/3433, 1987.
25. Lebow, S. T.; Foster, D. O.; Lebow, P. K. *For. Prod. J.* **2004**, *54* (2), 81–88.
26. Chung, P. A.; Ruddick, J. N. R. IRG/WP 04-50219, 2004.
27. García-Valcárcel, A. I.; Bravo, I.; Jiménez, C.; Tadeo, J. L. *Environ. Toxicol. Chem.* **2004**, *23*, 2682–2688.
28. Hasan, A. R.; Hu, L.; Solo-Gabriele, H. M.; Fieber, L.; Cai, Y.; Townsend, T. G. *Environ. Pollut.* **2010**, *158* (5), 1479–1486.
29. Tao, W.; Shi, S.; Kroll, C. N. *J. Hazard. Mater.* **2013**, *260* (15), 296–304.
30. Waldron, L.; Cooper, P. A. *Wood Sci. Technol.* **2010**, *44*, 129–147.

31. Belford, D. S.; Nicholson, J. *Proc. Am. Wood-Preserv. Assoc.* **1969**, *65*, 38–46.
32. Edlund, M. L.; Sundman, C. E. *International Res. Group on Wood Preservation*; Doc No. IRG/WP 3533, 1989.
33. Hedley, M.; Durbin, G.; Wichmann-Hansen, L.; Knowles, L. *International Res. Group on Wood Preservation*; Doc No. IRG/WP 04-20285, 2004.
34. Lindegaard, B.; Morsing, N. *International Res. Group on Wood Preservation*; Doc No. IRG/WP 03-10499, 2003.
35. Militz, H. M.; Broertjes, M; Bloom, C. J. *International Res. Group on Wood Preservation*; Doc No. IRG/WP 98-20143, 1998.
36. Rapp, A. O.; Peek, R. D.; Sailer, M. *Holzforschung.* **2000**, *54*, 111–118.
37. Rydell, A.; Bergstrom, M.; Elowson, T. *Holzforschung.* **2005**, *59*, 183–189.
38. Saladis, J.; Rapp. *International Res. Group on Wood Preservation*; Doc No. IRG/WP 04-20299, 2004.
39. Brischke, C.; Rapp, A. O. *Wood Sci. Technol.* **2008**, *42* (8), 663–677.
40. Isaksson, T.; Thelandersson, S. *Build. Environ.* **2013**, *59* (1), 239–249.
41. Cooper, P. A.; Ung, Y. T. *International Res. Group on Wood Preservation*; Doc No. IRG/WP 97-50086, 1997.
42. Lebow, S. T. *Coatings minimize leaching from treated wood*; Durability Techline; U.S. Department of Agriculture, Forest Service, Forest Products Laboratory, Madison, WI, 2002; pp 1–2.
43. Lebow, S. T.; Williams, R. S.; Lebow, P. K. *Environ. Sci. Technol.* **2003**, *37*, 4077–4082.
44. Stilwell, D. E.; Sawhney, B. L.; Musante, C. L. *Proceedings of the Annual International Conference on Soils, Sediments, Water and Energy*; 2006, Volume 11, Article 10.
45. Nejad, M.; Cooper, P. *Environ. Sci. Technol.* **2010**, *44* (16), 6162–6166.
46. Nejad, M.; Ung, T.; Cooper, P. *Wood Sci.Technol.* **2012**, *46* (6), 1169–1180.
47. Plackett, D. V. *International Res. Group on Wood Preservation*; Doc No. IRG/WP/3310, 1984.
48. Ruddick, J. N. R. *Mater. Org.* **1993**, *27* (2), 135–144.
49. Kartal, S. N.; Dorau, B. F.; Lebow, S. T.; Green, F., III *For. Prod. J.* **2004**, *54* (1), 80–84.
50. Kartal, S. N.; Hwang, W. J.; Imamura, Y. *Build. Environ.* **2007**, *42* (3), 1188–1193.
51. Hingston, J. A.; Bacon, A.; Moore, J.; Collins, C. D.; Murphy, R. J.; Lester, J. N. *Bull. Environ. Contam. Toxicol.* **2002**, *689* (1), 118–125.
52. Lebow, S. T.; Foster, D. O.; Lebow, P. K. *For. Prod. J.* **1999**, *49* (7/8), 80–89.
53. Cooper, P. A. *For. Prod. J.* **1991**, *41* (1), 30–32.
54. Kim, J. J.; Kim, G. H *International Res. Group on Wood Preservation.* Doc No. IRG/WP/93-50004, 1993.
55. Murphy, R. J.; Dickinson, D. J. *International Res. Group on Wood Preservation.* Doc No. IRG/WP/3579, 1990.
56. Brooks, K. M. In *Managing Treated Wood in Aquatic Environments*; Morrell, J. J., Brooks, K. M., Davis, C. M., Eds.; Forest Products Society: Madison, WI, 2011; pp 163–207.

57. Becker, L.; Matuschek, G.; Lenoir, D.; Kettrup, A. Leaching behaviour of wood treated with creosote. *Chemosphere.* **2001**, *42*, 301–308.
58. Brooks, K. M. *Proceedings, Enhancing the Durability of Lumber and Engineered Wood Products*; Forest Products Society: Madison, WI, 2002; pp 59–72.
59. Xiao, Y.; Simonsen, J.; Morrell, J. J. Res. Note FPL–RN–0286. U.S. Department of Agriculture, Forest Service, Forest Products Laboratory: Madison, Wisconsin. 2002
60. Bergholm, J. Rep. 166. Swedish Wood Preservation Institute: Stockholm, Sweden, 1992
61. Crawford, D.; Fox, R.; Kamden, P.; Lebow, S.; Nicholas, D.; Pettry, D.; Schultz, T.; Sites, L.; Ziobro, R. *International Res. Group on Wood Preservation*; Doc No. IRG/WP 02-50186, 2002
62. Lebow, P.; Ziobro, R.; Sites, L.; Schultz, T.; Pettry, D.; Nicholas, D. *Wood Fiber Sci.* **2006**, *38* (3), 439–449.
63. Schultz, T. P.; Nicholas, D. D; Pettry, D. *Holzforschung.* **2002**, *56*, 125–129.
64. Wang, J. H.; Nicholas, D. D.; Sites, L. S.; Pettry, D. E. *International Res. Group on Wood Preservation*; Doc No. IRG/WP/98-50111, 1998.
65. Wang, L.; Kamdem, D. P. *International Res. Group on Wood Preservation*; Doc No. IRG/WP 11-50280, 2011.
66. Cooper, P. A.; Ung, Y. T. *For. Prod. J.* **1992**, *42* (9), 57–59.
67. Evans, F. G.; Nossen, B.; Edlund, M. L. *International Res. Group on Wood Preservation*; Doc No. IRG/WP 94-50026, 1994
68. Haloui, A.; Vergnaud, J. M. *Wood Sci.Technol.* **1997**, *31*, 51–62.
69. Lebow, S; Lebow, P.; Foster, D. Estimating Preservative Release From Treated wood Exposed to Precipitation. *Wood Fiber Sci.* **2008**, *40* (4), 562–571.
70. Anon. JIS K 1571. Japan Standards Association: Tokyo, 2004.
71. Anon. CNS 6717. Bureau of Standards, Metrology and Inspection, Ministry of Economic Affairs (MOEA): Taipei, Taiwan, 2000.
72. Anon. BS EN 84. British Standards Institute: London, 1997.
73. OECD, Test No. 313; OECD Environment Directorate: Paris, France, 2007.
74. OECD, Series on Testing and Assessment No. 107; OECD Environment, Health and Safety Publications. OECD Environment Directorate: Paris, France.
75. Baines, E. F. *International Res. Group on Wood Preservation*; Doc No. IRG/WP 05-50224-7, 2005.
76. Morsing, N.; Lindegaard, B. *International Res. Group on Wood Preservation*; Doc No. IRG/WP 04-20303, 2004.
77. Baldwin, W. J.; Pasek, E. A. *Proc. Am. Wood-Preserv. Assoc.* **1994**, *90*, 300–316.
78. Kang, S. M.; Morrell, J. J.; Simonsen, J.; Lebow, S. T. *For. Prod. J.* **2005**, *55* (12), 42–46.
79. Cooper, P. A.; Ung, Y. T. *Wood Fiber Sci.* **2009**, *41* (3), 229–235.
80. Garcia-Valcarcel, A. I.; Tadeo, J. L. *Environ. Toxicol. Chem.* **2006**, *25* (9), 2342–2348.

81. Shibata, T.; Solo-Gabriele, H. M.; Fleming, L. E.; Cai, Y.; Townsend, T. G. *Sci. Total Environ.* **2007**, *372* (2-3), 624–635.
82. Stefanovic, S.; Cooper, P. In *Environmental Impacts of Treated Wood*; Townsend, T., Solo-Gabriel, H., Eds.;CRC Press, Taylor and Francis Group: Boca Raton, FL, 2006; pp 101–116.
83. Cooper, P. A., MacVicar, R. *International Res. Group on Wood Preservation*; Doc No. IRG/WP 95-50044, 1995.
84. Choi, S.; Ruddick, J. R.; Morris, P. *For. Prod. J.* **2004**, *54* (3), 33–37.
85. Cui, F.; Walcheski, P. *International Res. Group on Wood Preservation*; Doc No. Document No. IRG/WP 00-50158, 2000.
86. Kennedy, M. J.; Collins, P. A. *International Res. Group on Wood Preservation*; Doc No. IRG/WP 01-50171, 2001.
87. Khan, B. I.; Solo-Gabriele, H. M.; Townsend, T. G.; Cai, Y. *Environ. Sci. Technol.* **2006**, *40* (3), 988–993.
88. Mitsuhashi, J.; Love, C. S.; Freitag, C.; Morrell, J. J. *For. Prod. J.* **2007**, *57* (12), 52–57.
89. Morrell, J. J.; Chen, H.; Simonsen, J. *Proc. Am. Wood-Preserv. Assoc.* **2004**, *100*, 158–161.
90. Waldron, L.; Cooper, P. A.; Ung, Y. T. *Holzforschung* **2005**, *59* (5), 581–588.
91. Waldron, L.; Cooper, PA. *Wood Sci. Technol.* **2007**, *42*, 299–312.
92. Barna, L.; Schiopu, N. *J. Hazard. Mater.* **2011**, *192* (3), 1476–1483.
93. Lupsea, M.; Mathies, H.; Schoknecht, U.; Tiruta-Barna, L.; Schiopu, N. *Sci. Total Environ.* **2013**, *444*, 522–530.
94. Lupsea, M.; Tiruta-Barna, L.; Schiopu, N.; Schoknecht, U. *Sci. Total Environ.* **2013**, *461–462*, 645–654.
95. Schultz, T. P.; Nicholas, D. D.; Dalton, T. D.; Keefe, D. *For. Prod. J.* **2004**, *54* (3), 85–90.
96. Lebow, S.; Foster, D. *For. Prod. J.* **2010**, *60* (2), 183–189.
97. Cooper, P. A.;Ung, Y. T. *International Res. Group on Wood Preservation*; Doc No. IRG/WP/97-50087, 1997.
98. Cserjrsi, A. J. *For. Prod. J.* **1976**, *26* (12), 34–39.

Chapter 15

Discussion on Prior Commercial Wood Preservation Systems That Performed Less Well Than Expected

Tor P. Schultz,*,[1] Darrel D. Nicholas,[2] and Patti Lebow[3]

[1]Silvaware, Inc., 303 Mangrove Palm, Starkville, Mississippi 39759
[2]Department of Forest Products/FWRC, Mississippi State University, Mississippi State, Mississippi 39762
[3]Forest Service, Forest Products Laboratory, U.S. Department of Agriculture, Madison, Wisconsin 53726
*E-mail: tschultz.silvaware@bellsouth.net.

This article reviews five prior commercial wood preservatives that had efficacy concerns. A common factor among all five systems was minimal or no field testing of the proposed system prior to commercialization. Also, the formulation of a successful preservative was twice changed, and one successful system was employed with a new wood species. There is no intent to hold responsible any individual(s), company(ies), or organization(s) for these failures; the purpose is to simply discuss systems which performed less well than expected and report on failure factors. Because it appears that longer field evaluations prior to commercialization might have identified poor performance, a preliminary study of the effect of exposure time was performed. Data from three ground-contact studies of the fungal efficacy of experimental systems run at two sites were compared to the efficacies of positive control biocides which have long provided adequate commercial wood protection. Systems that were "poor" because of low initial biocide treatment required an exposure of three or fewer years to detect differences from a positive control. However, some systems which were treated to moderate biocide levels, and initially

© 2014 American Chemical Society

performed adequately, suffered greater fungal degradation than the positive controls after four or more years. This study indicates that exposure times longer than the currently-required three years may be needed to determine if a system treated to the proposed commercial retentions will perform adequately for the long service life expected by consumers.

Introduction

Because some untreated wood products may perform favorably for years in outdoor exposure, especially when used above ground, predicting the long-term durability of a wood preservative system requires years of field tests at multiple sites (*1*). Similarly, it may be years before consumers become aware that a wood product is not performing as expected, such as a structural member failing below the ground level where decay or termite degradation is not visible. No company that has been in business for any length of time wants, or even considers, the possibility of providing a product with the potential to fail with the resulting financial loss and negative publicity. However, failures do happen. While many types of failure can occur with treated wood products, in this chapter we focus only on biodegradation of commercially-treated wood where deterioration occurred with a frequency that caused concern within the forest products industry and was likely due to poor performance of the wood preservative system.

A system must protect a treated wood product against a wide variety of fungi and/or insects, of which any single species can invade and degrade the product over its expected long service life. It is impossible to test a system in the laboratory against all possible organisms which can attack wood, or to conduct ground-contact field studies in the many soil environments to which a treated wood product will be exposed. Furthermore, biocides can be depleted over time and depletion is best measured by long-term field studies, but outdoor ground-contact depletion data among replicate samples are typically very erratic (*e.g.* (*2*, *3*)). Finally, many technically complex factors can be involved and more than one cause can lead to failure under different circumstances – and human factors at the treating facility can also be involved in some circumstances.

For this chapter, we choose five major prior commercial preservatives which had efficacy concerns after commercialization and discuss each in chronological order in a balanced and factual manner. We then list the common factors among the five systems discussed. Finally, we examine preliminary data on the effect of outdoor ground-contact exposure test time and site on the significance of the fungal efficacy of experimental compared to well-recognized commercial systems.

Prior Commercial Systems with Efficacy Concerns

Volatile Solvent Pentachlorophenol Treatment

The first/earliest system discussed is pentachlorophenol (penta)-treated utility poles. Penta in heavy oil carriers, which gives the wood an oily surface, have been employed to treat utility poles and crossarms for at least 60 years. In the early

1960's volatile solvents, specifically liquefied propane or butane (the CELLON™ process) or methylene chloride, were employed to formulate penta with the volatile solvent recycled after use (*4–8*). The treated poles had a dry and visually-appealing surface that could be stained or painted. This also eliminated the heavy oil carrier and enabled recycling of the volatile solvent which reduced costs. Poles with these treatments were extensively used. However, in a relatively short time soft-rot decay problems were noticed in both poles and research stakes (*9*). Much resulting discussion ensued on whether a problem existed and, if so, possible causes considered such as the penta retention gradient in small research stakes versus commercial-sized poles or interpretation of the rating index of research stakes.

Eventually, it was generally agreed that serious problems indeed existed and a number of factors were identified. One major problem was that the poles emerged so dry and uncolored after treatment that any untreated or partially treated poles, especially poles on the top layer in cylinders which were only partially filled, could not be readily identified. When employed, the entire pole or untreated portion quickly experienced decay or termite attack. By contrast, poles treated with a heavy oil carrier that were only partially treated could be visually observed and set aside for retreatment.

Another problem was penta migrating to the surface of the poles as the volatile solvent was removed, causing blooming [crystallizing] on the wood surface. The pole surface was sometimes washed with aqueous alkali to remove the surface residue (*10*), but this wash also removed some of the penta in the thin outermost layer of the pole. The lower penta retention in the vulnerable surface layer afforded an opportunity for soft-rot fungi to become established, as soft-rot fungi are more resistant to penta than brown- or white-rot fungi. It was also reported that various additives may have been employed to reduce blooming, and a few of these may have negatively affected treatment efficacy and/or enhanced penta leaching (*11*).

The heavy oil carrier was originally assumed to be biologically inert. However, extensive studies have now conclusively shown that stakes treated with heavy oils alone are more durable than untreated stakes (*12*). Today, most professionals agree that heavy oil carriers impart some biological efficacy against decay and termites (*13*) and enhance the activity of penta as a wood preservative. Further, heavy oils impart some water repellency which reduces decay potential.

Another possible problem was penta leaching. Penta is a highly acidic phenol with a pKa of about 5, which meant that this biocide readily forms the ionized, water soluble salt form when exposed to water at normal pHs. Thus, some researchers suspected that any crystallized penta without a heavy oil carrier might be easily solubilized and leached by free water in the ground-contact area of the pole. While some penta in the outer zone of the pole initially depleted relatively rapidly, the depletion rate decreased over time and only slightly higher penta levels were found to be retained in poles with heavy oils (*5, 8*). The only poles which had severe penta depletion were those installed in alkaline and/or water-saturated soils, and reportedly even penta in heavy oil will leach from utility poles in highly alkaline soils (*14*).

CCA Treatment of Eucalypts Utility Poles

CCA is a waterborne preservative that was extensively employed in both residential and industrial applications and is still employed in industrial and agricultural uses where it provides excellent and cost-effective service for a wide variety of applications. The vast majority of treated wood is cut from gymnosperms (softwoods) that include pines, spruces, and firs. Thus, the wood treating profession has had a long and positive experience with CCA in protecting softwood products.

Softwood pole availability was limited in Australia, however, so poles from certain hardwood (angiosperm) Eucalyptus species were employed with the assumption that the highly effective CCA would continue to perform well with hardwoods to give a durable product. This assumption quickly proved wrong, however, as initial reports of soft-rot attack on CCA-treated Eucalyptus poles was later verified by extensive field research (*15, 16*). In an exhaustive study, slight to severe fungal deterioration was observed in all 1,000 CCA-treated Eucalyptus poles examined (*16*). The problem was successfully addressed by inspection of standing poles and a remedial treatment applied where necessary (*17*).

The reasons identified appear rather subtle at first but, perhaps, with hindsight should have been considered before large volumes of Eucalyptus poles were treated. Specifically, hardwoods have a more complex anatomy, and different lignin chemical structure and lower lignin levels, than softwoods. As CCA components are fixed within softwood, often to lignin, the relatively simple anatomy and uniform lignin structure and distribution in softwoods means that the CCA is relatively homogenously distributed. However, hardwoods have a more complex anatomy which results in hardwood fibers sometimes being poorly treated, and pockets of untreated fiber cells existed with CCA-treated hardwoods (*18, 19*), Further, as the lignin structure in hardwoods differ among the various cell types, and the metallic biocides preferentially fix to lignin, CCA microdistribution within hardwoods was more variable than in softwoods. Finally, hardwoods are generally more susceptible than softwoods to soft-rot fungi (*10*).

To summarize the possible reasons for the failure observed with hardwoods treated with CCA; 1) hardwoods are more susceptible to soft- and white-rot fungi than softwoods; 2) CCA fixes to the wood, especially the lignin component, and hardwoods have a more complex lignin structure than softwoods that varies among the cell types and the different layers of the cell wall; 3) hardwood fibers are more difficult to treat than softwood fibers; so 4) due to 2 and 3 above a non-uniform CCA microdistribution exists in hardwoods; and 5) metallic biocides have smaller zone of inhibition compared to organic biocides, so that decay fungi can grow in untreated cells that border cells treated with copper.

Quaternary Ammonium Compounds in Australia and Asia

A variety of quats, or quaternary ammonium compounds, are employed in wood preservatives. Quats have many positive properties including being extremely economical per unit weight and a relatively broad efficacy against a wide variety of fungi and insects. They are also waterborne but fix in wood upon treatment by ion-exchange reactions. Quats were tested as a wood preservative using approved laboratory tests and one outdoor above-ground test with painted wood (*20*). After favorable test results (*21*), quats were employed to treat wood for above ground applications in Australia and New Zealand. An above-ground test run shortly thereafter in North America with unpainted wood suggested some concerns, however, and decay problems later surfaced in applications with high decay potential.

Relatively little data was made publicly available from studies on the causes. One cause was likely poor penetration of quats in some cases. Under alkaline conditions quats sometimes rapidly fixed within the wood so that only the outer shell of commercial-sized lumber was treated, leaving the center core untreated as observed by one author, DDN, in field trials in Houston, TX. This effect would not necessarily have been observed with the small samples typically employed in laboratory efficacy tests. Secondly, a supposedly less active benzalkonium chloride (BAC) quat was used rather than the more active dimethyl didecylammonium chloride (DDAC) (*22*), although a later field study found little efficacy differences between BAC and DDAC (*23*). Some bacterial degradation of the quat may have also occurred (*24, 25*). Overall, it appears that a large number of factors and special circumstances occurred in the various failures, and hindsight suggested that more and longer field tests should have been conducted (*20*).

Companies in Japan later examined quats and, upon favorable results (*26, 27*), employed them. A few decay problems later appeared in certain high-decay hazard applications (*20*).

Quats have many advantages, as listed above, and when combined with other biocide(s) to ensure greater and/or broader activity they provide good protection against a wide variety of fungi and insects in some current systems.

Waterborne Pentachlorophenol

In the late 1970's waterborne penta [the salt of pentachlorophenol] was employed to treat wood (*28*). While the salts were water soluble, upon treatment and exposure to the naturally acidic wood the salt was protonated and the penta precipitated in the wood. As with the volatile solvent penta-treated poles discussed above, the surface of the resulting product was clean and could be painted, individually-treated wood could be glued together to form composites, and the cost of the process was reduced by not employing a heavy oil carrier. However, in a relatively short time concern was expressed over decay seen in some research stakes that further exposure time confirmed in commercially treated wood products (*9, 12, 29*).

Many of the problems identified were similar to the volatile solvent-formulated penta poles discussed earlier, including the lack of a heavy oil carrier making the wood more susceptible to decay, especially by soft-rot fungi. Another factor included the possible use of a surfactant in the formulation which may have increased penta leaching (*30*). Further, the aqueous alkali treating solution solubilized wood extractives such as sugars and resin acids which, along with any remaining soluble penta salts, crystallized on the surface. The treated poles were washed with an alkaline solution to remove the unsightly surface deposits, but this also resulted in reduced penta levels in the outer pole surface to make it more vulnerable to initial fungal colonization and subsequent fungal movement into the inner/deeper zones (*10*). Finally, soft-rot fungi appeared to "grow around" areas with penta crystals.

Tributytin Oxide

Tributyltin oxide (TBTO), a colorless biocide with low mammalian toxicity, was not patented and so was freely available for commercial use. A number of factors suggested that TBTO would be a good wood preservative, including laboratory decay tests which indicated it would be effective in above-ground applications and a long and positive history in other biocidal applications. Also, the treated wood had good appearance and no strength loss, and TBTO has low volatility and good water leach resistance. Based on the promising laboratory fungal efficacy tests and the good TBTO properties (*31*), it was extensively used in millwork and other low-deterioration hazard applications in North America and Europe. Unfortunately, initial reports of deterioration appeared within a few years, which were later confirmed as the TBTO-treated products aged.

After numerous studies in different laboratories it was generally acknowledged that with long-term exposure TBTO underwent chemical and/or biological dealkylation to form dibutyl and monobutyl compounds that had reduced fungal activity (*32–38*). While the initial laboratory efficacy tests were encouraging, the abiotic and/or biotic transformation of the biocide which resulted in reduced fungal protection were relatively slow reactions - so that the resulting lower efficacy was not observed in the relatively short-duration laboratory tests employed to test TBTO prior to commercialization. Thus, the fundamental cause for the efficacy problem with TBTO was the lack of long-term outdoor above-ground efficacy and depletion tests prior to commercialization. Further, retention of TBTO in millwork and other wood products was determined by analysis of only the tin, rather than determining the amount of the actual organotin biocide (*39*). Consequently, the tin retention deceivingly remained almost constant while the organotin biocide underwent dealkylation reactions in the treated wood to give an altered, less biocidal compound.

Summary of Common Factors

The mostly frequently occurring factors associated with the five systems discussed above are:

- Minimal or no valid outdoor, long-term efficacy and/or depletion trials prior to commercialization, including the need to employ multiple test sites for systems intended for ground-exposure applications. This apparently occurred with all five systems discussed.

- A change in the formulation of an established system with the assumption that the biocide would continue to be effective even when formulated in a different carrier (*e.g.* the volatile solvents and waterborne penta systems), or using a successful system with a new wood species (*e.g.* CCA/Eucalypts poles).

Preliminary Study on the Effect of Field Ground-Contact Exposure Time

One common factor which occurred in all five of the systems discussed above was minimal or no valid outdoor efficacy and depletion tests. For ground-contact applications, systems proposed for American Wood Protection Association (AWPA) standardization are required to be exposed for a minimum of three years at two different sites with high or severe deterioration hazard along with depletion analyses ((*40*), Appendix A). Similar requirements exist for systems being submitted for International Code Council-Evaluation Service (ICC-ES) Standardization (*41*). Recently, much discussion has occurred on the sufficiency of three years of field exposure at two sites to determine if a proposed system will be effective for the many years of service expected by consumers.

To address the question of exposure time, long-term field AWPA E7 test data were obtained from multiple studies conducted by the USDA-Forest Products Laboratory and Mississippi State University. Results from a preliminary analysis of three field data sets established at the Dorman (AWPA Deterioration Hazard Zone 4/High, with copper tolerant fungi present) and Saucier (AWPA Deterioration Hazard Zone 5/Severe) test sites in Mississippi were conducted. All three field stake tests included positive control samples of a long-established commercial biocide. One set consists of the synergistic copper/Cu-8 combination (*42, 43*) with CCA as the positive controls. The other two sets examined the organic PXTS (polymeric xylenol polysulfide) system, a potential creosote substitute which tests showed to be about twice as effective as creosote (*44*). The two PXTS sets employed different formulations, with one set having CCA and the other creosote positive controls. As this is a preliminary analysis only the fungal decay results were examined. The results were modeled over a 12-year period. Fungal efficacies were statistically compared at the 0.05 significance level, adjusted for multiple time comparisons, using as a comparison a commercial preservative system at the specified retention for residential (Cu/Cu8) or industrial applications (PXTS). The objective was to determine whether three years of field exposure is sufficient to decide if a proposed ground-contact system is comparable to a commercial system and, if three years is insufficient, how long a test exposure period is necessary.

The first system examined was PXTS with creosote as the positive control. The results were statistically different between Saucier and Dorman sites. At Saucier, the positive control was 144.5 kg/m^3 creosote, and at Dorman the positive control was 113.1 kg/m^3 creosote. What would be anticipated as poor systems, e.g. low retentions of creosote [19.7 to 75.4 kg/m^3] or PXTS [18.9 and 39.8 kg/m^3], gave statistically poorer results compared to the positive control after only one year of exposure at Saucier. PXTS at levels of 112.5 and 159.2 kg/m^3 at the same site required seven and eight years of exposure before they performed significantly better than creosote at 144.5 kg/m^3. Similarly, at Dorman PXTS with relatively high levels of 112.5 and 159.2 kg/m^3 performed better than the creosote positive control after only four years of exposure. For low biocide retention levels which would be expected to have "poor" performance, the two lowest creosote levels of 19.7 and 38.7 kg/m^3 took only one year of exposure at Dorman to show they had poor decay performance. However, the efficacy of stakes treated with the relatively low PXTS retention of 39.8 kg/m^3 at Dorman was statistically similar to the positive control, and after five years of exposure PXTS at 76.6 kg/m^3 performed statistically better than the positive control.

The second set was PXTS, with CCA at 10.1 kg/m^3 as the positive control. The results were statistically different at the two sites, and differences with the positive control were detected more often and sooner at Saucier. PXTS at the relatively low level of 46.6 kg/m^3 required only three years of exposure at Saucier and four years at Dorman to show significantly poorer efficacies compared to the positive control. The moderate PXTS level of 70.9 kg/m^3 required four years of exposure at Saucier before it performed poorer than the positive control, while at Dorman this set performed equally to the positive control for all 12 years. PXTS at 94.2 kg/m^3 also performed equally to the positive control at Dorman for 12 years but statistically lower at Saucier after only two years of exposure; however, at the end of the study (12 years), the performance matched the controls at Saucier. Finally, CCA at a level of 3.4 kg/m^3, or about half of the UC4A retention and so would be considered to be a "fair" system, required nine and eight years of exposure at Dorman and Saucier, respectively, before it performed significantly poorer than the positive control.

The third set examined, copper/Cu-8, proposed to be a possible copper-based residential system so CCA at 6.1 kg/m^3 was employed as the positive control. Not unexpectedly, significant fungal efficacy differences were noted between Saucier and Dorman, with Dorman having lower ratings sooner. This was most likely due to copper-tolerant fungi present at Dorman. However, unlike the creosote studies an interaction between site and exposure time was not detected, indicating that although fungal attack occurred sooner at Dorman the ability to detect the differences between treatments did not occur any sooner at Dorman than Saucier. The systems which would be expected to be "poor", those treated with Cu8 alone (0.3, 1.0, and 1.9 kg/m^3) or relatively low levels of Cu/Cu8 mixtures (0.6/0.3, 1.6/0.2, and 1.3/0.3 kg/m^3), required only one to two years of exposure to show poor fungal performance. A moderate Cu/Cu8 level of 3.2/0.5 kg/m^3 required six years of exposure. Relatively high levels of Cu/Cu8, 3.7/0.6 to 5.0/1.1 kg/m^3, showed equal performance to the positive control of CCA at 6.1 kg/m^3 for all 12 years.

Overall, in this preliminary analysis it appears that three years of exposure at two sites is sufficient to show that treatments with relatively low retentions, which would be expected to give poor performance, did indeed perform poorly relatively to the positive controls. However, systems that would be considered as "fair" required longer exposure periods. For example, CCA at about half the UC4A retention required exposures of about eight years, and moderate Cu/Cu8 retentions of 3.2/0.5 kg/m^3 required six years of exposure, to show significantly poorer fungal efficacies than the positive controls. Conversely, PXTS at 76.6 kg/m^3 at five years of exposure performed better than the positive control. Thus, we conclude that exposure times longer than the currently-required three years may be necessary to determine if a system treated to a potential commercial retention would likely perform adequately in commercial service for the relatively long duration expected by consumers. However, longer field test exposure time will make it more difficult and expensive to develop new systems as quickly as the public currently demands.

This study is only a preliminary analysis of fungal efficacy data from three research studies. We hope to report more fully on further statistical analyses employing these and additional field stake data in the future.

Conclusions

A review of five prior commercial systems which had experienced poor service efficacy found that all systems underwent limited or no valid long-term outdoor efficacy and depletion testing prior to commercialization. In addition, formulations were changed in two systems which had a previously successful biocide, and one effective system was employed with a new wood species. A preliminary analysis of three AWPA E7 field test studies conducted at two sites suggests that a longer exposure period than the three years currently required may be necessary to adequately test the fungal efficacy of a proposed ground-contact preservative prior to commercialization.

References

1. Schultz, T. P.; Nicholas, D. D. *For. Prod. J.* **2008**, *58* (5), 73–76.
2. Lebow, S.; Halverson, S. *AWPA Proc.* **2008**, *104*, 55–60.
3. Schultz, T. P.; Nicholas, D. D.; Pettry, D. E. *Holzforschung* **2002**, *56*, 125–129.
4. Boyce, J. D. *AWPA Proc.* **1967**, *63*, 55–56.
5. Arsenault, R. D. *AWPA Proc.* **1970**, *66*, 197–211.
6. Ochrymowych, J.; McOrmond, R. R., III *The Western Elec. Engineer* **1976**, *XX* (1), 24–33.
7. Marouchoc, S. R. *AWPA Proc.* **1972**, *68*, 148–153.
8. McOrmond, R. R., III)Ochrymowych, J.; Arsenault, R. D. *AWPA Proc.* **1978**, *74*, 64–79.
9. *1965 Progress Report, Comparison of Wood Preservatives in Stake Tests*; USDA Forest Products Laboratory: Madison, WI.
10. Goodell, B.; Qian, Y.; Jellison, J. In *Development of Commercial Wood Preservatives: Efficacy, Environmental, and Health Issues*; Schultz, T. P., Militz, H., Freeman, M. H., Boodell, B., Nicholas, D. D., Eds.; ACS Symposium Series 982; American Chemical Society: Washington, DC, 2008; Chapter 2.
11. Arsenault, R. D.; Ochrymowych, J.; Kressback, J. N. *AWPA Proc.* **1984**, *80*, 140–168.

12. *2000 Progress Report. Comparison of Wood Preservatives in Stake Tests*; USDA Forest Products Laboratory: Madison, WI.
13. Nicholas, D. D.; Sites, L.; Barnes, H. M.; Ng, H. *AWPA Proc.* **1994**, *66*, 44–45.
14. Barnes, H. M.; Amburgey, T. L.; Sanders, M. G. *For. Prod. J.* **2006**, *56* (6), 43–47.
15. Leightely, L. E.; Norton, J. *Int. Res. Group/Wood Preserv.* **1983** IRG Doc. 3226.
16. Leightley, L. E. *Int. Res. Group/Wood Preserv.* **1986** IRG Doc. 1301.
17. Ziobro, R. J.; McNamara, W. S.; Triana, J. F. *For. Prod. J.* **1987**, *37* (3), 42–45.
18. Levy, C. R. *AWPA Proc.* **1978**, *74*, 145–164.
19. Greaves, H. *Holzforschung* **1974**, *28*, 193–200.
20. Preston, A. F. Personal communication, 2008.
21. Preston, A. F.; Chittenden, C. M. *N. Z. J. For. Sci.* **1982**, *12* (1), 102–106.
22. Preston, A. F.; Nicholas, D. D. *Wood Fiber* **1982**, *14*, 37–42.
23. Hedley, M.; Tsunoda, K.; Suzuki, K. *Int. Res. Group/Wood Preserv.* **1995** IRG Doc. 95-30083.
24. Dubois, J. W.; Ruddick, J. N. R. *Int. Res. Group/Wood Preserv.* **1998** IRG Doc. 98-10263.
25. Wallace, D. F.; Cook, S. R.; Dickinson, D. J. In *Development of Commercial Wood Preservatives: Efficacy, Environmental, and Health Issues*; Schultz, T. P., Militz, H., Freeman, M. H., Boodell, B., Nicholas, D. D., Eds.; ACS Symposium Series 982; American Chemical Society: Washington, DC, 2008; Chapter 18.
26. Hedley, M.; Tsunoda, K.; Nishimoto, K. *Wood Res.* **1982**, *68*, 37–46.
27. Tsunoda, K.; Nishimoto, K. *Mokuzai Gakkaishi* **1987**, *33*, 589–595.
28. Hatcher, D. B. *AWPA Proc.* **1981**, *77*, 89–98.
29. *1973 Progress Report. Comparison of Wood Preservatives in Stake Tests*; USDA Forest Products Laboratory: Madison, WI.
30. Freeman, M. H. Personal communication, 2008.
31. Nicholas, D. D. *AWPA Proc.* **1973**, *69*, 65–66.
32. Dudley-Brendell, T. E.; Dickinson, D. J. *Int. Res. Group/Wood Preserv.* **1982** IRG Doc. 1156.
33. Orsler, R. J.; Holland, G. E. *Int. Res. Group/Wood Preserv.* **1984** IRG Doc. 3287.
34. Belford, P. S.; Dickinson, D. J. *Int. Res. Group/Wood Preserv.* **1985** IRG Doc. 1258.
35. Beiter, C. B.; Arsensault, R. D. *AWPA Proc.* **1981**, *77*, 58–63.
36. Edlund, M.-E.; Jenningsson, B.; Jensen, B.; Sundman, C.-E. *Int. Res. Group/Wood Preserv.* **1988** IRG Doc. 3476.
37. Blunden, S. J.; Hill, R. *Int. Res. Group/Wood Preserv.* **1989** IRG Doc. 3508.
38. Reinprecht, L. *Int. Res. Group/Wood Preserv.* **1998** IRG Doc. 98-30185.
39. Schultz, T. P.; Nicholas, D. D.; Henry, W. *Forest Products Society Proc. On Enhancing the Durability of Lumber and Engineered Wood Products*; Forest Products Soc.: Madison, WI, 2002; pp 273–280.

40. *2013 AWPA Book of Standards*; American Wood Protection Association: Birmingham, AL, 2013.
41. ICC-ES Doc. AC326, Approved Feb. 2013. Available at www.icc-es.org.
42. Schultz, T. P.; Nicholas, D. D.; Henry, W. P. *Holzforschung* **2005**, *59*, 370–373.
43. Schultz, T. P.; Nicholas, D. D. *For. Prod. J.* **2008**, *58* (5), 73–76.
44. Freeman, M. H.; Nicholas, D. D.; Renz, D.; Buff, R. *Int. Res. Group/Wood Preservs.* **2004** IRG Doc. 04-30350.

Nonbiocidal Modification

Chapter 16

Processes and Properties of Thermally Modified Wood Manufactured in Europe

H. Militz* and M. Altgen

Wood Biology and Wood Products, Georg-August-University Goettingen,
Buesgenweg 4, 37077 Goettingen, Germany
*E-mail: hmilitz@gwdg.de.

Several processes to thermally modify wood have been commercialised in Europe in the past decades. Due to the high temperatures, 180 - 220 °C, used in most processes, the chemical structure of the wood components are greatly changed. Heat treated lumber has altered biological and physical properties. The wood is more resistant against basidiomycetes and soft rot fungi, and has a lower equilibrium moisture content and fibre saturation point. Consequently, the dimensional stability is improved. Because of the increased brittleness of the wood, some strength properties are greatly decreased. Due to the enhanced durability, dimensional stability, and good appearance, thermally-treated wood is currently used in Europe in many indoor and outdoor applications.

Introduction

Research efforts have long examined processes to chemically modify wood. Because of the availability of tropical timbers with high natural quality and cheap effective wood preservatives, however, only a few wood modification processes were commercialized in the past. This has changed in the last few years, with increased interest in alternatives for tropical timbers and preservative-treated wood leading to several new wood treatments that have recently been commercialized in Europe. Acetylation with acetic anhydride, furfurylation with furfural alcohol, or treating wood with modifying resins are examples of the processes that have been commercialized in Europe (*1, 2*). Most of these new treatments are non-biocidal alternatives to conventional wood treatment with biocidal preservatives, and provide wood with improved dimensional stability and a pleasant appearance for interior or exterior use.

It has long been known that wood properties can be altered when wood is heated at elevated temperatures. However, only recently has this knowledge led to the development of commercial processes. Today, many production units with various production capacities exist in several European countries (*3*). This article will give an overview of the existing technology and the material properties of thermally treated wood.

Treatment Processes

Initial attempts to use the scientific knowledge of Stamm *et al.* (*4*) and Burmester (*5*) to develop a commercial heat-treatment process for mid-European wood species were made by Giebeler (*6*) in Germany. For more than 20 years knife handles were produced in a small scale production plant. The original goal, introduction of a large scale process for exterior wood, was not reached because of the lack of interest from the wood industry in the 1980's. About 10 years later the idea of thermally treating wood was taken up by several research groups and the industry in Europe. More or less independently from each other, several processes were developed and taken from the laboratory to commercial production. Nowadays, thermally modified wood is produced in more than 90 production facilities all over Europe, featuring an European production capacity of approximately 280,000 m^3 per year in 2010 (*3*).

All of the applied processes have in common a thermal treatment at elevated temperatures (160 – 240 °C) greater than that normally used to dry lumber (50 – 120 °C) and a minimization of the residual oxygen content. The main differences between the various processes are the process conditions that are reached with different treatment technologies. Besides treatment temperature and duration, the initial moisture content of the wood, the pressure conditions and the heat transferring media are some of the key parameters to produce wood with good decay resistance and physical properties. In the following, the characteristics of some of the existing process technologies are described.

An industrial scale wood heat treatment process, under the trade name of ThermoWood was developed in Finland. The process is licensed to members of the International ThermoWood Association. It is carried out at atmospheric pressure with a constant steam flow throughout the process that removes volatile degradation products and acts as a heat transferring media. The temperature inside the wood is used to regulate the temperature rise in the kiln. In the first step, the temperature is increased steadily to 130 °C, during which time high temperature drying takes place. In the second step, the temperature is raised to 185 – 230 °Cand then held for 2-3 hours, depending on the end-use applications. The third step is a cooling and conditioning step. This final stage lowers the temperature using a water spray system, and at a temperature of 80 – 90 °C re-moisturising and conditioning takes place to bring the wood moisture content to 4-6 %. The wood employed can be freshly sawn or kiln dried (*7*, *8*).

The PLATO-process (*9*, *10*) operated by PLATO BV in the Nethderlands combines a hydrothermolysis step with a dry curing step. During the hydrothermolysis step, the wood is treated in an aqueous environment with saturated steam as the heating media (1-2 hours at 160 – 190 °C). After an intermediate drying step (3-5 days) the final curing step at atmospheric pressure is applied (8-12 hours at 170 – 190 °C). In some cases a conditioning step (2-3 days) is needed. Depending on the wood species and the thickness of the material, these times can be shorter. The heating medium can be steam or heated air (*11*).

In France, there are several companies that produce thermally modified wood using the Retification-process. In this one step process, pre-dried wood (approx. 12 % MC) is heated to 200 – 240 °C in a nitrogen atmosphere that has limited oxygen content of less than 2 % (v/v). The total duration of the process is approx. 9 - 12 hours, depending on the wood dimensions and wood species (*12*). A similar process technology which also uses a nitrogen atmosphere was developed by a Swiss company named Balz Holz.

The main characteristic of the OHT (oil heat treatment) process, operated by Menz Holz (Germany), is the use of linseed oil as the drying medium and to improve heat flow into the timber. At the same time, the oxygen level in the vessel is low due to the oil. Fresh or pre-dried timber can be used in this process. The heat treatment is performed at 180 to 220 °C for 2 - 4 hours in a closed vacuum-pressure process vessel. Additional treating steps include heating up and cooling down, with various time depending on the wood dimension. Typical process duration for a whole treatment cycle (including heating up and cooling down) for logs with a cross section of 100 mm x 100 mm and length of 4 meters is 18 hours (*13*).

Based on a method for vacuum kiln drying, Opel Therm GmbH developed a process without any addition of steam, nitrogen or oil. A German company (Timura) produces thermally modified wood using this process under the name of "vacuum press dewatering method" (Vacu^3). The method is based on applying a vacuum of 150 mbar that decreases the boiling point of water and thus accelerates the drying of the wood. During the process, the heat is transfered to the wood by heating plates between the boards while pressure is applied from the top of each stack to reduce deformation. Byproducts from the wood thermal degradation taking place during the process are continuously discharged from the process. A similar process technology is SmartHeat, which also uses heating plates in combination with an optional vacuum (*14*) and is orperated by Lignius (the Netherlands).

WTT (Wood Treatment Technology, Denmark) introduced a process similar to Burmester's technique to thermally modify wood under elevated pressure (*15*). FirmoLin Technologies (the Netherlands) further enhanced the process (*16*). This process is based on using superheated steam at elevated pressure in an autoclave equipped with a water reservoir. By controlling the temperature of the vessel and the water reservoir as well as the steam pressure, the relative humidity can be regulated and the dry state of the wood is avoided. The treatment requires pre-dried material with a typical moisture content of approximately 12 % (*17*). A very similar process technology is provided by Moldrup Systems Pte. Ltd. (Denmark).

A more complete list of the existing process techniques and manufacturers in Europe is given by EUWID (*3*).

Products and Production of Heat-Treated Wood

Depending on the wood species and the production process, several biological and technological properties of the wood are changed by the treatment. The colour of the wood turns brownish, which is used by some companies to give local wood the appearance of exclusive tropical wood. Because no chemicals are used, heat treated wood can be used in both exterior (with increased resistance against wood degrading organisms) and interior applications. At the present time thermally treated wood is used in many applications, including windows, claddings, play ground equipment, sauna interiors, bath rooms, parquet flooring, decking, etc.

The production capacity of thermally modified wood in Europe increased considerably in recent years. While Militz (*18*) estimated the production capacity in 2001 at approximately 165,000 m^3, it was reported that the capacity reached 280,000 m^3 in 2010 (*3*). In addition to the extension of already existing capacities, this increase can be explained by the construction of new manufacturing facilities (*19*). Scheiding (*20*) and Welzbacher (*21*) reported that more than 90 manufacturing facilities spread in at least 25 countries in Europe. The production capacities for thermal modification are, however, very diverse. While some manufacturers feature annual production capacities of only 1,000 m^3, others exceed a capacity of 30,000 m^3 (*3*). The capacity could easily be further increased, because the equipment for thermally modifying wood is relatively simple and has a low capital cost.

Anatomical and Chemical Changes

During the thermal treatment of wood there is a risk of surface cracking as well as internal fissuring (*7*). Recent investigations (*22–26*) on anatomical changes in thermally treated wood by optical and scanning electron microscopies revealed damages to the wood structure such as radial and tangential cracks, collaps of vessels or the destruction of the ray tissue as shown in Figure 1. These damages are strongly dependent on the process conditions and the wood species. With optimized processes, however, these damages can be reduced to a minimum (*7, 22, 23*).

Even though anatomical changes might be contributing to the alteration of wood properties during thermal treatments of wood, their impact is believed to be outweighed by the chemical changes that occur. Chemical changes to the wood structural polymers caused by high treatment temperatures, lead to altered wood properties, such as increased resistance against wood degrading organisms, altered physical/strength properties, darker colour, etc. Intensive studies (*18–42*) of this aspect have shown that many different chemical transformations occur during the thermal treatment.

Within the temperature range usually applied during the thermal treatment, hemicelluloses are the first cell wall components that are affected. Their degradation starts with deacetylation resulting in the formation of acetic acid which catalyses the further degradation of the carbohydrates (*27–31*). This acid-catalyzed degradation results in the formation of aldehydes, with furfural and hydroxymethylfurfural as the main degradation products of pentoses and hexoses, respectively (*27, 32, 33*). Increasing the severity of the treatment process by using an elevated treatment temperature or using an extended treatment duration facilitates the degradation of the carbohydrates (*34, 35*).

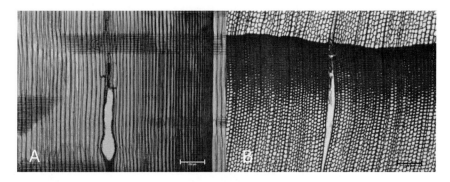

Figure 1. Destruction of axial resin canals (A) and radial crack following a ray (B) as typical defects in thermally modified spruce due to the harsh process conditions.

The sensitivity towards thermal degradation depends on the chemical structure of the carbohydrates. Pentoses are known to be less resistant against a temperature-induced degradation than hexoses (*36, 37*). Kotilainen (*35*) found a decrease in the mass ratio of pentoses to hexoses with increasing treatment severity for thermally treated Norway spruce (*Picea abies*). Consequently, hardwood carbohydrates (mainly xylan) degraded under milder treatment conditions than softwood carbohydrates (mainly galactoglucomannans). Furthermore, cellulose is less affected by a thermal treatment than hemicelluloses (*34, 35, 38*). Investigations by Kotilainen (*35*) revealed that the ratio of glucose to the total amount of monosaccharides in the hydrolysate of thermally treated pine (*Pinus sylvestris*) and birch (*Betula pendula*) increased upon heating. A similar observation was reported by Esteves *et al.* (*39*) for thermally treated eucalypt wood (*Eucalyptus globulus*). Degradation of cellulose mainly occurs in the amorphous regions while the crystalline regions of the cellulose remain unaffected. As a result, the degree of polymerization decreases (*27*) while the relative amount of crystalline cellulose increases (*28, 40, 41*).

Compared to the carbohydrates, lignin is more stable against thermal degradation. Consequently, the lignin content in the wood increases upon heating. Bourgois and Guyonnet (*31*) observed an increase in the lignin content in pine (*Pinus pinaster*) from 28 % to 41 %, 54 % and 84 % caused by heating at 260 °C for 0.5, 1 and 4 hours, respectively. Similar results were reported by Zaman (*42*) in Scots pine and birch and by Esteves *et al.* (*39*) as well in eucalypt wood. Despite the increase in the lignin content during thermal treatments, there is evidence of degradation reactions taking place in the lignin as well. Ahajji *et al.* (*43*), Niemz *et al.* (*44*) and Hofmann *et al.* (*45*) observed an increase in the total phenol content in extracts of thermally treated wood. This increase was explained as the degradation of lignin resulting in the formation of low molecular weight products. Furthermore, Sivonen et al. (28), Ahajji et al. (43) and Willems et al. (46) used ESR-spectroscopy to detect stable free radicals in wood after the thermal treatment process and suggested that they were formed within the lignin network.

By cleavage of ether linkages, especially the β-O-4 links, in the lignin, free phenolic as well as carbonyl groups are formed, which enables cross-linking reactions via methylene bridges (27, 30, 32, 40, 47). Furthermore, demethoxylation at the aromatic ring of guaiacyl and syringyl units of the lignin creates new reactive sites, leading to further condensation reactions (28, 40). Condensation reactions may also be involved in other cell wall components or the degradation products of carbohydrates, which contributes to the increased lignin content after thermal treatements (27, 30, 48).

The majority of the extractives presented in untreated wood are degraded or leached out during the process. Using ATR and reflection FTIR microscopy, Nuopponen et al. (49) detected the movement of fats and wax along the axial parenchyma cells to the surface of pine sapwood edges during the treatment between 100 and 160 °C. Above 180 °C they disappeared from the sapwood surfaces, whereas resin acids were detactable until the temperature became greater than 200 °C. While native extractives disappear, new extractable compounds are formed during the process, mainly as a result of the carbohydrate degradation (39, 50). Poncsak et al. (51) discovered for Jack pine (*Pinus banksiana*) that most of the original extractives left the wood below 200 °C, while most of the new extractives appear only above 200 °C. Furthermore, extractives of untreated Jack pine were dominated by non-polar components, whereas a thermal treatment mainly produced polar compounds.

Properties

In the last decades many publications have studied the material properties of heat treated timber. Overview articles are given by Rapp (52), Militz (18), Ewert and Scheiding (53) as well as by Esteves and Pereira (54). In general, as was shown earlier by Stamm et al. (2), Burmester (3) and Giebeler (4), the durability, sorption, shrinkage and swelling, and strength properties are changed by a heat treatment. The level of change depends on the wood species and process conditions, in which the temperature, the duration of treatment, the wood moisture content and the oxygen level are the most critical process factors.

Sorption and Dimensional Stability

Because of the chemical modification of the wood cell wall structural polymers, the sorption behaviour of the thermally treated wood is altered (55). Tjeerdsma et al. (11) measured the hygroscopicity of PLATO-treated wood. The strong impact of the treatment on the hygroscopicity of softwood and hardwood was illustrated by the reduced sorption curves of the treated samples compared to the unmodified wood. Reduced hygroscopicity was most pronounced at higher relative humidity (R.H. > 70 %). The hysteresis effect between sorption and desorption was found to be unchanged by the heat treatment of wood. It is known that the hygroscopicity of heat-treated wood can vary considerably with varying process time and temperature in the second treatment step of the PLATO-process (*11*).

Popper *et al.* (*56*) investigated the influence of temperature between 100 and 200 °C on sorption and swelling properties of several wood species (*Pinus radiata*, *Pseudotsuga menziesii*, *Laurelia sempervirens*, *Castanea sativa* and *Quercus robur*). They noticed that even low temperature treatments resulted in a lower equilibrium moisture content, with the effect being greater with increasing temperature. The sorption analysis, according to the Hailwood-Horrobin model, suggested that changes in the void volume and cross linking of the holocellulose could be responsible for this change.

Esteves *et al.* (*57*) investigated the properties of pine (*Pinus pinaster*) and eucalypt (*Eucalyptus globulus*) wood after thermal treatment. Besides reduction in the bending strength, they reported a decrease in the equilibrium moisture content and the dimensional stability with increasing treatment intensity. This decrease already occured at low treatment intensities up to a mass loss between 6 and 8 % during the process due to carbohydrate degradation. A further increase in the treatment intensity did not affect the hygroscopicity of the wood considerably. Similar results were observed by Viitaniemi *et al.* (*58*) who reported a maximum reduction in the equilibrium moisture content for a mass loss of 6 % during the process.

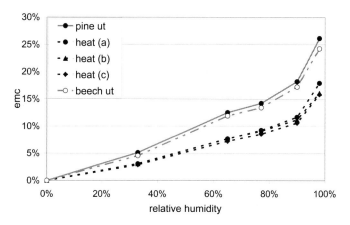

Figure 2. Equilibrium moisture content of untreated (ut) and heat treated Scots pine (Pinus sylvestris) and beech (Fagus sylvatica) from three commercially available processes (heat a-c). (Reproduced with permission from reference (59). Copyright 2004.)

In a joint research programme with the German window industry, the physical properties of several wood samples treated using commercial heat treatment processes were compared (*59*). All processes lowered the equilibrium moisture content in the examined range of relative humidities (Figure 2), as well as the volumetric swelling. At higher humidity, the volumetric swelling was reduced to approx. 50 - 60 % of its original values (Figure 3).

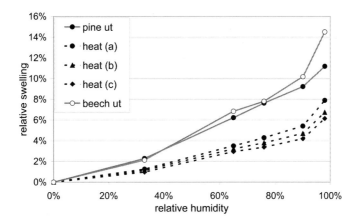

Figure 3. Maximal volumetric swelling of untreated Scots pine (ut) (Pinus sylvestris) and beech (Fagus sylvatica) and heat treated Scots pine from three commercially available processes (heat a-c). (Reproduced with permission from reference (59). Copyright 2004.)

Metsä-Kortelainen et al. (60) investigated the water absorption of the sap- and heartwood of pine and spruce after thermal treatment at different temperatures. The thermal treatment reduced the water absorption in a floating test for spruce sap- and heartwood as well as for pine heartwood. With increasing treatment temperature less water was absorbed. In contrast, the water absorption of pine sapwood increased for treatment temperatures up to 210 °C compared to the untreated control. The water absorption did not decrease for samples treated at temperatures lower than 230 °C.

Resistance against Fungi and Insects

Many authors have shown that the durability of wood against decay can be improved considerably by a thermal treatment of the wood (61–65). This improvement is characterized by a decrease in the mass loss caused by fungal degradation during laboratory monoculture tests on malt-agar medium as shown in Figure 4. The efficacy strongly depends on the wood species, the test fungus and the process conditions.

For PLATO-treated wood, Tjeerdsma et al. (65) found an improved resistance against all of the examined fungi. They showed that the effectiveness against decay was improved by employing a hydrothermal step prior to the dry heat treatment step. However, while the white rot decay was more affected by the hydrothermolysis step, the process conditions in the curing step had the largest effect on the resistance against soft and brown rot decay.

For the OHT-process, Rapp and Sailer (*13*) and Sailer et al. (*66*) studied the resistance of heat treated wood to *Coniophora puteana* with different oil loadings. With increasing temperatures in the range of 180 – 220 °C, the resistance of heat-treated spruce and pine to the brown rot fungus *C. puteana* improved considerably. Mass loss of less than 2 % was found in pine sapwood treated in oil at 200 °C. With spruce, a decisive increase in resistance was only obtained at 220 °C.

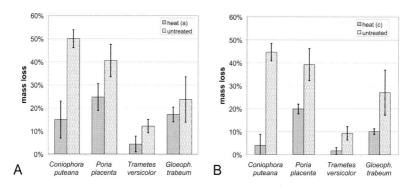

Figure 4. Mass loss of untreated and heat treated Scots pine (Pinus sylvestris) from two commercially available processes (heat a, c) in a 6 week monoculture test following a modified EN 113. (Reproduced with permission from reference (59). Copyright 2004.)

Metsä-Kortelainen and Viitanen (*67*) investigated the durability of Scots pine (*Pinus sylvestris*) and Norway spruce (*Picea abies*) thermally treated according to the ThermoWood process. Resistance of soft and brown rot decay increased with the treatment temperature. However, while a treatment temperature of 210 °C was sufficient to increase the durability against brown rot decay to class 1-2, a treatment temperature greater than 230 °C was needed to increase the durability against soft rot decay to a similar level. The effect of the thermal treatment on the durability appeared to be most effective on pine heartwood. Consistently, Boonstra et al. (*61*) reported that Scots pine heartwood was more resistant than sapwood even after the thermal modification process.

Welzbacher and Rapp (*68*) investigated the durability in laboratory and field tests using material from several commercial treatment batches of Scots pine sapwood and Norway spruce. Improved durability of thermally treated wood was found in laboratory tests and confirmed in field tests in above ground conditions after 5.5 years' exposure. In contrast, thermally treated wood from field tests in soil contact was only rated as slightly durable to non-durable (class 4-5). Thus, they concluded that thermally modified wood is unsuitable for the application in ground contact, irrespective of the treatment process. A non-sufficient decay resistance of thermally modified wood in ground contact coincides with previous investigations (*64, 69*).

Investigations by Boonstra et al. (61) on the growth of non-decaying fungi did not reveal an effect of thermal treatment on mold and sap stain fungi. They concluded that changes in the main components of wood did not affect the growth of such fungi. However, degradation products of the carbohydrates appeared to accelerate the mold growth, as Radiata pine sapwood (*Pinus radiata*) was very sensitive to mold especially after the hydrothermolysis step.

Ewert and Scheiding (53) tested resistance against blue stain fungi with *Aureobasidium pullulans* and *Sclerophoma pithyophila*, and reported no difference between treated wood and controls in the colonisation at the surface of the samples. However, penetration of the hyphae into the wood was only seen with non treated (control) pine, whereas the heat treated wood was only superficially colonised.

Research performed at the University of Kuopio (Finland) and at the French institute CTBA showed a higher resistance of thermally treated wood against longhorn beetles, *Anobium punctatum* and *Lyctus brunneus*. However, preliminary trials with termites showed no improved resistance (8). This result was later confirmed by Surini et al. (70). A non-choice feeding test using Maritime pine (*Pinus pinaster*) showed no effect on the durability against termites for all applied treatment conditions. Due to the removal of termite inhibiting compounds that are present in some untreated wood species, the wood can even become less durable against termites after a thermal treatment. Consequently, Shi et al. (71) reported for Scots pine an increase in the mass loss caused by termite attack from 10.3 to 33.6 % after a treatment at 215 °C.

Mechanical Properties

The changes in the cell wall chemistry (changes in the hemicellulose and lignin structures, cellulose depolymerisation and increased crystallinity, etc.) affect the mechanical properties of heat treated wood (11, 72–76). The extent of the changes in the mechanical properties strongly depends on the process conditions and the the properties of the raw material.

Strength results are often based on small wood samples free of defects and planks treated under mild conditions. Boonstra et al. (22) reported that during the process, high tension can occur in the wood as it is exposed to high temperatures and rapid evaporation of water. In their study, some of the wood species were found difficult to treat and showed a number of defects, mainly cracks, if not treated carefully. Several softwood species are known to have a high resistance against liquid impregnation. These wood species were difficult to be heat treated and showed a relatively high strength loss.

In static bending tests, modulus of elasticity (MOE) is normally less affected than the bending strength. For mild treatment conditions the MOE often increases, whereas it decreases for severe treatment conditions. Boonstra et al. (72) reported for Scots pine (*Pinus sylvestris*) that was PLATO treated using mild conditions (< 200 °C) an increase in the MOE of 10 % while the bending strength showed a small reduction of 3 %. For Radiata pine (*Pinus radiata*) an increase in the MOE of 13 % and a reduction in the bending strength of 9 % was observed at a hydrothermolysis temperature of 165 °C. Bengtsson *et al.* (*77*) tested the strength of heat treated beams (45 x 145 mm) of spruce (*Picea abies*) and pine (*Pinus sylvestris*) from higher temperature ranges (200 – 220 °C) and found reduced bending strengths of up to 50 %, but only minor MOE changes. Similar results have also been reported by other authors (*13*, *57*, *73*).

The impact bending strength is probably the most critical mechanical property for all heat treatment processes. It decreases considerably because the wood becomes brittle. Even a mild PLATO-treatment (< 200°C) lead to a severe decrease for Scots pine (56 %), Norway spruce (79 %) and Radiata pine (80 %) (*72*). Rapp *et al.* (*52*) found that Scots pine treated at 200 °C in oil and air achieved only 51 and 37 % of the impact bending strength of unmodified pine, respectively. For several soft- and hardwood species that were treated in the temperature range between 185 and 220 °C, Hanger *et al.* (*78*) found reductions in the static bending strength up to 50 %, while impact bending strength decreased up to 80 %. They also reported a change in the breaking behaviour. For thermally treated samples, the fracture surfaces were mainly brittle and abrupt (Figure 5). This has also been reported by Boonstra *et al.* (*22*).

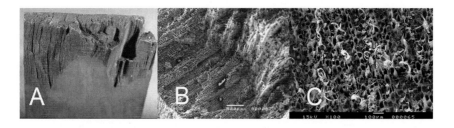

Figure 5. The typical fracture of a heat treated poplar specimen after a bending test (A). Microscopical photo of a heat treated (B) and non-treated (C) Radiata pine, fracture surface after bending test. (Reproduced with permission of references (22) and (23). Copyright 2006 Maderas: Cienc Tecnol.)

Further strength properties have been measured by Boonstra *et al.* (*72*). They measured a strong reduction of the tensile strength (39 %) and the radial compression strength (53 %) for a mild PLATO-treatment of Scots pine. An increase was found for the compressive strength parallel to the grain (28 %), the tangential compression strength (8%) and the Brinell hardness parallel to the grain (48 %). Somewhat different results for heat treated Scots pine were reported by Korkut *et al.* (*79*). In addition to a reduction in impact bending strength, bending strength and MOE, they also found a decrease in compression strength parallel to the grain, Janka-hardness and tension strength perpendicular to the grain. Similar results obtained by Korkut *et al.* (*80*) for Red-bud maple (*Acer trautvetteri*).

Colour and Odor

Due to the high temperatures employed, all heat-treated wood species show a characteristic brownish colour (comparable to the natural colour of *Thuja plicata*, Western Red Cedar). The colour is affected by the intensity of the treatment process, i.e. by the treatment temperature and the duration of the processes. The higher the temperature and the longer the duration the darker the colour.

After treatment, the wood has a characteristic caramellish smell, likely due to furfural formation. Measurements by Manninen *et al.* (*81*) also showed that emission of monoterpenes from treated pine is considerably reduced, but heat-treated wood does emit acetic acid, furancarboxaldehyde and 2-propanone as major components in the VOC.

Coating Performance

Without coating, the original brown colour of thermally modified wood is not stable during weathering and thus turns grey, similar to unmodified wood (*82*). Furthermore, the crack formation of thermally modified wood without coating is at the same level as unmodified wood, despite the lower moisture contents and the higher dimensional stability of thermally modified wood (*83*). Consequently, a surface treatment with oils or paints can be required.

In a field test, Jämsä *et al.* (*83*) showed that unpigmented or low build stains and oils did not prevent cracking of thermally modified wood during natural weathering. However, weather resistance was improved by water- or solvent-borne paints. They report that thermally modified wood as a substrate for the tested coating systems is comparable to unmodified wood and no alterations of the coating recommendations are needed. The results coincide with the investigation performed by Krause and Militz (*84*) who found no changes in the painatbility of thermally modified wood with water borne acrylic or solvent borne alkyds. Due to its UV-degradability, opaque systems are recommended over priming oils and stains (*11*).

Gluability

Schmid *et al.* (*85*) evaluated glued wood following different German and European standards for strength and moisture performance. They showed that thermally modified wood can be glued with many industrial adhesives (polyvinyl alcohol and other polyvinylic glues, polyurethane, isocyanate, and resorcinol-phenolic glues). Due to the lower shear strength and tension strength perpendicular to grain with heat-treated lumber, a higher wood failure was found. Furthermore, the hydrophobic wood surface caused a slower penetration of the solvents from the glue to the surrounding wood, thus indicating that it is necessary to modify the gluing process.

Sernek *et al.* (*86*) investigated the bonding of wood after the PLATO-treatment using melamine-urea-formaldehyde (MUF), phenol-resorcinol-formaldehyde (PRF) and polyurethane (PUR) adhesives. Although they concluded that PLATO-treated wood generally can be bonded with structural adhesives, they found that the thermal modification process affected the shear strength and the delamination of the laminated wood depending on the applied adhesive system. Unmodified wood appeared to perform better than thermally modified wood, especially in case of water-borne adhesives. They suggested that the low pH (PRF) and the low wettability (PRF and MUF) of thermally modified wood have a negative effect on the bonding. Furthermore, MUF and PUR systems performed better than the PRF adhesive. They stated that the bonding performance of thermally modified wood might be improved by alterations of the adhesive composition and/or the bonding process.

Quality Assessment

Despite the fact that the commercial acceptance of thermally modified wood in Europe is leading to an increase in the production capacity and the number of manufacturers in recent years, it is impossible to define characteristic attributes that describe the whole range of thermally modified wood products. The properties of thermally modified wood can vary greatly depending on the process conditions, the raw material properties or the process technology that is used. Variation can even occur within one treatment batch, if the intial moisture content of the wood or the process conditions are not optimally steered. Consequently, a quality assement is required for a reliable end-use application of thermally modified wood.

Efforts for a transparent quality assurance system have been made in different European countries separately. In Finland, several manufacturers formed the International ThermoWood Association that defines two standard treatment classes (Thermo-S and Thermo-D) and also specifies the requirements for the raw material. Further approaches exist in the Netherlands by the KOMO® certificate "Timber modification" or in Germany where several manufacturers of thermally modified wood founded a working group within the Federal Association of Sawmill and Woodworking Industry (BSHD e.V.) and developed a quality assurance concept. More detailed information about the implematation of quality assurance systems for thermally modified wood are given by Scheiding (*87*) or Welzbacher and Scheiding (*88*).

Recent scientific research activities focus on the development of methods for a fast and reliable prediction of the properties of thermally modified wood. These methods could be used for off- or on-line process control and the quality approval for different thermally modified wood products. Different approaches exist that are for example based on color measurement (*89, 90*), infrared-spectroscopy (*91–93*), electronspin-resonance spectroscopy (*46, 94*) or elemental composition (*95, 96*). One of the upcoming challenges is the implementation of these methods in the industrial production process.

References

1. Militz, H.; Hill, C. A. S. *Proceedings of the 2nd European Conference on Wood Modification*; Göttingen, Germany, 2005.
2. Hill, C. A. S. *Wood modification: Chemical, thermal and other processes*; John Wiley & Sons Ltd.: Chichester, 2006; Vol 1, pp 99−126.
3. *EUWID* (42/2010); p 80−82
4. Stamm, A. J.; Burr, H. K.; Kline, A. A. *Ind. Eng. Chem.* **1946**, *38*, 630–634.
5. Burmester, A. *Eur. J. Wood Prod.* **1973**, *31*, 237–243.
6. Giebeler, E. *Holz Roh- Werkst.* **1983**, *41*, 87–94.
7. Syrjänen, T.; Kangas, E. *Int. Res. Group Wood Protect.*; Doc. no IRG/WP 00-40158; 2000.
8. Mayes, D.; Oksanen, O. *ThermoWood Handbook*; Finnish ThermoWood Association: Helsinki, 2003.
9. Boonstra, M. J.; Tjeerdsma, B.; Groeneveld, H. A. C. *Int. Res. Group Wood Preserv.*; Doc. no. IRG/WP 98-40123; 1998.
10. Ruyter, H. P. European Patent Appl. No. 89-203170.9, 1989.
11. Tjeerdsma, B.; Boonstra, M.; Militz, H. *Intern. Res. Group Wood Preserv.*; Doc. no. IRG/WP 98-40124; 1998.
12. Duchez, L. New Option Wood. Personal communication. 2002.
13. Rapp, A. O.; Sailer, M. *Proceedings of Seminar "Production and developement of heat treated wood in Europe"*. Helsinki, Stockholm, Oslo, 2000.
14. Michon, S. G. L. Patent Appl. No. WO 01/85410 A1, 2001.
15. Skovmand, E.; Christensen, P. Patent Appl. No.WO 2003106123 A2, 2003.
16. Willems, W. Patent Appl. No. WO 2008079000 A1, 2008.
17. Willems, W. Proceedings of the 4th European Conference on Wood Modification, Stockholm, Sweden, 2009.
18. Militz, H. *Enhancing the Durability of Lumber and Engineered Wood Products*; Forest Products Society: Madison, WI. 2002, pp. 239−249.
19. *EUWID* (23/2009), pp 1, 12.
20. Scheiding, W. *6. Europäischer TMT-Workshop*, Dresden, Germany, 2010.
21. Welzbacher, C. *7. Europäischer TMT-Workshop*, Dresden, Germany, 2012.
22. Boonstra, M.; Rijsdijk, J. F.; Sander, C.; Kegel, E.; Tjeerdsma, B.; Militz, H.; Van Acker, J.; Stevens, M. *Maderas: Cienc. Tecnol.* **2006**, *8*, 193–208.
23. Boonstra, M.; Rijsdijk, J. F.; Sander, C.; Kegel, E.; Tjeerdsma, B.; Militz, H.; Van Acker, J.; Stevens, M. *Maderas: Cienc. Tecnol.* **2006**, *8*, 209–217.
24. Awoyemi, L.; Jones, I. P. *Wood Sci. Technol.* **2011**, *45*, 261–267.
25. Biziks, V.; Andersons, B.; Belkova, L.; Kapaca, E.; Militz, H. *Wood Sci. Technol.* **2013**, *47*, 717–735.
26. Altgen, M.; Ala-Viikari, J.; Hukka, A.; Tetri, T.; Militz, H. Proceedings of the 6th European Conference on Wood Modification, Ljubljana, Slovenia, 2012.
27. Tjeerdsma, B. F.; Boonstra, M.; Pizzi, A.; Tekely, P.; Militz, H. *Holz Roh-Werkst.* **1998**, *56*, 149–153.
28. Sivonen, H.; Maunu, S. L.; Sundholm, F.; Jämsä, S.; Viitaniemi, P. *Holzforschung* **2002**, *56*, 648–654.
29. Wienhaus, O. *Wiss. Z. Tech. Univ. Dresden* **1999**, *48*, 17–22.

30. Tjeerdsma, B. F.; Militz, H. *Holz Roh- Werkst.* **2005**, *63*, 102–111.
31. Bourgois, J.; Guyonnet, R. *Wood Sci. Technol.* **1988**, *22*, 143–155.
32. Nuopponen, M.; Vuorinen, T.; Jämsä, S.; Viitaniemi, P. J. *Wood Chem. Technol.* **2005**, *24*, 13–26.
33. Sandermann, W.; Augustin, H. *Eur. J. Wood Prod.* **1963**, *21*, 256–265.
34. Alen, R.; Kotilainen, R.; Zaman, A. *Wood Sci. Technol.* **2002**, *36*, 163–171.
35. Kotilainen, R. Ph.D. Thesis, University of Jyväskylä, Finland, 2000
36. Runkel, R.; Wilke, K. *Eur. J. Wood Prod.* **1951**, *9*, 260–270.
37. Runkel, R.; Witt, H. *Eur. J. Wood Prod.* **1953**, *11*, 457–461.
38. Sandermann, W.; Augustin, H. *Eur. J. Wood Prod.* **1963**, *21*, 305–315.
39. Esteves, B.; Graça, J.; Pereira, H. *Holzforschung* **2008**, *62*, 344–351.
40. Wikberg, H.; Maunu, S. *Carbohydr. Polym.* **2004**, *58*, 461–466.
41. Bhuiyan, M. T. R.; Hirai, N.; Sobue, N. *J. Wood Sci.* **2000**, *46*, 431–436.
42. Zaman, A.; Alén, R.; Kotilainen, R. *Wood Fiber Sci.* **2000**, *32*, 138–143.
43. Ahajji, A.; Diouf, P. N.; Aloui, F.; Elbakali, I.; Perrin, D.; Merlin, A.; George, B. *Wood Sci. Technol.* **2009**, *43*, 69–83.
44. Niemz, P.; Hofmann, T.; Retfalvi, T. *Maderas: Cienc Tecnol* **2010**, *12*, 69–78.
45. Hofmann, T.; Wetzig, M.; Retfalvi, T.; Sieverts, T.; Bergemann, H.; Niemz, P. *Eur. J. Wood Prod.* **2013**, *71*, 121–127.
46. Willems, W.; Tausch, A.; Militz, H. *Int. Res. Group Wood Protect.*; Doc. no. IRG/WP 10-40508; 2010.
47. Rousset, P.; Lapierre, C.; Pollet, B.; Quirino, W.; Perre, P. *Ann. For. Sci.* **2009**, *66*, 110p1–110p8.
48. Kosikova, B.; Hricovini, M.; Cosentino, C. *Wood Sci. Technol.* **1999**, *33*, 373–380.
49. Nuopponen, M.; Vuorinen, T.; Jämsä, S.; Viitaniemi, P. *Wood Sci. Technol.* **2003**, *37*, 109–115.
50. Esteves, B.; Videira, R.; Pereira, H. *Wood Sci. Technol.* **2011**, *45*, 661–676.
51. Poncsak, S.; Kocaefe, D.; Simard, F.; Pichette, A. *J. Wood Chem. Technol.* **2009**, *29*, 251–264.
52. Rapp, A. O. Review on heat treatments of wood; Proceedings of a special seminar held in Antibes, France, COST Action E22, Brussels, 2001.
53. Ewert, M; Scheiding, W. *Holztechnologie* **2005**, *46*, 22–29.
54. Esteves, B. M.; Pereira, H. *BioResources* **2009**, *4*, 370–404.
55. Kollmann, F; Schneider, A. *Eur. J. Wood Prod.* **1963**, *21*, 77–85.
56. Popper, R; Niemz, P.; Eberle, G. *Holz Roh- Werkst.* **2005**, *63*, 135–148.
57. Esteves, B.; Marques, A. V.; Domingos, I.; Pereira, H. *Wood Sci. Technol.* **2007**, *41*, 193–207.
58. Viitaniemi, P.; Jämsä, S.; Viitanen, H. U.S. Patent-No. 005678324, 1997.
59. Militz, H.; Krause, A. Modified wood for window and cladding products; Proceedings of the COST conference, Florence, 2004.
60. Metsa-Kortelainen, S.; Antikainen, T.; Viitaniemi, P. *Holz Roh- Werkst.* **2006**, *64*, 192–197.
61. Boonstra, M. J.; van Acker, J.; Kegel, E.; Stevens, M. *Wood Sci. Technol.* **2007**, *41*, 31–57.
62. Hakkou, M.; Petrissans, M.; Gerardin, P.; Zoulalian, A. *Polym. Degrad. Stab.* **2006**, *91*, 393–397.

63. Dirol, D.; Guyonnet, R. *Int. Res. Group Wood Protect.*; Doc. no. IRG/WP 93-40015; 1993.
64. Kamdem, D. P.; Pizzi, A.; Jermannaud, A. *Eur. J. Wood Prod.* **2002**, *60*, 1–6.
65. Tjeerdsma, B.; Stevens, M.; Militz, H. *Int. Res. Group Wood Preserv.* Doc. no. IRG/WP 00-40160; 2000.
66. Sailer, M.; Rapp, A. O.; Leithoff, H. *Int. Res. Group Wood Preserv.* Doc. no. IRG/WP 00-40162; 2000.
67. Metsä-Kortelainen, S.; Viitanen, H. *Wood Mater. Sci. Eng.* **2009**, *4*, 105–114.
68. Welzbacher, C. R.; Rapp, A. O. *Wood Mater. Sci. Eng.* **2007**, *2*, 4–14.
69. Westin, M.; Rapp, A. O.; Nilsson, T. *Int. Res. Group Wood Preserv.* Doc. no. 04-40288; 2004.
70. Surini, T.; Charrier, F.; Malvestio, J.; Charrier, B.; Moubarik, A.; Castera, P.; Grelier, S. *Wood Sci. Technol.* **2012**, *46*, 487–501.
71. Shi, J.; Kocaefe, D.; Amburgey, T.; Zhang, J. *Holz Roh- Werkst*.**2007**, *65*, 353–358.
72. Boonstra, M. J.; van Acker, J.; Tjeerdsma, B.; Kegel, E. *Ann. For. Sci.* **2007**, *64*, 679–690.
73. Kubojima, Y.; Okano, T.; Ohta, M. *J. Wood Sci.* **2000**, *46*, 8–15.
74. Epmeier, H.; Westin, M.; Rapp, A. *Scand. J. For. Res.* **2004**, *19*, 31–37.
75. Schneider, A. *Holz Roh- Werkst.* **1971**, *29*, 431–440.
76. Windeisen, E.; Bächle, H.; Zimmer, B.; Wegener, G. *Holzforschung* **2009**, *63*, 773–778.
77. Bengtsson, C.; Jermer, J.; Brem, F. *Int. Res. Group Wood Preserv.*; Doc. no. IRG/WP 02-40242; 2002.
78. Hanger, J.; Huber, H.; Lackner, R.; Wimmer, R. *Holzforsch. Holzverwert.* **2002**, *54*, 111–113.
79. Korkut, S.; Akgül, M.; Dündar, T. *Bioresour. Technol.* **2008**, *99*, 1861–1868.
80. Korkut, S.; Kök, M. S.; Korkut, D. S.; Gürleyen, T. *Bioresour. Technol.* **2008**, *99*, 1538–1543.
81. Manninen, A. M.; Pasanen, P.; Holopainen, J. K. *Atmos. Environ.* **2002**, *36*, 1763–1768.
82. Huang, X. A.; Kocaefe, D.; Kocaefe, Y.; Boluk, Y.; Pichette, A. *Appl. Surf. Sci.* **2012**, *258*, 5360–5369.
83. Jämsä, S.; Ahola, P.; Viitaniemi, P. *Pigm. Resin. Technol.* **2000**, *29*, 68–74.
84. Krause, A.; Militz, H. Proceedings of the 2nd European Conference on Wood Modification, Göttingen, Germany, 2005.
85. Schmid, J.; Illner, M.; Schwarz, B.; Stetter, K.; Militz, H. Einheimisches dimensions stabilisiertes Holz für den Fenster- und Fassadenbau; Abschlussbericht, DGFH München, 2004.
86. Sernek, M.; Boonstra, M.; Pizzi, A.; Despres, A.; Gérardin, P. *Holz Roh-Werkst.* **2008**, *66*, 173–180.
87. Scheiding, W. *Holztechnologie* **2007**, *48*, 50–51.
88. Welzbacher, C.; Scheiding, W. *Int. Res. Group Wood Protect.*; Doc. no. IRG/WP 11-40558; 2011.
89. Gonzalez-Pena, M. M.; Hale, M. D. C. *Holzforschung* **2009**, *63*, 394–401.

90. Brischke, C.; Welzbacher, C.; Brandt, K.; Rapp, A. O. *Holzforschung* **2007**, *61*, 19–22.
91. Altgen, M.; Welzbacher, C.; Militz, H. *Holztechnologie* **2013**, *54*, 40–44.
92. Bächle, H.; Zimmer, B.; Windeisen, E.; Wegener, G. *Wood Sci. Technol.* **2010**, *44*, 421–433.
93. Bächle, H.; Zimmer, B.; Wegener, G. *Wood Sci. Technol.* **2012**, *46*, 1181–1192.
94. Altgen, M.; Welzbacher, C.; Humar, M.; Militz, H. COST-Action FP0904 - Thermo-Hydro-Mechanical Wood Behaviour and Processing, Nancy, France, 2012.
95. Sustersic, Z.; Mohareb, A.; Chaouch, M.; Petrissans, M.; Petric, M.; Gérardin, P. *Polym. Degrad. Stab.* **2010**, *95*, 94–97.
96. Inari, G. N.; Petrissans, M.; Petrissans, A.; Gérardin, P. *Polym. Degrad. Stab.* **2009**, *94*, 365–368.

Chapter 17

Wood Protection with Dimethyloldihydroxy-Ethyleneurea and Its Derivatives

Yanjun Xie,[1,2] Andreas Krause,[3] and Holger Militz[*,1]

[1]Department of Wood Biology and Wood Products,
Georg August University Göttingen, Buesgenweg 4,
D-37077 Goettingen, Germany
[2]Key Laboratory of Bio-based Material Science and Technology,
Northeast Forestry University, 26 Hexing Road,
Harbin 150040, People's Republic of China
[3]Mechanical Wood Technology, University Hamburg,
Leuschnerstr. 91, D-21031 Hamburg, Germany
*E-mail: andreas.krause@uni-hamburg.de.

The low-molecular-weight N-methylol compounds, dimethyloldihydroxyethyleneurea (DMDHEU) and its derivatives, have been successfully used to modify wood. The N-methylol compounds can penetrate and react in wood cell walls. The reaction modes may be crosslinking of cell wall polymers by DMDHEU and/or self-condensation of DMDHEU within the cell wall. As a result, the modified wood exhibits a permanent cell wall bulking; the swelling and shrinkage is reduced, depending on the modification levels. This causes an improved anti-swelling efficiency of up to 70%. Modification does not substantially influence the equilibrium moisture content of wood but improves the durability against white, brown, and soft rot fungi. The treatments also enhance the wood's surface hardness and compression strength, but do not change its flexural properties. The adhesion of coatings on the modified wood is greater than on the untreated wood and the weathering properties of both uncoated and coated wood are improved. The simple processing, enhanced material properties, and acceptable production cost make this modification technique applicable to industry.

© 2014 American Chemical Society

Introduction

Chemical modification can improve the properties of wood and impart a protection efficacy comparable to that given by preservatives (*1*). The chemicals used do not contain heavy metal elements and are able to react with wood cell wall polymers or condense in wood micro-structures. As a result, there is little risk of chemical leaching and, therefore, protection to wood can last for a long-term service period. Consequently, chemical modification has been recognized as an important alternative to the use of tropical hardwood species or preservatives in the wood protection industry. Chemical modification can generally be classified as either cell wall modification, or filling of large cell cavities, or a combination of both (*2*). Cell wall modification refers to the process whereby wood cell wall constituents are altered through reactions with reactive low molecular weight monomers or oligomers, or by heating under high temperature conditions. Filling of cell cavities is the process by which chemicals are deposited in the large cell cavities such as lumens to block the physical passages thereby reducing water/moisture access to wood cell walls. The protection mechanisms of wood modification are mainly proposed effected through bulking of cell walls, reduction of moisture content, and/or changes of molecular structure of the cell wall polymers (*3*).

Various chemical modification techniques have been investigated for many years, particularly acetylation (*4*) and treatments with melamine (*5*). Among the most promising chemicals used for wood modification are *N*-methylol compounds, which are widely used in the textile industry to improve cotton or other cellulose-based fabrics. They enhance wash- and wear-properties and help fix color or other agents to fibers (*6*). Dimethyloldihydroxyethyleneurea (DMDHEU) was the most widely used *N*-methylol compound in the textile industry, but due to formaldehyde emissions in the process and from the textiles, low-formaldehyde containing agents were developed (*7*). Modification of wood with DMDHEU or its derivatives could be applicable to both solid lumber and wood based composites (*8, 9*). The mode of action is based on DMDHEU cross-linking with wood compounds and self poly-condensation within the cell wall. Technically, the modified material is a wood polymer composite with the appearance and texture of solid wood (*10*).

Chemical Agents and Reactions

Chemical Agents

Various *N*-methylol compounds have been developed by the textile industry over the past 40 years (*11*), but only DMDHEU and its derivatives were widely accepted. The reactive functional groups in the molecule are the two *N*-methylol groups (Figure 1). The molecule is also partially methylolated to mDMDHEU to reduce the formaldehyde emissions from DMDHEU, but this reduces the reactivity.

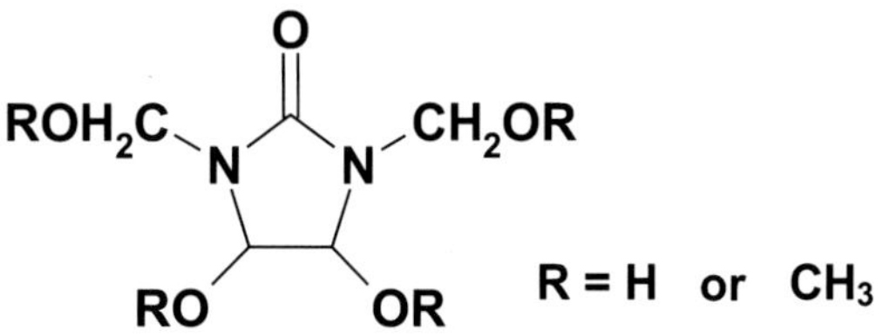

$$\text{ROH}_2\text{C}-\underset{\underset{\text{RO}}{|}}{\text{N}}-\overset{\overset{\text{O}}{\|}}{\text{C}}-\underset{\underset{\text{OR}}{|}}{\text{N}}-\text{CH}_2\text{OR} \qquad R = H \text{ or } CH_3$$

Figure 1. The chemical structure of dimethyloldihydroxy-ethyleneurea (DMDHEU as R=H) or its methylolated derivative (mDMDHEU as R=CH₃).

DMDHEU can be modified by an N,O-acetalization with an alcoholic compound to prevent hydrolytic release of formaldehyde (*12*). Formaldehyde emissions can be further reduced by adding formaldehyde scavengers, such as citric acid, chitosan or glyoxal (*13*). Formaldehyde free finishing agents, such as dihydroxydimethylimidazolidinone (DHDMI), are also used for finishing of textiles. However, this compound has a low reactivity and is thus not suitable for wood modification (*10*). Various catalysts are used to enhance the reactivity of cross-linking agents (*10, 14*). One of the best catalysts is magnesium chloride $MgCl_2$ which is used in the reported results below.

Mechanism of Reaction and Treatment of Wood

The chemical reaction mechanism has been extensively investigated by textile researchers (*6*). The *N*-methylol group reacts with hydroxyl groups to form acetal bonds. The following reactions can occur:

- Cross-linking with hydroxyl groups of wood
- Hydrolysis of *N*-methylol groups to formaldehyde and NH-groups
- Condensation with NH groups to form methylene bonds
- Condensation with hydroxyl groups of alcohols to form ether bonds

Reactions of *N*-alkoxymethyl compounds are subjected to a general acid catalysis (*6*). The main goal in modifying wood with *N*-methylol compounds is to achieve both a high extent of cross-linking with wood components coupled with self-condensation in the wood cell wall.

The treatment procedures for textiles and solid wood are different. Wood tends to form cracks after treatment due to drying stresses. Also, since wood will undergo structural changes at temperatures above 130°C, relatively mild reaction processes are necessary. The typical treatment consists of following steps:

- Impregnation of wood with an aqueous solution containing agent and catalyst
- Drying the wood to below fiber saturation point (optional)
- Curing at temperatures above 90°C and below 130°C
- Conditioning the modified wood to a final equilibrium moisture content (optional)

Dry wood is normally impregnated. During impregnation, the agent is incorporated into the wood cell wall. The curing at high temperatures leads to the formation of cross-links between wood hydroxyl groups and *N*-methylol groups, and to poly-condensation between *N*-methylol groups (Figure 2).

$$NCH_2OH + HO\text{-}WOOD \xrightarrow{crosslink} NCH_2\text{-}O\text{-}WOOD$$

$$NCH_2OH + HOCH_2N \xrightarrow{condense} NCH_2\text{-}O\text{-}CH_2N$$

Figure 2. Schematic reaction of DMDHEU with hydroxyl groups of wood cell walls and condensation of DMDHEU.

Uniform distribution of DMDHEU within the wood is required when treating large size wood since uneven distribution will lead to heavy cracking of the treated wood when it is dried after treatment. Therefore, a novel curing process which employs superheated steam was developed (*15*). Wood with large pores and low extractive content are suitable for treatment with this modification technique.

Properties of Treated Wood

Moisture Content and Dimensional Stability

Incorporation of *N*-methylol compounds in the cell walls influences the moisture sorption behavior of wood. At 20 °C and 65% relative humidity, beech wood (*Fagus sylvatica*) treated with 22.5% DMDHEU and 1.5% $MgCl_2*6H_2O$, had an equilibrium moisture content (EMC) of 9.3%, lower than the EMC of untreated wood (13%), as calculated based on the dry mass of modified wood. The reduced EMC can be attributed to incorporation of resin molecules into the wood cell walls (bulking). Increasing amounts of *N*-methylol resin reduced the pore size and numbers in the cell walls of beechwood (*16*), which caused a reduced free space for water accessibility. In addition, the reactive *N*-methylol groups of DMDHEU may also react with the OH groups of wood cell walls, thereby blocking the water absorption sites of cell walls (*17*).

The calculation of EMC for modified wood has been the subject of some debate. Some recommend that EMC is calculated from the mass of wood before treatment to exclude the effect of chemicals deposited in the wood (*18, 19*). DMDHEU-treated wood exhibits comparable, sometimes even higher equilibrium moisture contents than the untreated wood using this calculation method (*20*). This is mainly because each incorporated DMDHEU molecule also contains two non-reactive OH groups that are accessible to water (Figure 1). The trace, but highly hygroscopic magnesium chloride (as catalyst) in wood can also adsorb water, thereby offsetting the reduction in the moisture sorption of cell wall substances (*10*). The moisture sorption behavior of DMDHEU-modified wood is also supported by the production of more energy during the sorption process compared to the untreated controls, as determined by isosteric method and solution calorimetry (*20*).

Swelling behavior of treated wood differs considerably from untreated wood compared to the minor effect on the EMC. Monomeric *N*-methylol compounds are able to penetrate the cell wall and bulk the wood cell wall in a permanently swollen state. This bulking effect can increase the volume of treated beech wood up to 10% compared to the volume of untreated wood. Consequently, the swelling and shrinking of wood is considerably reduced. The complementary effects of bulking and cross-linking result in an anti-swelling efficiency (ASE) of up to 70%. The correlation between EMC and the swelling/shrinking of treated beech wood is not linear, however, unlike untreated wood (*21*). EMC increases more than swelling with increasing relative humidity (*22, 23*).

Durability against Biological Decay

Protection of wood against biological decay is one of the main objectives of *N*-methylol modification. While DMDHEU-modified wood has enhanced fungal resistance, the protective effect by DMDHEU is not based on a biocidal effect but on wood modification (*24, 25*). The mechanism of protection against biological decay is generally assumed to be based on: (1) the EMC in modified wood is below the value required by fungi; (2) deposition of DMDHEU in the cell wall causes a reduction in pore diameter of cell wall smaller than the diameter of decay enzyme, which cannot access to interior of cell wall; and (3) grafting of DMDHEU onto cell wall polymers makes the modified wood non-recognizable by fungal enzymes.

Brown and White Rot Decay Resistance

Durability tests against brown and white rot fungi was done according to EN113 using beech and pine sapwood impregnated with mDMDHEU and diethyleneglycol (DEG) with magnesium chloride as the catalyst. A negative relationship between the chemical loading with DMDHEU/DEG and wood mass loss was observed; weight percent gains of more than 15% to 20% assure complete protection against decay by four fungi species (Figure 3). Consequently, this study confirmed that DMDHEU modification improves wood durability against basidiomycetes.

Soft-Rot Decay Resistance

The durability of treated pine sapwood against soft rot decay was investigated in laboratory (ENv807) and field tests (EN252). As expected, the laboratory results showed that the resistance of modified beech wood in soil contact depended on chemical loading (WPG). The difference in decay resistance between wood treated with DMDHEU vs. mDMDHEU was minor. The laboratory tests indicated that beech wood treated with DMDHEU or mDMDHEU was classified as highly durable.

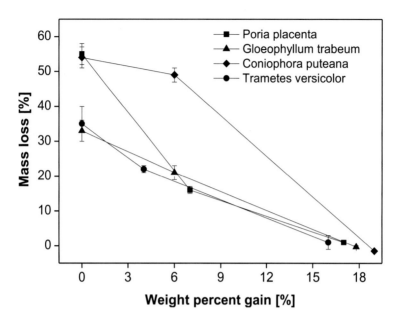

Figure 3. Mass loss of mDMDHEU/DEG-treated wood due to various fungi after 16 weeks incubation in an EN113 decay test.

Figure 4. Rating of treated pine after 3 years of ground contact in field exposure according to EN252. DMD and DMD/DEG content is expressed as WPG. DMD = DMDHEU. Column = mean value, line = maximal and minimal rating.

Reliable conclusions about the durability of modified wood can be only attained after ground-contact outdoor tests. Those tests, however, require considerable exposure time to obtain meaningful results. The results presented below have only been exposed for the relatively short period of three years and, thus, the results are only preliminary (Figure 4). The failure-rate of untreated pine sapwood samples in the test indicated a normal infestation by fungi in the test field. In contrast, only a few of the DMDHEU/DEG treated samples exhibited minor evidence of decay, and DMDHEU-treated samples at a 24% WPG had no decay. Based on these results, modified pine sapwood at WPG above 15% can possibly be classified as highly durable, independent of the specific DMDHEU-agent employed.

Mechanical Properties

Modification with DMDHEU increased hardness up to four times at a high-level, compared to untreated wood (Figure 5). The combined treatments of thin veneer strips of Scots pine with DMDHEU and magnesium chloride caused a strength loss of up to 50% in zero-span (mainly measuring the intra-fibre strength) and up to 70% in finite-span test mode (mainly measuring the inter-fibre shear strength). Acid-catalyzed hydrolysis of polysaccharides and crosslinking were the main reasons for the losses (26). DMDHEU-treated beech wood blocks exhibited an increase in compression strength of up to 65% (Figure 6a), but a decrease in tensile strength of up to 40% by increasing the DMDHEU concentration using magnesium chloride as the reaction catalyst (27). Only a slight decrease of MOR in bending was observed. This minor change in MOR can be explained by the increased compression strength on the top layer of wood sample, which compensated the loss of tensile strength on the bottom layer during the static bending test (28).

The most adverse effects of DMDHEU treatments may be a reduction in the dynamic mechanical properties. Impact strength decreased with increasing DMHDEU concentration; at the highest concentration of 2.3 mol l^{-1}, treated beech wood showed up to 80% loss in impact strength (Figure 6b). Scanning electron microscope (SEM) observation revealed that some microfibril bundles were pulled out on the fractured surface of the untreated control (Figure 7a), while samples treated to 28% weight percent gain (WPG) exhibited a regular fractural surface with some fragments (Figure 7b). The results show that DMDHEU-treated wood is brittle.

Surface Properties

Wood used outdoors can be susceptible to biotic and abiotic damage. One target of chemical modification is to create durable wood. For example, the acetylation of wood enhances the weathering resistance of wood compared to untreated controls (29). Acetylation is compatible with coatings and improves coating properties, such as adhesion or drying rate (30).

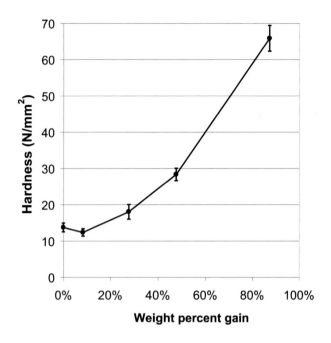

Figure 5. Brinell hardness perpendicular to grain of pine sapwood treated to increasing levels of DMDHEU.

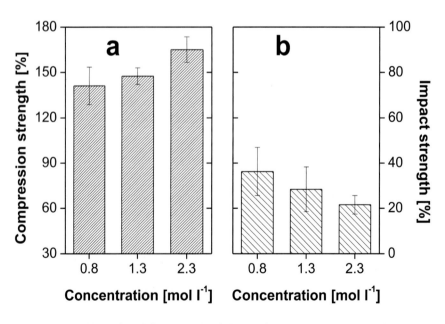

Figure 6. Effect of modification of beech wood with DMDHEU/$MgCl_2 \cdot 6H_2O$ (5wt% based on DMDHEU) on increase of compression strength (a) and retention of impact strength (b), respectively.

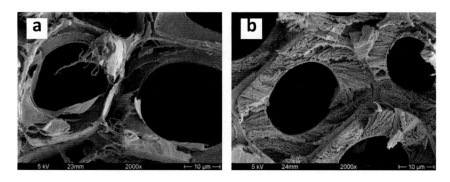

Figure 7. Micrographs of Scots pine latewood after impact fracture. The untreated wood exhibits a tough fracture surface (a) but a brittle fracture surface for the wood treated with DMDHEU to WPG of 20% (b).

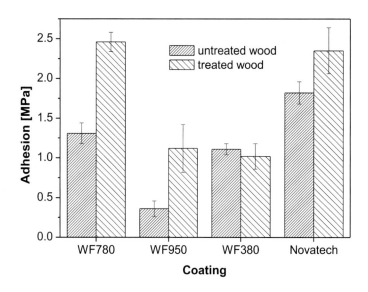

Figure 8. Wet adhesion of coatings on wood determined by the pull-off method (PrENV 927-8). WF780, WF 950, and WF380 are coatings containing core-shell type acrylic binders and Novatech is an alkyd-based solvent-borne stain. The wood substrates are untreated controls and treated with 50% DMDHEU, respectively.

Coating Performance

DMDHEU-treated wood exhibit similar surface wettability with several waterborne acrylic and solvent-borne alkyd coatings to untreated wood (*31*). Both the drying rates of various coatings on the wood and the blocking effect between coated wood were unaffected by the DMDHEU treatment, but by the types of coatings (*32*). The wet adhesion was considerably improved due to treatment, which may extend the service life of coatings on the treated wood surface (Figure 8).

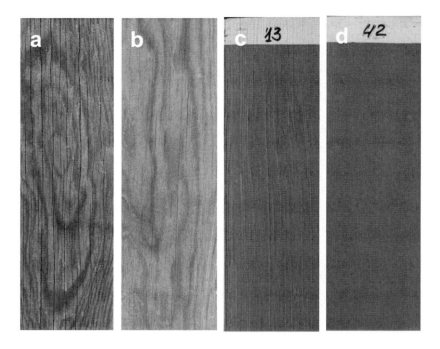

Figure 9. The appearance of untreated (a) and treated (b) pine sapwood after 18 months of natural weathering; the appearance of untreated (c) and treated (d) pine sapwood which were coated with water-borne acrylic stain and weathered outdoors for seven years in Goettingen, Germany. The wood treated with DMDHEU had a WPG of 22%.

Weathering Resistance

Thin veneers of pine sapwood treated with DMDHEU to 48% weight gain and artificially weathered for 72hs experienced tensile strength losses that were lower than that of untreated veneers, likely due to reduced cellulose degradation (*33*). FT-IR spectroscopy suggested that DMDHEU slow the lignin photodegradation, but did not inhibit color change. SEM examination also revealed that DMDHEU treatment is highly effective at reducing the degradation of wood cell wall during weathering. Specifically, the tracheids in untreated veneers became distorted within 48hs of artificial weathering, whereas tracheids in modified veneers retained their shape even after 144 hours of weathering. The stabilization effect increased with increasing DMDHEU-content within the veneers (*33*).

Flat-sawn panels of pine sapwood modified with DMDHEU to 22% weight gain and naturally weathered for 18 months experienced reduction of discoloration and cracking on the wood surface compared to untreated wood (Figure 9a, b). The surface erosion caused by weathering, especially in the less dense earlywood, was lower in the treated wood. The modified panels also had less colonization by blue stain or molds. This latter observation may be due to a reduced hydroscopicity of modified wood, rather than a biocidal effect. Coatings on modified pine sapwood with waterborne stains or oils experienced significantly less cracking after weathering for 18 months than those on untreated wood (*34*). The appearance of modified and coated wood did not obviously change even after a seven-year outdoor weathering; however, the surface of untreated and coated wood had a large amount of cracking and staining (Figure 9c, d).

Conclusion and Outlook

Treatments of permeable wood with low-molecular-weight *N*-methylol compounds, such as DMDHEU and its derivatives, can improve wood properties; the degree of improvement highly depends on the modifying levels. The swelling and shrinking of wood can be reduced up to 70% due to treatments. Mean ASE of 50% in an industrial process may be achievable when wood is treated at the proper chemical concentration. The treated wood exhibited resistance to decay fungi and can be graded as naturally durable material. The treatment did not prevent surface growth of molds and stains, but it reduced the growth of non-wood destroying molds to a high extent in exterior exposure. Surface hardness of wood was increased by several fold through treatment, and this characteristic makes it suitable for flooring material. Modification with DMDHEU did not influence the wettability or drying rate of coatings on the wood substrate, but did improve their wet adhesion. The modified wood exhibited less deformation, cracking and erosion on the surface during exposure outdoor. Synergistically protecting wood with surface coating, DMDHEU modification can extend the service life of wood during outdoor use.

Modification with *N*-methylol resin has good commercial potential. The chemicals used are easily available industrial products and their cost is acceptable compared to the added value to wood. The treatment uses aqueous solution, thereby avoiding the pollution problem caused by solvent evaporation. Although there is an issue of formaldehyde release during the curing and use, this problem can be properly controlled by using modified chemicals and adjusting the curing parameters. Modification with *N*-methylol resins could be used to improve the quality of fast-growing wood species that have been extensively used in the wood industry in the countries such as China.

References

1. Rowell, R. M. *For. Prod. Abstr.* **1983**, *6*, 363–382.
2. Xie, Y.; Fu, Q.; Wang, Q.; Xiao, Z.; Militz, H. *Eur. J. Wood Prod.* **2013**, *71*, 401–416.
3. Hill, C. *Wood Modification, Chemical, Thermal and Other Processes*; Wiley: Chichester, 2006.
4. Beckers, E. P. J.; Militz, H. In Second Pacific Rim Bio-Based Composites Symposium, Vancouver, Canada, 1994; pp 125−135.
5. Lukowsky, D. *Holz Roh- Werkst.* **2002**, *60*, 349–355.
6. Petersen, H. In *Chemical Processing of Fibers and Fabrics. Funktional Finishes Part A*; Lewin, M., Sello, S. B., Eds.; Marcel Dekker, Inc.: New York, 1983; pp 47−327.
7. Reeves, W. A.; Day, M. O. *J. Coat. Fabrics.* **1983**, *13*, 50–58.
8. Weaver, J. W.; Nielson, J. F.; Goldstein, I. S. *For. Prod. J.* **1960**, *10*, 306–310.
9. Militz, H. *Wood Sci. Technol.* **1993**, *27*, 347–355.
10. Krause, A.; Jones, D.; van der Zee, M.; Militz, H. In *European Conference on Wood Modification*; Van Acker, J., Hill, C.; Eds.; Ghent, Belgium, 2003; pp 317−327.
11. Petersen, H. *Textilveredlung* **1968**, *3*, 160–179.
12. Vieweg, R.; Becker, E. *Kunststoffhandbuch Band X*; Carl Hanser Verlag: Muenchen, 1968.
13. Bhattacharyya, N.; Doshi, B. A.; Sahasrabudhe, A. S.; Mistry, P. R. *Ameri. Dyestuff Rep.* **1993**, *82*, 96–103.
14. Zee Van der, M.; Beckers, E. P. J; Militz, H.; *International Research Group on Wood Preservation*; 1998; Doc. 98-40119.
15. Schaffert, S.; Krause, A.; Militz, H. In *European Conference on Wood Modification;* Militz, H., Hill, C., Eds.; Goettingen, Germany, 2005.
16. Dieste, A.; Krause, A.; Mai, C.; Sèbe, G.; Grelier, S.; Militz, H. *Holzforschung* **2009**, *63*, 89–93.
17. Kollmann, F. *Technologie des Holzes und der Holzwerkstoffe*; Springer Verlag: Berlin, 1951.
18. Dieste, A.; Krause, A.; Mai, C.; Militz, H. *Wood Sci. Technol.* **2010**, *44*, 597–606.
19. Xie, Y.; Hill, C.; Xiao, Z.; Jalaludin, Z.; Militz, H.; Mai, C. *J. Appl. Polym. Sci.* **2010**, *117*, 1674–1682.
20. Dieste, A.; Krause, A.; Militz, H. *Holzforschung* **2008**, *62*, 577–583.
21. Niemz, P. *Physik des Holzes und der Holzwerkstoffe*; DRW-Verlag Weinbrenner DmbH & Co.: Leinfelden-Echterdingen, 1993.
22. Krause, A. *Holzmodifizierung mit N-Methylolvernetzern*; Sierke Verlag: Goettingen, 2006.
23. Wepner, F.; Militz, H. In *European Conference on Wood Modification*; Goetingen, Germany, 2005; pp 169−177.
24. Ritschkoff, A.-C.; Ratto, M.; Nurmi, A. J.; Kokko, H.; Rapp, A. O.; Militz, H. *International Research Group on Wood Preservation*; 1999; Doc. 99-10318.
25. Verma, P.; Junga, U.; Militz, H.; Mai, C. *Holzforschung* **2009**, *63*, 371–378.

26. Xie, Y.; Krause, A.; Militz, H.; Turkulin, H.; Richter, K.; Mai, C. *Holzforschung* **2007**, *61*, 43–50.
27. Bollmus, S.; Rademacher, P.; Krause, A.; Militz, H. In *The Fifth European Conference on Wood Modification*; Riga, Latvia, 2010.
28. Winandy, J. E.; Rowell, R. M. In *Handbook of Wood Chemistry and Wood Composites*; Rowell, R. M., Ed.; CRC Press: Boca Raton, 2005; pp 303–347.
29. Feist, W. C.; Rowell, R. M.; Ellis, W. D. *Wood Fiber Sci.* **1991**, *23*, 128–136.
30. Beckers, E. P. J.; de Meijer, M.; Militz, H.; Stevens, M. *J. Coat. Technol.* **1998**, *70*, 59–67.
31. Tomazic, M.; Kricej, B.; Pavlic, M.; Petric, M.; Krause, A.; Militz, H. In *Woodcoatings-Developments for a Sustainable Future*; The Hague, The Netherlands, 2004.
32. Xie, Y.; Krause, A.; Militz, H.; Mai, C. *Prog. Org. Coat.* **2006**, *57*, 291–300.
33. Xie, Y.; Krause, A.; Mai, C.; Militz, H.; Richter, K.; Urban, K.; Evans, P. D. *Polym. Degrad. Stab.* **2005**, *89*, 189–199.
34. Xie, Y.; Krause, A.; Militz, H.; Mai, C. *Holz Roh- Werkst.* **2008**, *66*, 455–464.

Chapter 18

Acetylation of Wood

Roger M. Rowell[1] and James P. Dickerson[*,2]

[1]Professor Emeritus, University of Wisconsin, Madison, Wisconsin 53705
[2]Eastman Chemical Company, 200 South Wilcox Drive, Kingsport, Tennessee 37662
[*]E-mail: jpd@eastman.com.

Wood is a porous three dimensional, hydroscopic, viscoelastic, anisotropic bio-polymer composite composed of an interconnecting matrix of cellulose, hemicelluloses and lignin with minor amounts of inorganic elements and organic extractives. Some, but not all, of the cell wall polymer hydroxyl groups are accessible to moisture and these accessible hydroxyls form hydrogen bonds with water. As the water layers build up, the cell wall expands to accommodate the water resulting in an increase in wood/water volume up to the fiber saturation point. Increased moisture levels also offer a large variety of micro-organisms the opportunity to colonize and begin the process of decay. If these accessible hydroxyl groups are chemically substituted with a larger and more hydrophobic chemical groups, the bonded chemical can expand the cell wall until it reaches its elastic limit. And if the hydrophobic nature of substituted groups sufficiently reduces the cell wall moisture levels, the wood will no longer support the colonization of micro-organisms. This modified wood then achieves a high level of dimensional stability and durability. One technology that has now been commercialized to achieve these properties is acetylation: a reaction between the hydroxyl groups on the wood cell wall polymers with acetic anhydride. While all woods contain a low level of acetyl groups, increasing this acetyl content changes the properties and performance of the

© 2014 American Chemical Society

reacted wood. When a substantial number of the accessible hydroxyl groups are acetylated consistently across the entire cell wall, the wood reaches its highest level of dimensional stability and durability.

Introduction

Wood has been used by humans since the first humans walked the earth. They used it for fuel, shelter, weapons, tools and for decoration. They found it easy to work, renewable, sustainable and available. For the most part, it has been used without modification. Solid timbers and lumber were treated for decay and fire resistance as recorded in ancient accounts; however, most applications for wood today have little treatment other than a coating or finish. We learned to use wood accepting that it changes dimensions with changing moisture content, decomposes by a wide variety of organisms, burns and is degraded by ultraviolet energy. Wood has mainly been used as a construction material because it is widely available, renewable, sustainable and cheap. The average person does not consider wood as a high performance material.

In ancient Africa, natives hardened wood spears by placing a sharpened straight wooden stick in the bottom of glowing coals following by pounding the burned end with a rock and repeating this process many times until the end was sharp and hard. It would be many hundreds of years before we understood that pyrolysis of mainly hemicelluloses produced furan resins which when combined with carbon and compressed, results in an extremely hard resin-impregnated carbon composite.

The ancient Egyptians used dry to wet wooden wedges to split giant granite obelisks from the side of a quarry long before we discovered the tremendously large swelling pressure that is exerted when wood swells. The Egyptians also bent wood for furniture using hot water long before we understood the hydro-thermal glass transition of lignin. The Bible records a message to Noah to build an ark using a wood known to resist decay long before we understood how microorganisms recognized wood as a food source. Finally, the Vikings burned the outside of their ships to make them water and flame resistant without knowing anything about hydrophobicity or the insulating properties of pyrolignas char.

Connecting studies on the chemistry of wood with observations on properties and performance, it became clear that cell wall chemistry and properties were, for the most part, responsible for the observed performance. These early observations led researchers to connect chemistry, property and performance.

Combining all of the art and science of wood recorded from ancient times to the present, we have discovered that if you change the chemistry of wood, you change it properties and that leads to a change in performance. From this foundation, the science of chemical modification of wood was born.

While there are many early references to reacting chemicals with wood, the term "chemical modification of wood" was first used in 1946 by Tarkow (*1*). Chemical modification of wood is defined as covalently bonding a chemical group to some reactive part of the cell wall polymers.

Many chemical reaction systems have been published for the modification of wood and these systems have been reviewed in the literature several times in the past (*2–8*). These chemicals include anhydrides such as, acetic, butyric, phthalic, succinic, maleic, propionic and butyric anhydride, acid chlorides, ketene carboxylic acids, many different types of isocyanates, formaldehyde, acetaldehyde, di-functional aldehydes, chloral, phthaldehydic acid, dimethyl sulfate, alkyl chlorides, β-propiolactone, acrylonitrile, epoxides, such as, ethylene, propylene, and butylene oxide, and di-functional epoxides.

Most of the research in chemical modification of wood has been focused on improving dimensional instability and resistance to biological degradation. And of all of the chemistries published on the chemical modification of wood, the reaction of wood with acetic anhydride (acetylation) has been studied the most.

Acetylation of Wood

The acetylation of wood was first performed in Germany by Fuchs in 1928 using acetic anhydride and sulfuric acid as a catalyst (*9*). Fuchs found an acetyl weight gain of over 40 percent, which meant that he decrystalized the cellulose in the process. He used the reaction to isolate lignin from pine wood. In the same year, Horn acetylated beech wood to remove hemicelluloses in a similar lignin isolation procedure (*10*). Also in 1928, Suida and Titsch acetylated powdered beech and pine using pyridine or dimethylaniline as a catalyst to yield an acetyl weight gain of 30 to 35 percent after 15 to 35 days at 100°C (*11*).

In 1945, Tarkow first demonstrated that acetylated balsa was resistant to decay (*12*). Tarkow also was first to described the use of wood acetylation to stabilize wood from swelling in water (*1, 13*). The first patent on wood acetylation was filed by Suida in Austria in 1930 (*14*). Nearly two decades later in 1947, Stamm and Tarkow filed a patent on the acetylation of wood and boards using pyridine as a catalyst (*15*). In 1961, the Koppers Company published a technical bulletin on the acetylation of wood using no catalyst but an organic co-solvent.

In 1973, Rowell started a research program on chemical modification that included new ideas for wood acetylation (*3*). Several new ideas for wood acetylation were developed in this program depending on the size of the wood to be acetylated. For fiber, flakes, chips and wood up to about 3 cm in thickness, uncatalyzed hot acetic anhydride with a small amount of acetic acid was used. Wood does not float in acetic anhydride so the fiber, flakes or chips can be removed from the bottom of soaking container by a screw or belt for continuous removal of the saturated wood. Technology was also developed to use a minimal amount of anhydride based on the over dry weight of the wood to reduce the energy required to heat the reaction mixture. For wood thicker than about 3 cm, the wood is first soaked in a vacuum/pressure process with cold anhydride to saturate the wood and then heated. If this is not done, as the reaction starts on the outside of the thick wood, the by-product acetic acid builds up and will either slow the reaction or stop it (*16*).

In 1977 in Russia, Otlesnov and Nikitina came close to commercializing acetylation but the process was discontinued, presumably because it was not cost effective. In the late 1980s in Japan, Daiken started commercial production of acetylated wood for flooring called alpha-wood. Much more will be written about commercialization of acetylated wood later in this chapter.

The reaction of acetic anhydride with wood results in esterification of the accessible hydroxyl groups in the cell wall, with the formation of by-product acetic acid. The by-product acid must be removed to low levels from the product as the human nose is quite sensitive to the odor of acetic acid.

Chemistry

Acetylation is a single-addition reaction, which means that one acetyl group is on one hydroxyl group with no polymerization:

$$WOOD\text{-}OH + (CH_3CO)_2O \rightarrow WOOD\text{-}O\text{-}COCH_3 + CH_3COOH$$
$$\text{acetic anhydride} \qquad \text{acetylated wood} \qquad \text{acetic acid}$$

Thus, all the weight gain in acetyl can be directly converted into units of hydroxyl groups blocked. This is not true for a reaction where polymer chains are formed (epoxides and isocyanates, for example). In these cases, the weight gain cannot be converted into units of blocked hydroxyl groups. Table 1 shows that when green wood is dried, it shrinks about 10% in volume. When that wood is acetylated to about 20 weight percent gain (WPG), the new dry modified volume has increased about 10% and the wood is now approximately the same size as it was when green (Figure 1).

Figure 1. Video image capture from laboratory acetylation of southern pine. A: anhydride wet board prior to acetylation, B: acetylated, acid wet, C: acetylated, after acid removal (17).

Table 1. Change in Volume in Wood from Green to Dry to Acetylated

Green Volume (cm2)	Oven dry Volume (cm2)	Change (%)	Ac (%)	Ac Volume (cm2)	Change (%)
38.84	34.90	-10.1	22.8	38.84	+10.1

Acetylation has also been done using ketene gas (*12, 18–20*). In this case, esterification of the cell wall hydroxyl groups takes place but there is no formation of byproduct acetic acid. While this is interesting chemistry and eliminates a

$$WOOD\text{-}OH + CH_2=C=O \rightarrow WOOD\text{-}O\text{-}COCH_3$$
$$\text{ketene} \qquad \text{acetylated wood}$$

byproduct, it has been shown that reactions with ketene gas results in poor penetration of reactive chemical and it has been shown that the rate of the reaction is determined by the rate of diffusion of the vapor into the wood. Since the rate of diffusion into a porous solid varies inversely with the square of its thickness, reaction of wood with ketene has been restricted to a maximum wood thickness of about 3 mm if the reaction is to be carried out in a reasonable length of time. The properties of the reacted wood are less desirable than those of wood reacted with acetic anhydride (*20*).

Through the years, many catalysts have been tried for acetylation, both with liquid and vapor systems. These include zinc chloride, urea-ammonium sulphate, dimethylformamide, sodium acetate, magnesium persulfate, trifluoroacetic acid, boron trifluoride, and γ-rays. In the early 1980's, the technology changed to using a controlled and limited amount of acetic anhydride, no catalyst or cosolvent; a small amount of acetic acid; and a reaction temperature of between 120° and 130° C for solid wood (after soaking in cold anhydride) and 120° to 165° C for fibers, particles, chips, veneer and thin solid wood (*21*). The rate-controlling step in the chemical modification of solid wood is the rate of penetration of the reagent into the cell wall.

In the reaction of liquid acetic anhydride with wood, at an acetyl weight percent gain of about 4, there is more bonded acetyl in the S2 layer than in the middle lamella. At a WPG of about 10, acetyl is equally distributed throughout the S2 layer and middle lamella. At a WPG over 20, there is a slightly higher concentration of acetyl in the middle lamella than in the rest of the cell wall. These results were found using chloroacetic anhydride and following the fate of the chlorine by energy-dispersive x-ray analysis (*22*).

Questions have been raised about the long-term stability of the acetate group in wood. Table 2 shows the stability of acetyl groups in pine an aspen flakes to cyclic exposure to 30 and 90 percent relative humidity (RH) (3 months at 30% RH, followed by 3 months at 90% RH). Within experimental error, no loss of acetyl occurred over 41 cycles. This experiment has been ongoing for more than 20 years; results continue to show little or no loss of acetyl from humidity cycling (*23*).

Table 2. Stability of Acetyl Groups in Pine and Aspen Flakes after Cyclic Exposure between 90% Relative Humidity (RH) and 30% RH

	Acetyl Content (%) after cycle (number)				
	0	13	21	33	41
Pine	18.6%	18.2%	16.2%	18.0%	16.5%
Aspen	17.9%	18.1%	17.1%	17.8%	17.1%

The mass balance in the acetylation reaction shows that all the acetic anhydride going into the acetylation of hardwood and softwood could be accounted for as increased acetyl content in the wood, acetic acid resulting from hydrolysis by moisture in the wood, or as unreacted acetic anhydride. The consumption of acetic anhydride can be calculated stoichiometrically based on the degree of acetylation and the moisture concentration (MC) of the wood (24).

Table 3 shows the distribution of acetyl groups in southern yellow pine reacted with acetic anhydride to three levels of WPG at 120°C. Assuming that part of the cellulose is not accessible due to its crystallinity and the fact that cellulose did not react as seen in Figure 1, at a WPG of 23.6, all of the theoretical hydroxyl groups on lignin are acetylated (assuming approximately 1 hydroxyl group per C-9 unit) and over 60% of the hemicellulose hydroxyls are acetylated (25).

Table 3. Distribution of Acetyl Group in Southern Yellow Pine

Wpg	Total Accessibility*	Limited Accessibility**	Degree Of Substitution In Lignin
8.5	0.12	0.28	0.78
18.5	0.19	0.46	1.10
23.6	0.26	0.63	1.15

* Assuming accessibility of all cell wall hydroxyl groups   ** Assuming 100% accessibility of hemicelluloses and lignin hydroxyl groups but no accessibility of cellulose hydroxyl groups.

Figure 2 shows the rate and extent of each isolated cell wall polymer in the wood along with whole wood. Reacting wood with acetic anhydride reacts first with the more acidic phenolic hydroxyl on lignin but the bulk of the bonded acetyl is on the hemicelluloses (22).

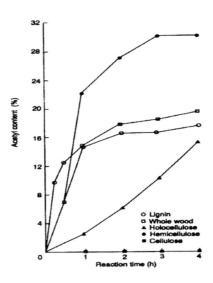

Figure 2. Reaction of acetic anhydride with isolated cell wall polymers and whole wood.

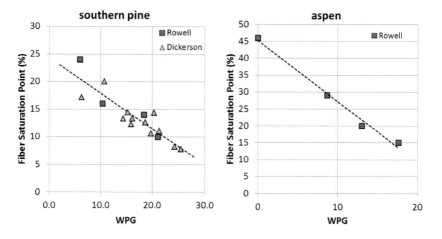

Figure 3. Fiber saturation point depression with increasing WPG in acetylated southern pine (26, 27) and aspen. WPG estimated from bound acetyl for reference (27).

Properties of Acetylated Wood

Moisture and Water Sorption

The replacement of some hydroxyl groups on the cell wall polymers with bonded acetyl groups reduces the hygroscopicity of wood. One measurement of level of this change is the reduction in fiber saturation point. Figure 3 shows the fiber saturation point for acetylated pine and aspen. As the level of acetylation increases, the fiber saturation point decreases, in both softwood and hardwood.

Table 4 shows the equilibrium MC (EMC) of control and acetylated pine and aspen at several levels of acetylation and three levels of RH. In all cases, as the level of chemical weight gain increases, EMC is reduced in the modified wood. Figure 4 shows the sorption/desorption isotherm for acetylated and control spruce fiber (28). Moisture is presumed to be sorbed either as primary or secondary water. Primary water is water sorbed to primary sites with high binding energy, such as the hydroxyl groups. Secondary water is water sorbed to sites with less binding energy; water molecules are sorbed on top of the primary layer. Since some hydroxyl sites are esterified with acetyl groups, there are fewer primary sites to which water sorbs. And since the fiber is more hydrophobic as a result of acetylation, there may also be fewer secondary binding sites.

Changes in dimensions are a great problem in wood composites as compared to solid wood. Composites undergo not only normal bulk wood swelling (reversible swelling) hut also swelling caused by the release of residual compressive stresses imparted to the board during the composite pressing process (irreversible swelling) (29, 30). Water sorption causes both reversible and irreversible swelling; some reversible shrinkage occurs when the board dries.

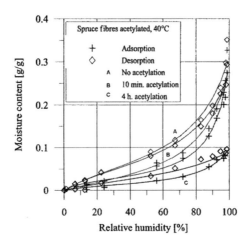

Figure 4. Sorption/desorption isotherms for control and acetylated spruce fiber.

Table 4. Equilibrium Moisture Content of Acetylated Pine and Aspen

Specimen	WPG (%)	Equilibrium Moisture Content (%) at 27°C		
		30%RH	*65%RH*	*90%RH*
acetylated pine	0	5.8	12.0	21.7
	6.0	4.1	9.2	17.5
	10.4	3.3	7.5	14.4
	14.8	2.8	6.0	11.6
	18.4	2.3	5.0	9.2
	20.4	2.4	4.4	8.4
aspen	0	4.9	11.1	21.5
	7.3	3.2	7.8	15.
	11.5	2.7	6.9	12.9
	14.2	2.3	5.9	11.4
	17.9	1.6	4.8	9.4

The rate of swelling is much slower in acetylated wood as compared to controls. Table 5 shows the rate of swelling of fiberboards made from control and acetylated fiber. The control board has increased more than 25% in thickness after just 15 minutes and has increased over 35% after 5 days. The board made from acetylated fibers only increased less than 5% after 5 days.

Table 5. Rate and Extent of Thickness Swelling in Liquid Water of Pine Fiberboards Made from Control and Acetylated Fiber (8% Phenolic Resin)

	Minutes			Hours			Days
	15	30	60	3	6	24	5
Control	25.7	29.8	3.5	33.8	34	34	36.2
Acetylated 21.6 WPG	0.6	0.9	1.2	1.9	2.5	3.7	4.5

The dimensional stability, as measured as anti-shrink efficiency (ASE) of acetylated solid wood varies depending on the species acetylated. Table 6 shows the ASE of two soft woods and two hardwoods. As the percent acetyl content increases, dimensional stability increases (*31*). The data indicates shows that softwoods achieve a higher weight gain in acetyl as compared to hardwoods. This may be because hardwoods contain a higher content of xylans which do not have a primary hydroxyl group in which to react. One hundred percent dimensional stability is not achieved by acetylation since water molecules will still interact with the wood structure even in "completely acetylated wood;" however, the swelling does not exceed the elastic limit of the cell wall.

Table 6. Dimensional Stability of Solid Wood As Measured as Antishrink Efficiency (ASE)

Wood	WPG	ASE	Wood	WPG	ASE
Ponderosa Pine	0	---	Beech	0	---
	12.0	39.4		10.6	43.8
	13.6	56.3		11.9	56.4
	17.1	64.8		15.4	66.0
	20.8	78.9		17.5	75.6
Sitka Spruce	0	---	Oak	0	---
	13.5	34.1		11.9	46.8
	14.8	46.3		13.9	65.5
	18.7	54.9		17.2	73.4
	24.1	69.5		17.8	84.9

Table 7 shows that repeated wetting and drying of acetylated pine does not decrease the dimensional stability and further indicates the stability of the acetyl group (*32*).

Table 7. Repeated Antishrink Efficiency (ASE) of Acetylated Solid Pine

	WPG	ASE1	ASE2	ASE3	ASE4	Weight Loss After Test
Acetic Anhydride	22.5	70.3	71.4	70.6	69.2	<0.2

Table 8 shows that a lower level of equilibrium moisture content and a higher level of dimensional stability is achieved in fiberboards made from acetylated fiber as compared to solid wood. This is due to more accessible reaction sites available at the fiber lever as compared to solid wood (*33*).

Table 8. Equilibrium Moisture Content and Antishrink Efficiency

	WPG	EMC	ASE
	0	19.6	---
Fiberboard made from Acetylated Pine fiber (5% phenolic resin)	12.3	10.8	61.9
	15.8	8.9	77.1
	18.9	5.3	86.3
	20.8	3.8	94.7

Resistance to Biological Attack

Fungi – Lab Tests

Various types of solid wood, particleboards, and flakeboards made from acetylated wood have been tested for resistance to different types of organisms. Acetylated wood has been tested with several types of decay fungi in an ASTM standard 12-week soil block test using the brown-rot fungus Gloeophyllum trabeum or the white-rot fungus Trametes versicolor. Table 9 shows the resistance of pine acetylated to several levels of chemical modification to attack by brown-and white-rot fungi. As the level of acetylation rises, the resistance to attack increases. Weight loss resulting from fungal attack is the method most frequently used to determine the effectiveness of a preservative treatment to protect wood from decay. In some cases, especially for brown-rot fungal attack, strength loss may be a more important measure of attack since large strength losses are known to occur in solid wood at very low wood weight loss.

Figure 5 shows the control pine sample before and after attack by the brown-rot fungus *Gloeophyllum trabeum* (*34–36*). The wood is badly deteriorated with major damage to the cell wall structure. The figure also shows the acetylated wood after the same test. The fungal hyphae can be seen but there is no visible attack on the wood.

Weight loss resulting from fungal attack is the method most frequently used to determine the effectiveness of a preservative treatment to protect wood from decay. In some cases, especially for brown-rot fungal attack, strength loss may be a more important measure of attack since large strength losses are known to occur in solid wood at very low wood weight loss. A dynamic bending creep test (Figure 6) has been developed to determine strength loss when wood composites are exposed to a brown or white-rot fungus (*37*).

Table 9. Weight Loss of Acetylated Southern Pine in a Soil Block Test *(ASTM D14113 – 07e1. Standard Test method for Wood Preservatives by Laboratory Soil-Block Cultures)*

Acetyl Weight Gain (%)	Weight Loss After 12 Weeks (%)	
	Brown-rot Fungus	White-rot fungus
0	61.3	7.8
6.0	34.6	4.2
10.4	6.7	2.6
14.8	3.4	<2
17.8	<2	<2

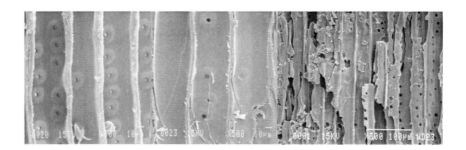

Figure 5. SEM of brown-rot fungal attack on wood. Left = control, Center = acetylated wood after 12 weeks in the ASTM soil block test, Right = after 51.1% weight loss.

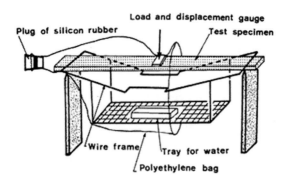

Figure 6. Test Equipment for strength loss in control and acetylated wood (32).

In this bending creep test of aspen flakeboards, control boards made with phenol-formaldehyde adhesive failed in an average of 71 days when exposed to the brown-rot fungus *Tyromyces palustris* and in 212 days when exposed to the white-rot fungus *Trametes versicolor* (*35, 38, 39*). At failure, weight loss averaged 7.8 percent for *T. palustris* and 31.6 percent for *T.versicolor*. Isocyanate-bonded control flakeboards failed in an average of 20 days with *T. palustris* and 118 days with *T. versicolor*, with an average weight loss at failure of 5.5 and 34.4 percent, respectively. Very little or no weight loss occurred with both fungi in flakeboards made using either phenol-formaldehyde or isocyanate adhesive with acetylated flakes. None of these specimens failed during the 300-day test period. Mycelium fully covered the surfaces of isocyanate-bonded control flakeboards within 1 week, but mycelial development was considerably slower in control flakeboards bonded with phenol-formaldehyde. Acetylated flakeboards bonded with both isocyanate- and phenol-formaldehyde showed surface mycelium colonization during the test, but no strength loss (Figure 7).

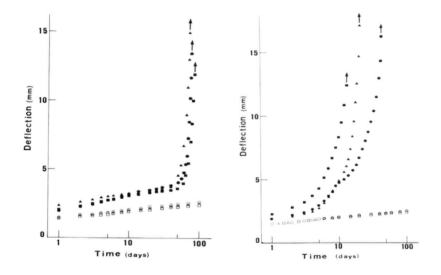

Figure 7. Deflection-time curve for control and acetylated flakeboards: phenol-formaldehyde bonded (left), isocyanate bonded (right). [Control = solid dots, acetylated = hollow dots].

In similar bending creep test, both control and acetylated pine particleboard made using melamine-ureaformaldehyde adhesive failed because *T. palustris* attacked the adhesive in the glueline. Mycelium invaded the inner part of all boards, colonizing in both the wood and glue line in control boards but only in the glueline in acetylated boards. These results show that the glueline is also important in protecting composites from biological attack. After 16 weeks of exposure to *T. palustris*, the internal bond strength (IBS) of control aspen flakeboards made using a phenol-formaldehyde resin was reduced more than 90 percent; IBS of flakeboards made using an isocyanate resin was reduced 85 percent (*39*). After 6 months of exposure in moist unsterile soil, the same control flakeboards made using a phenol-formaldehyde resin lost 65 percent IBS and those made using an isocyanate resin lost 64 percent IBS. Failure was due mainly to strength reduction in the wood resulting from fungal attack. Acetylated aspen flakeboards lost much less IBS during the 16-week exposure to *T. palustris* or 6-month soil exposure.

The mechanism of resistance to fungal attack by chemical modification is hypothesized to be related to low moisture sorption in the cell wall, below that needed for biological attack. Other theories suggest the mechanism may be from blocking of specific enzymatic reactions as a result of changes in configuration and conformation of the polymers in the cell wall of the modified wood. In the case of brown-rot fungal attack, researchers have suggested that the reduced moisture of acetylated wood prevents the fungus from initiating the breakdown of the hemicelluloses as an energy source. This mechanism is consistent with the data from soil block weight loss tests and strength loss tests.

Table 10. Sugar Analysis on a Pine Sample after the 12 Week ASTM Soil Block Test with the Brown-Rot Fungus *Gloeophyllum trabeum*

Wood Fiber WPG	% Lost							
	Weight	Total Carbo	Araban	Galac-tan	Rham-nan	Glucan	Xylan	Mannan
0	51.7	85.8	87.9	71.9	90.0	83.8	90.6	92.5
13	1.4	13.2	89.0	55.2	70.0	0	38.3	42.0

Table 10 shows the sugar analysis after the 12 week soil block test with a brown-rot fungi on a control fiberboard and an acetylated fiber with a WPG of 13. The control sample lost 51.7 percent in total weight and a carbohydrate loss of 86% while the acetylated sample only lost 1.4% in total weight and 13.2% carbohydrate. Almost all of the araban, rhamnan, mannan and xylan were lost in the control samples as well as most of the galactan and glucan. No glucan was lost in the acetylated sample showing that in the early stages of brown-rot decay, the cellulose is not attacked. A large portion of the araban and the rhamnan were lost in the early stages of attack on the acetylated board. There is also a significant loss of galactan, xylan and mannan during the early stages of attack. All of these sugars are in the hemicellulose polymers which is even more evidence that it is the hemicelluloses that need to be protected for fungal resistance, at least, for brown-rot fungal attack. Arabinose is the only sugar in wood that is in a strained five membered ring. It is possible that this easily hydrolyzed sugar is the recognition site for the fungal enzymes that starts the entire decay process in brown-rot fungi.

Fungi – Fungal Cellar Test

Another test to determine the fungal and bacterial resistance of acetylated composites is a fungal cellar containing brown-, white-, and soft-rot fungi and tunneling bacteria. Control blocks were destroyed in less than 6 months while flake-boards made from acetylated furnish above 16 WPG showed no attack after 3 years (Table 11). These data show that no attack occurs until swelling of the wood occurs. This is additional evidence that the moisture content of the cell wall is critical to fungal attack.

Table 11. Fungal Cellar Tests of Aspen Flakeboards Made from Control and Acetylated Flakes

WPG	*Rating Interval (months)*							
	2	3	4	5	6	12	24	36
0	S/2	S/3	S/3	S/3	S/4	--	--	--
7.3	S/0	S/1	S/1	S/2	S/3	S/4	--	--
11.5	0	0	S/0	S/1	S/2	S/3	S/4	--
13.6	0	0	0	0	S/0	S/1	S/2	S/3
16.3	0	0	0	0	0	0	0	0
17.9	0	0	0	0	0	0	0	0

- Nonsterile soil containing brown-, white-, and soft-rot fungi and tunneling bacteria.
- Flakeboards bonded with 5% phenol-formaldehyde adhesive.
- Rating system: 0 = no attack; 1 = slight attack; 2 = moderate attack; 3 = heavy attack; 4 = destroyed; S = swollen.

Fungi – In-Ground Tests

Acetylated solid wood and flakeboards have been subjected to in-ground tests in the United States and New Zealand (*40*), and Sweden (*41*), with specimens showing little or no attack after 10 years of exposure. In Indonesia (*42*) specimens failed in less than 3 years, mainly as a result termite attack. In Sweden, acetylated pine at a WPG of 21.2 has been outperforming wood treated with copper chromium arsenic at 10.3 kg/m3 after 15 years of exposure (*41*).

Termites – Lab Tests

Table 12 shows the results of a 2-week termite test using Reticulitermes flavipes (subterranean termites) on control and acetylated pine. While protection was afforded to samples with higher levels of acetylation, the lack of complete resistance to attack may be attributed to the low mortality rate in termites during the test. Termites can live on acetic acid and decompose cellulose to mainly acetate.

Table 12. Wood Weight Loss in Control and Acetylated after a Two-Week Exposure to *Reticulitermes flavipes*

	WPG	Wood Weight Loss (%)
Control	0	31
Acetylated	10.4	9
	17.8	6
	21.6	5

Additional tests were conducted using with dry wood and subterranean termites. The two different woods were placed in test using dry wood termites (*Cryptotermes cynocephalus*). Fifty healthy and active nymphae placed in each box and the boxes were put in a dark room at an average temperature of 20 to 32°C and 81 to 89 percent relative humidity (RH) for 10 weeks. At the end of the test nymphae mortality and wood weight loss were determined (Table 13).

Table 13. Acetylated Pine and Jabon Exposed to *Cryptotermes cynocephalus*

Wood Species	Modification	Weight Loss (%)	Mortality (%)
Indonesian pine	Control	9.3	42
	Acetylated	1.9	99
Indonesian jabon	Control	16.5	31
	Acetylated	1.9	95

The two different woods (19 mm x 19 mm x 10 mm) were placed in test using subterranean termites *Coptotermes gestroi*. (Table 14). Each wood specimen was put in an acrylic cylindrical tube (sized 60 mm height and 80 mm diameter), and to each tube was put 150 workers and 15 soldiers of nymphae. A wet tissue was placed in each tube to maintain humidity. The tubes were put in a dark room at an average temperature of 20 to 32 °C and 81 to 89 percent relative humidity (RH) for 5 weeks. At the end of the test the percentage weight loss of each specimen was determined as well as nymphae mortality. The results of this test are shown below (*43*).

Table 14. Acetylated Pine and Jabon Exposed to *Coptotermes gestroi*

Wood Species	Modification	Weight Loss (%)	Mortality (%)
Indonesian pine	Control	9.9	70
	Acetylated	1.7	100
Indonesian jabon	Control	12.5	83
	Acetylated	3.5	100

Termites – In-Ground Tests

Termite tests were run on acetylated wood at several test sites (*40*). The first sites were in Indonesia: Bogor and Bandung (Table 15). Rubber wood, spruce and aspen particles were acetylated to two levels of acetyl content: low 8 – 12 WPG and high 20 WPG and made into particleboards. The rubber wood particle board contained 9% phenolic resin with a target density of 750 kg/m^3, the spruce fiber board contained 8% phenolic resin with a target density of 750 kg/m^3 and the aspen fiber board contained 8% phenolic resin with a target density of 800 kg/m^3 (*43*).

From the data collected to date, the mechanism of resistance of acetylated wood to termites may be due to several factors, including: moisture content: EMC reduced below that needed for attack, increased hardness, modification of typical nutrients (such as hemicellulose acetate, lignin acetate, i.e. no recognition message), and since the actual digestion of the wood is done by bacteria living inside the termite and acetylated wood is stable to attack by microorganisms. The termite may graze but do not attack.

Marine Organisms

Acetylated wood is somewhat resistant to attack by marine organisms (Table 16). In Florida, control specimens were destroyed in 6 months to 1 year, mainly because of attack by *Limnoria tripunctata*, while acetylated wood showed good resistance. In similar tests in Sweden, acetylated wood failed after 2 years of exposure and control specimens failed in less than 1 year. For both control and acetylated specimens in Sweden, failure was due to attack by crustaceans and mollusks (*44*).

Table 15. Termite Ratings for Wood Species in Indonesia

species	time (mo.)	Bandung rating			Bogor rating		
		control	acetyl level low	acetyl level high	control	acetyl level low	acetyl level high
Rubber wood	1	7	7	10	0	4	10
	3	7	7	10	--	4	10
	12	0	0	10	--	0	9
	36	--	--	0	--	--	0
Spruce	1	7	9	10	7	9	10
	3	4	9	10	0	9	10
	12	4	7	10	--	9	9
	36	0	0	0	--	0	0
Aspen	1	7	7	10	4	0	7
	3	4	0	9	0	--	7
	12	0	--	4	--	--	7
	36	--	--	0	--	--	0

	Termite rating criteria
10	No attack or a few nibbles present
9	Small tunnels on surface less than 3% of cross section area affected at any location
7	Termite attack affects 10-25% of cross section area at any location
4	Termite attack affects more than 50% of cros sectional area at one
0	Failure

Thermal Properties

Figure 8 and Table 17 show the results of thermogravimetric and evolved gas analysis of control and acetylated pine. The control and acetylated samples show two peaks in the thermogravimetric runs and the lower temperature peak represents the hemicellulose fraction and the higher peak represents the cellulose. Acetylated pine pyrolyze at about the same temperature and rate as controls (*45*). The heat of combustion and rate of oxygen consumption are approximately the same for control and acetylated wood which means that the acetyl groups added have approximately the same carbon, hydrogen and oxygen content as the cell wall polymers. Acetylated wood has essentially the same thermal properties as unmodified wood.

Table 16. Ratings of Acetylated Southern Pine Exposed to a Marine Environment

	WPG (%)	Exposure (yrs)	Mean rating due to attack by	
			Limnoriid and Teredinid Borers (Key West, FL) 1975-1987	Shaeroma terebrans (Tarpon Springs, FL) 1984-1987
Control	0	1	2-4	3.4
Acetylated	22	3	8	8.8

Rating system - 10 = no attack; 9 = slight attack; 7 = some attack; 4 = heavy attack; 0 = destroyed

Table 17. Thermal Properties of Control and Acetylated Pine Fiber

	WPG (%)	Temp of maximum weight loss (°C)	Heat of combustion (Kcal/g)	Rate of oxygen consumption (mm/g sec)
Control	0	335/375	2.9	0.06/0.13
Acetylated	21.1	338/375	3.1	0.08/0.14

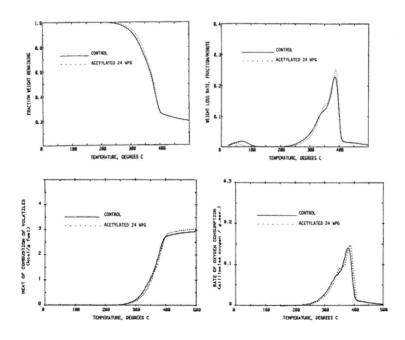

Figure 8. Thermogravimetric and evolved gas analysis of acetylated pine.

Weathering

Reaction of wood with acetic anhydride has also been reported to improve the ultraviolet resistance of wood (46). After 700 hours of accelerated weathering, controls eroded at a rate of about 0.12 µm/h or about 0.02%/h (Table 18). Acetylation reduced surface erosion by 50 percent. The depth of penetration as a result of weathering was about 200 µm for unmodified boards and half that for acetylated boards. In outdoor tests, the color of acetylated pine remained unchanged after 1 year while that of control boards turned from dark orange to light gray. After 3 years, the control wood was dark orange and parts starting to turn gray while the acetylated wood had just started turning darker. Acetylated pine exposed behind glass retained its bright color for 10 years.

Table 18. Weight Loss and Erosion of Control and Acetylated Aspen after 700 Hours of Accelerated Weathering

	WPG	Weight Loss in Erosion (%/hr)	Erosion Rate (µm/hr)	Reduction in Erosion (%)	Depth of Weathering (µm)
Control	0	0.019	0.121	---	199-210
Acetylated	21.2	0.010	0.059	51	85-105

Table 19 shows the acetyl and lignin analysis before and after 700 hours of artificial weathering. It can be seen that the acetyl level before and after weathering in the surface and interior of the acetylated wood is about the same however the amount of lignin in the surface in both the control and acetylated samples is greatly reduced.

Mechanical Properties

Table 20 shows the strength and stiffness properties of control and acetylated pine. There is considerable loss of wet strength and wet stiffness in non-acetylated wood as compared to acetylated wood. There is a loss of over 60% in wet strength in non-acetylated wood while acetylated only drops 10%. There is a loss of over 35% in wet stiffness of non-acetylated wood while acetylated wood only loses less than 9% in wet stiffness. Acetylation has been shown to slightly increase the strength properties of fiberboards and flakeboards. Strength properties of wood are very dependent on the moisture content of the cell wall. The mechanical properties of fiber stress at proportional limit, work to proportional limit, and maximum crushing strength are most affected by changing moisture content by only ±1 percent below the fiber saturation point. The fact that the EMC and fiber saturation point of acetylated wood are much lower than that of unmodified wood alone accounts for their difference in strength properties.

Table 19. Acetyl and Lignin Analysis before and after 700 Hours of Accelerated Weathering of Aspen Fiberboards Made from Control and Acetylated Fiber

		WPG	Before surface weathering (%)	Remainder (%)	After surface weathering (%)	Remainder (%)
Acetyl	Control	0	4.5	4.5	1.9	3.9
	Acetyl'd	19.7	17.5	18.5	12.8	18.3
Lignin	Control	0	19.8	20.5	1.9	17.9
	Acetyl'd	1.7	18.5	19.2	5.5	18.1

Table 20. Dry and Wet Strength and Stiffness of Pine Control and Acetylated

Sample	MOR Strength (N/mm²)			MOE Stiffness (N/mm²)		
	Dry	Wet	%Diff.	Dry	Wet	%Diff.
Pine	63.6	39.4	-62	10,540	6,760	-36
acetyl'd (19WPG)	64.4	58.0	-10	10,602	9,690	-8.6

%Diff is determined between the wet and dry states.

Commercialization of Acetylated Wood

While laboratory acetylation of wood has been practiced for nearly a century, the commercialization of acetylated wood has been met with several challenges. Koppers Company may have made the first earnest, albeit short-lived, attempt at entry into the acetylated wood market in the 1960's. This was followed by efforts in Russia and Japan (Diaken) in the 1970's and 1980's. Before the turn of the century efforts began in northern Europe and elsewhere to develop a cost effective commercial process for acetylation (47). In the late 1980's and early 1990's, ACell, in Sweden, were granted many patents and built two pilot plants: One for solid wood using microwave technology and one for acetylating fibers. In 2005, Accsys Technologies PLC, which had acquired technologies developed earlier at Stichting Hout Research (the Netherlands) and Scion (New Zealand), launched trial quantities of Accoya®, an acetylated pinus radiata, into the market and began full commercial scale production in Arnhem, the Netherlands, in March 2007. This was followed in 2012 with Eastman Chemical Company introducing Perennial Wood™ using acetylated southern pine produced at its pilot facility in Kingsport, Tennessee.

The long latency between the first commercial activities of the 1960's and now may be partially explained as much more favorable markets that exist now driven by a focus on sustainable products and the increased popularity of outdoor living. Advances have also been made in the process technology to reduce the capital expenses, improve cycle time, and decrease the manufacturing costs (*42*). Still, challenges face the commercial scale-up of acetylated wood manufacturing facility and can generally be attributed to the following areas:

- The inherent variability of wood as a raw material in a chemical manufacturing process
- Efficient removal and recapture of byproducts from acetylation
- Developing application expertise to enter the market with, what is essentially, a new species of wood.

Wood as a Raw Material

Each piece of wood is unique. The grain direction, pattern, texture, and natural variations are aesthetically appealing and shaped by the environmental exposure unique to each tree. Even so, this uniqueness also creates manufacturing difficulties, especially in the chemical industry, from wood's inhomogeneous and anisotropic nature. This is further complicated by the variations in saw patterns from milling operations. Operations that use wood as a raw material can partially address this issue by reducing wood to smaller form factors, such as chips and flour, and blending to produce a more homogenous material, but at the expense of the loss of the aesthetic appeal and functional properties of solid lumber. As lumber typically represents the higher value markets, the choice to acetylated solid wood greatly complicates both the experimental and manufacturing processes.

As the extent of reaction can be limited by the amount of reactant introduced into the wood, wood acetylation is dependent on the void volume of the wood in which to introduce the reactant. Even carefully sourced material may have large variations in wood density (inversely related to void volume), hence there are large variations in the level of impregnation into the wood and the amount of woody mass to acetylate. This variation occurs between boards and within different positions in the board. To minimize this variation in a batch process, a robust process and/or careful material screening may be required to achieve high levels of acetylation (*48*).

Removal and Use of Byproduct Acid

The acetylation of cellulose using acetic anhydride has been practiced commercially since the 1920's. Although the initial market for cellulose acetate was Safety Film (used as a replacement for the highly flammable cellulose nitrate used in movie film), the material can be today found in numerous application including filter media, fiber, plastics, and coatings additives. Chemical companies have built infrastructure not only for the production of acetic anhydride, but also to handle the byproduct acetic acid generated from this process. One option to manage the byproduct acid created by acetylation is to chemically reproduce acetic anhydride. This can be done by dehydrating one molecule of acetic acid to produce ketene, and subsequently reacting ketene with another acetic acid molecule to produce acetic anhydride.

$$CH_3COOH \rightarrow H_2C=C=O + H_2O$$
$$\text{acetic acid} \quad\quad \text{ketene}$$

$$H_2C=C=O + CH_3COOH \rightarrow (CH_3CO)_2O$$
$$\text{ketene} \quad\quad \text{acetic acid} \quad\quad \text{acetic anhydride}$$

The acetylation of solid wood faces the same industrial challenges as initial acetylation of cellulose – the liberation of large amounts of byproduct acid. A commercially viable processes require a rapid and thorough removal of the acid both to recover the acid for future use, and to reduce the acid odor in the final product (the human nose can detect very small amounts of acetic acid). This operation can be the slow step in the acetylation process and can limit production rates.

Application Expertise

Wood fabrication techniques have been around for centuries. Indeed, wood's workability has been one of the compelling reasons for its use in building and construction markets. Because of the change in hydrophilic nature, acetylated wood has a lower equilibrium moisture content than its unacetylated counterpart and as a result, a small increase in surface hardness is also realized. Fabrication with acetylated wood is subtly different from its unacetylated counterpart and understanding how these differences impact historical construction techniques requires reinvestigating the wood, essentially treating it as a new species. A small amount of acetic acid will typically remain in the wood at low levels and can lead to material compatibilities issues. Depending upon the manufacturer's process and quality control standards for residual acid, the acidity of acetylated wood may be comparable to or less than that of widely used hardwoods, including oak. However, because of the combination of long service life and the presence of some acetic acid, stainless steel fasteners are recommended with acetylated wood. Testing with different materials of construction for fasteners, flashing, adhesives, coatings, and fillers is required for new market areas.

In addition to the application development effort, Life Cycle Assessment (LCA) of the product and process has become an industry necessity for addressing the needs of the architectural community. Processes development effort must now consider impacts to the environment, such as energy use, in the overall product use, design and disposal.

Application Opportunities and Market Development

Any material which offers the potential performance of acetylated wood – durable, strong, stable, aesthetically pleasing, thermally resistive, readily machined, and so forth – will have multiple potential applications. However, potential customers are inevitably wary, as history is littered with examples of materials being launched to great fanfare, promising much and, unfortunately, failing to deliver upon their promises. Complicating matters further, the building trades are notoriously (no doubt justifiably) conservative.

There are many ways that the introduction of a material can be effected and can influence the overall rate of adoption in the market.

Expectations and Market Risk

While efforts to differentiate acetylated wood from other materials provide strong evidence that will support market acceptance, other challenges remain. Critical among these will be the establishment of standards to foster customer confidence in acetylated wood for new applications. Established wood regulatory agencies are actively working to understand the modified wood category. Standards, long established for pressure treated materials, are being tailored to ensure that modified products will meet the requirements for its intended uses and ensure that consumers are getting a high quality product. This also ensures that new entrants into this space will be pressed into meeting these standards as well. While acetylation has shown to improve the properties of most species tested, it is notable that not all acetylated wood will perform the same either because of differences in the underlying species, difference in the wood homogeneity, or differences in the acetylation process and technology used. Continued use testing and qualification will be a requirement as knowledge about these materials builds and acetylated wood is used in new and more diverse applications (*49*).

Concluding Remarks

The commercial deployment of acetylated wood, after several false starts, has finally begun in earnest. Consumers in North America have the choice of two branded acetylated wood products. In Europe, the Accoya, solid acetylated wood process technology has been licensed to Solvay, a large chemicals company with interests in cellulose acetylation, whose headquarters are located in Belgium (*50*).

In addition to solid wood acetylation, Accsys Technologies has formed a joint venture with Ineos, a global chemicals company, to develop and commercialize its proprietary acetylation technology for the production and licensing of acetylated wood elements (under the brand name Tricoya®) for use within medium density fiberboard (MDF), particle board and wood plastic composites. The Tricoya process technology been licensed to Medite Europe Limited, a subsidiary of Ireland's state-owned Coillte group (*51*).

With the existing commercial and pilot production at Accsys' and Eastman's European and North American sites respectively, together with potential production of acetylated wood being actively considered in other locations, acetylated wood has moved from the sole preserve of scientists in laboratories into the maelstrom of the commercial world. No doubt there will be challenges ahead. But, as the world seeks more sustainable building materials, acetylated wood looks set to become a material of choice.

Note

On January 31, 2014 Eastman Chemical Company announced it will discontinue the Perennial Wood™ product line and business operations.

References

1. Tarkow, H. *A New Approach to the Acetylation of Wood*; USDA Forest Service, Forest Products Laboratory: Madison, WI, 1946; p 9.
2. Jahn, E. C. Chemical modified wood. *Sven. Papperstidn.* **1947**, *17*, 393–4012.
3. Rowell, R. M. *Chemical Modification of Wood: Advantages and Disadvantages*; Proceedings of American Wood Preservers' Assoc.; Am. Wood Preservers' Assoc.: 1975; pp. 1–10.
4. Rowell, R. M. *Chemical Modification of Wood: A Review*; Commonwealth Forestry Bureau: Oxford, England, 1983; Vol. 6, pp 363–382.
5. Rowell, R. M. *Handbook of Wood Chemistry and Wood Composites*; Taylor and Francis: Boca Raton, FL, 2013; Vol. 15, pp 537–598.
6. Kumar, S. Chemical modification of wood. *Wood Fiber Sci.* **1994**, *26*, 270–2802.
7. Hon, D. N.-S. *Chemical Modification of Wood Materials.*; Marcel Dekker: New York, 1996; p 370.
8. Hill, C. *Wood modification: Chemical, Thermal and Other Processes*; John Wiley and Sons: Chichester, England, 2006; p 239.
9. Fuchs, W. Genuine lignin. I. Acetylation of pine wood. *Ber. Dtsch. Chem. Ges.* **1928**, *61B*, 948–951.
10. Horn, O. Acetylation of beech wood. *Ber. Dtsch. Chem. Ges.* **1928**, *61B*, 2542–45.
11. Suida, H.; Titsch, H. Chemistry of beech wood: Acetylation of beech wood and cleavage of the acetyl-beech wood. *Ber. Dtsch. Chem. Ges.* **1929**, *61B*, 1599–1604.
12. Tarkow, H. *Decay resistance of acetylated balsa*; USDA Forest Service, Forest Products Laboratory: Madison, WI, 1945; p 4.
13. Tarkow, H., Stamm, A. J.; Erickson, E. C. O. *Acetylated Wood. Rep. 1593*. USDA Forest Service, Forest Products Laboratory: Madison, WI, 1946.
14. Suida, H. Acetylating wood. Austria Patent 122,499, 1930.
15. Stamm, A. J.; Tarkow, H. Acetylation of Lignocellulosic Board Materials. U.S. Patent 2,417,995, 1947.

16. Rowell, R. M. *Reaction conditions for acetylation of fibers, flakes, chips, thin and thick woods.* Chalmers University, 1986.
17. Dickerson, J.; Guinn, T.; Allen, J. Eastman Chemical Company, Kingsport, TN. Personal communication, 2011.
18. Karlson, I.; Svalbe, K. Method of acetylating wood with gaseous ketene. *Uchen. Zap. Latv. Univ.* **1972**, *166*, 98–104.
19. Karlson, I.; Svalbe, K. Method of acetylating wood with gaseous ketene. *Latv. Lauksaimn. Akad. Raksti.* **1977**, *130*, 10–21.
20. Rowell, R. M.; Wang, R. H. S.; Hyatt, J. A. Flakeboards make from aspen and southern pine wood flakes reacted with gaseous ketene. *J. Wood Chem. Technol.* **1986**, *6*, 449–471.
21. Rowell, R. M.; Tillman, A-M.; Simonson, R. A simplified procedure for the acetylation of hardwood and softwood flakes for flakeboard production. *J. Wood Chem. Technol.* **1986**, *6* (3), 427–48.
22. Rowell, R. M.; et al. Acetyl distribution in acetylated whole wood and reactivity of isolated wood cell wall components to acetic anhydride. *Wood Fiber Sci.* **1994**, *1*, 11–1826.
23. Rowell, R. M., Lichtenberg, R. S.; Larsson, P. Stability of acetyl groups in acetylated wood to changes in pH, temperature, and moisture. In *Pacific Rim Bio-Based Composites Symposium: Chemical Modification of Lignocellulosics*; Plackett, D. V., Dunningham, E. A., Eds.; FRI Bulletin, 1992; Vol. 176, pp 33−40.
24. Rowell, R. M.; Simonson, R.; Tillman, A.-M. Acetyl balance for the acetylation of wood particles by a simplified procedure. *Holzforschung.* **1990**, *44* (4), 263–269.
25. Rowell, R. M. Distribution of reacted chemicals in southern pine modified with acetic anhydride. *Wood Sci.* **1982**, *15* (3), 172–182.
26. Rowell, R. M. *Handbook of Wood Chemistry and Wood Composites*; Taylor and Francis: Boca Raton, FL, 2005; p 487.
27. Dickerson, J.; Cwirko, E.; Allen, J. *Correlation of %Acetyl and Fiber Saturation in Acetylated Southern Pine Boards*; Proceedings IRG Annual Meeting, 2012; IRG/WP 12-40598.
28. Stromdahl, K. Water sorption in wood and plant fibers. Ph.D. Thesis, Department of Structural Engineering and Materials, The Technical University of Demark, Copenhagen, Denmark, 2000.
29. Youngquist, J. A.; Rowell, R. M.; Krzysik, A. Dimensional stability of acetylated aspen flakeboard. *Wood Fiber Sci.* **1986**, *18*, 90–981.
30. Youngquist, J. A.; Rowell, R. M.; Krzysik, A. Mechanical properties and dimensional stability of acetylated aspen flakeboards. *Holz Roh- Werkst.* **1986**, *44*, 453–457.
31. Militz, H. *Improvements of stability and durability of beechwood (Fagus sylvatica) by means of treatment with acetic anhydride*; International Research Group on Wood Preservation, 1991; Doc. No. IRG/WP 3645.
32. Rowell, R. M.; Ellis, W. D. Determination of dimensional stabilization of wood using the water-soak method. *Wood Fiber* **1978**, *10*, 104–1112.
33. Rowell, R. M.; et al. Dimensional stability of aspen fiberboards made from acetylated fiber. *Wood Fiber Sci.* **1991**, *23*, 558–566.

34. Rowell, R. M.; et al. Biological resistance of southern pine and aspen flakeboards made from acetylated flakes. *J. Wood Chem. Technol.* **1987**, *7*, 427–4403.
35. Rowell, R. M.; Youngquist, J. A.; Imamura, Y. Strength tests on acetylated flakeboards exposed to a brown rut fungus. *Wood Fiber Sci.* **1988**, *20*, 266–2712.
36. Rowell, R. M.; Bergman, O.; Nilsson, T. Resistance of acetylated wood to biological degradation. *Holz Roh- Werkst* **.2000**, 331–33758.
37. Imamura, Y.; Nishimoto, K. Bending creep test of wood-based materials under fungal attack. *J. Soc. Mater. Sci.* **1985**, *34*, 985–98938.
38. Imamura, Y. K.; et al. Bending-creep tests on acetylated pine and birch particleboards during white-and brown-rot fungal attack. *Pap. Puu* **1988**, *9*, 816–820.
39. Imamura, Y.; Nishimoto, K.; Rowell, R. M. Internal bond strength of acetylated flakeboard exposed to decay hazard. *Mokuzai Gakkaishi* **1987**, *33*, 986–99112.
40. Rowell, R. M. *Worldwide in-ground stake test of acetylated composite boards*; IRG Secretariat: Stockholm. Sweden, 1997; pp 1–7, IRGWP Section 4, Doc. No. IRG/WP 97-40088.
41. Larsson-Brelid, P.; et al. Resistance of acetylated wood to biological degradation. *Holz Roh- Werkst.* **2000**, 331–33758.
42. Hadi, Y. S.; Rowell, R. M.; Nelsson, T.; Plackett, D. V.; Simonson, R.; Dawson, B.; Qi, Z.-J. *In-ground testing of three acetylated wood composites in Indonesia*; Proc. 3rd Pacific Rim Bio-Based Composites Symposium, Kyoto, Japan; Marcel Dekker: New York, 1996; p 370.
43. Bonger, F.; et al. *The resistance of high performance acetylated wood to attack by wood-destroying fungi and termites*; 44th IRG Annual Meeting, Stockholm, Sweden, 2013.
44. Johnson, B. R; Rowell, R. M. *Resistance of chemically-modified wood to marine borers. Mater. Org.* **1988**, *23* (2), 147–156.
45. Rowell, R. M.; Susott, R. A.; De Groot, W. C.; Shafizadeh, F. Bonding fire retardants to wood. Part I. *Wood Fiber Sci.* **1984**, *16* (2), 214–223.
46. Feist, W. C.; Rowell, R. M.; Youngquist, J. A. Weathering and finish performance of acetylated aspen fiberboard. *Wood Fiber Sci.* **1991**, *23* (2), 260–272.
47. Rowell, R. M. Acetylation of Wood: Journey from Analytical Technique to Commercial Reality. *For. Prod. J.* **2006**, *56*, 4–12.
48. Allen, J.; Guinn, T; Dickerson, J. *The International Research Group on Wood Protection*, IRG/WP 11-40543, 2011.
49. Tullo, A. H. Making Wood Last Forever with Acetylation. *Chem. Eng. News* **2012**August6, 22–23.
50. reuters.com. *Reuters*. [Online] July 2, 2012; http://www.reuters.com/article/2012/07/02/idUS46627+02-Jul-2012+RNS20120702. RNS Number: 6037G.
51. reuters.com. *Reuters*. [Online] July 11, 2013; http://www.reuters.com/finance/stocks/ACCS.L/key-developments/article/2791950.

Approval Processes

Chapter 19

ICC-ES: The Alternate Path for Building Code Recognition

Craig R. McIntyre*

McIntyre Associates LLC, P.O. Box 220,
Dayton, Montana 59914-0220, United States
*E-mail: gooddrmc@gmail.com.

Building codes require that either specified products or their equivalents be used in a number of applications. Recognizing that not all products can be prescribed, the building codes allow for alternate materials. These alternates must be shown to be the equivalent to the prescribed material in a number of properties such as quality, strength, effectiveness and durability. This procedure allows for new products to enter the market and be used in building code applications. For preservative formulations, defined procedures discussed herein can be followed to obtain the designation of equivalency from the International Code Council-Evaluation Service.

Executive Summary

U.S. building codes require that either prescribed products or their equivalents be used in a number of applications. There are a number of products that are considered to be "alternates" to the prescribed material and these alternates must be shown to be the equivalent in a number of properties including among others quality, strength, effectiveness and durability.

For preservative formulations with promising efficacy, the first step is to propose a new Acceptance Criteria (AC) or amend an existing one that covers a similar product. Acceptance Criteria are developed for new and innovative products that are not recognized by the code thereby allowing for entrance of products into the marketplace. The AC is accepted at an open meeting of the International Code Council-Evaluation Service (ICC-ES) committee. This meeting is the only public opportunity that opponents or competitors have to

© 2014 American Chemical Society

comment on a preservative system. The AC specifies certain testing regimes which are common to preservative development to demonstrate efficacy and usefulness.

The testing must be done by an accredited institution or be accredited by an agency and once done, the test results are reviewed by independent experts and engineers hired by the proponent for completeness and correctness and that they comply with the relevant ACs. Then the reports are filed with the ICC-ES staff. The staff then reviews the data to ensure that it is satisfactory and that all issues are addressed. Upon approval, the staff issues an Evaluation Service Report (ESR) which indicates that the ICC-ES staff deems the material to be the equivalent of that prescribed in the code. Figure 1 summarizes the procedure.

Introduction

Once a new preservative system is under development, consideration must be given to evaluations with appropriate bodies to allow building code uses. Both the International Code Council-Evaluation Service (ICC- ES) and the American Wood Protection Association (AWPA) provide avenues to allow new preservative systems to be added to their respective domains. Both organizations require generally similar testing regimes but there are significant differences in the scope of and the time frames needed for such testing to be accomplished. Typically, the ICC-ES will accept accelerated tests or shorter time spans for the prescribed tests and the evaluation procedures are discussed fully in this paper.

One question that should be addressed though is "Why is product recognition by the codes important?" In today's open market, there are many treaters who sell wood treated with a non-evaluated preservative system. Yet almost all organizations pursuing new wood preservative systems choose to obtain evaluations from first the ICC-ES and then the AWPA. One must consider why the proponents of new systems pursue such evaluations if it is not absolutely necessary.

Accessing larger markets is the first reason. Historically, non-evaluated preservatives have had very limited distribution in the marketplace. The systems may have a small, regional market or perhaps a small niche market but the major retailers (such as Home Depot, Lowe's, etc.) typically consider non-evaluated systems as too "risky" and unvetted. The large retailers also prefer to have systems that have building code recognition which in turn requires evaluation.

Avoiding some risk is a second reason. The fear is that non-evaluated preservatives would be used in an application where they are not suitable. For example, if the treated wood was used for structural purposes even though structural uses are not recommended, there may be a catastrophic failure. Therefore, non-evaluated preservatives are further confined to non-structural uses such as fences, mailbox posts and the like. Another non-structural use is for decks built close to the ground but generally this application is considered too hard to control.

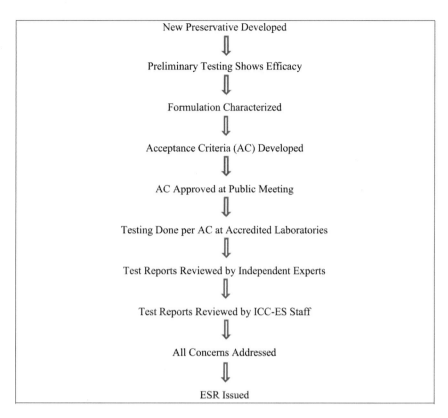

Figure 1. ICC-ES Procedure to Obtain Evaluation Service Report.

A third reason is that in today's litigious environment, producing or selling a non-evaluated product could leave one with a poor legal defense in the event of a catastrophic failure. The product could be deemed to be "not to industry standards" and the producer/seller would possibly have a variety of liability issues. Avoiding these liability issues is seemingly the major reason for evaluation.

The Building Codes

The International Code Council (ICC) is the dominant national code issuing body in the USA and the various ICC codes are essentially the building codes used throughout the USA. For preservative treated lumber, the two codes of interest are the International Residential Code (IRC) and the International Building Code (IBC). Both the IRC and the IBC have various structural applications where treated wood is required to be used. Local jurisdictions typically adopt these codes in whole or augment them with local requirements that address specific issues. For example, earthquake provisions are frequently incorporated in California jurisdictions.

Of major importance, the building codes prescribe materials for specific uses. For example, the codes state that any structural lumber that touches the ground must be preservative treated or a naturally durable species. Other provisions require structural lumber used above ground such as in decks to be preservative treated as well. Furthermore, the preservative system must be listed in the AWPA Book of Standards to meet the prescribed portions of the building codes.

However, the codes have provisions for alternate materials (i.e. alternates to the prescribed materials) to be used provided the alternates are deemed to be equivalent. For preservatives, the alternates must be of similar quality, effectiveness, durability and safety without meaningfully affecting the strength or fire properties. The issues then are what criteria are used to deem the alternate to be equivalent and what body considers the data to discern if the alternate is the equivalent. Fortunately, a well-organized, thorough process has evolved to address these issues within the ICC framework. This process is described further in the following sections.

ICC Organization

In addition to the parent organization, the ICC has three subsidiaries: the International Code Council Evaluation Service (ICC-ES) which will be discussed in detail, the International Accreditation Service (IAS) which will also be discussed and the ICC Foundation. The Foundation sponsors building products research programs and is of no further interest here. However, ICC-ES and the IAS play major roles in the evaluation of new preservative systems.

International Code Council-Evaluation Service (ICC-ES)

The ICC-ES is the organization that provides a mechanism for evaluating new products that are then deemed as equivalents for code uses. Within the ICC-ES procedures, the proponent develops an Acceptance Criteria (AC) that defines the product in a generic manner, specifies the necessary tests used to document the performance of the product and specifies the limitations as needed for the product. The proponent then gathers the necessary test data and has it reviewed by independent experts. The reports then go to the ICC-ES staff for review. Once all issues are resolved, an Evaluation Service Report (ESR) is released. Upon the issuance of the ESR, the product has been evaluated and found to be in compliance with the code. That is, the product has been deemed to be the equivalent of the code recognized product.

An important point is that a proponent has some choices in the performance tests for the product and can use tests that are already underway. For wood preservatives, the range of tests includes numerous efficacy tests as well as strength and fastener corrosion tests. The efficacy and corrosion tests are usually conducted by AWPA standard methods but similar test methods from other organizations such as ASTM or CEN can also be used. In-house testing is also acceptable if it is witnessed by an ISO 17025 accredited third-party agency.

ICC-ES Preservative Evaluation Procedures

Acceptance Criteria

The ICC-ES now has a parent acceptance criteria AC326 (*1*) for preservatives that lists the generic testing requirements in the body of the AC and then the specific preservative systems are listed as daughter appendices. AC326 is patterned after Appendix A of the AWPA Technical Committee Regulations. Both documents specify the extent of the testing necessary for various uses and basically, the body of AC326 is equivalent to the AWPA Appendix A. As mentioned, the various preservative systems are defined in appendices to AC326.

A draft of a new AC or a significant modification of an existing AC is typically submitted about three months prior to one of the three ICC-ES meetings per year which are held in February, June and October. An application fee is required at this time which covers most of the processing charges for an ESR but costs for testing and reviews are separate. The draft undergoes an internal ICC-ES review and then 30 days before the meeting, it is published on the ICC-ES web site (www.icc-es.org).

During the 30-day period and at the ICC-ES meeting, public comments are welcome on the proposed AC. Often these comments are critical of the proposed AC and suggest more strenuous testing or that more restrictions need to be placed on the product. Sometimes, organizations supply their own test data on similar products. The ACs can be and frequently are amended on the floor to accommodate any comments and issues that are raised.

It is absolutely critical that a representative of the proponent be at the meeting who understands the implications of changing the test or test criteria in terms of additional costs and time. The meeting debate is highly technical and there are many references to either code paragraphs or other ACs. The combination of test, code and AC references can sometimes be confusing to the uninitiated so it is best to have a representative who has weathered several AC meetings. Sometimes a seemingly simple change can radically increase the cost or lengthen the overall testing and such changes should be avoided if possible.

After the comments are heard, the ES committee votes on approving the AC. Most of the time, the AC is approved with modifications but there are a few that are approved without change. An additional option is to table the AC for review and at some meetings a significant number of the proposals are tabled. If the AC is approved, then the proponent sets about conducting the various testing specified in the AC.

Testing and Accreditation

The testing specified in the AC that was initiated after July 2004 must be done at a laboratory accredited by the International Accreditation Service (IAS). The inspection agency must also be accredited by the IAS. Table 1 lists a few of the laboratories and inspection agencies that are accredited by IAS and commonly used for preservative efficacy and other wood products testing. Other laboratories can be, and frequently are, used for certain tests.

Table 1. Accredited Testing and Inspection Organizations

Testing Organization	IAS Accreditation Number
Louisiana State University	TL-350
Michigan Technological University	TL-313
Mississippi State University	TL-301
Timber Products Inspection	TL-295

Inspection Organization	
Southern Pine Inspection Bureau	AA-680
Timber Products Inspection	AA-696

The IAS also has special provisions for one-time approvals for specific tests. This provision is usually reserved for accrediting long-term exposure tests that were started before the requirement for accreditation took effect in 2004. This special procedure requires an on-site inspection of the testing facility and the proponent is charged about $3000 to cover fees and travel costs. It may also be possible to accredit shorter term tests under certain circumstances.

Reviews

After the test data is collected, the test institution prepares a report that complies with AC 85 (2). Then the various test reports are submitted to critical reviews by wood preservation experts and professional engineers that are independent of the proponent organization. The reviews discuss the testing protocols, the data itself and the suitability of the data for the intended purpose. The reviewers must also state that it is their opinion that the proposed system would be suitable for the intended applications.

Once the critical reviews are done, the reports and reviews are sent to the ICC-ES for their review. The data is first reviewed by the lead engineer assigned to the file who is typically a senior staff member. For preservatives, most of the reviews are handled by the Birmingham office of the ICC-ES and the lead engineer there has over 25 years of experience with treated wood and actively participates in the AWPA. Once his review is complete, there is a peer review of all of the data and reports by two other ICC-ES engineers at other locations. Usually the same "team" reviews preservative packages and the peer engineers also have 20+ years of experience in preservative matters. After the ICC-ES review is complete, the proponent is provided with a written response that encompasses all of the issues raised in the review.

If there are any anomalies in the data, the proponent is asked for a plausible explanation and if satisfactory, the review is concluded. In some cases though, a test must be repeated or additional testing must be done. If field problems such as early deterioration have been reported with a reasonably similar product, the ICC-ES may request additional assurances that such problems will not occur with the current product. If there are perceptual differences such as attempting to use southern pine as representative of all softwood, then a discussion is needed. And lastly, if issues surfaced during the ICC-ES meeting to adopt the AC, then these may require further responses from the proponent. In short, the proponent must address any deficiencies and all questions and all issues regarding the system.

In general, the ICC-ES review process takes several months to complete after all the testing is submitted. The shortest review time span that is known was about six months. Other reviews have been ongoing for several years but no preservative systems have taken that long. A typical time frame would be 8-10 months.

Quality Control and Inspection

An integral part of the ESR process is that there must be a quality control manual submitted with the test reports and reviews that meets the criteria of AC10 (3). The quality documentation is also reviewed to ensure that the product produced will be essentially the same as that tested.

As well, there must be a third party inspection agency that monitors the production. In 2014, the inspection agencies will be contracted by the ICC-ES instead of to the production facility which will provide another degree of independence to the inspections. There have also been a number of changes to preservative inspecting and reporting procedures made in recent years.

Evaluation Service Report

Once all of the above is in place and the review is complete, the ICC-ES prepares and issues an ESR. When this occurs, the product is deemed to be an equivalent of products recognized in the IBC and IRC. Each of the required properties has been evaluated and the product is considered acceptable to that specified in the code. In short, the product is recognized for building code uses.

At the end of an ESR, there are typically several restrictions placed on the use of the product to prevent its misuse. For example, a product may have a restriction on the type of fasteners that can be used or on the particular use categories for which it is allowed.

The initial ESR is issued with a one year expiration date. After the one year period, the ESR is reexamined for any deficiencies and then reissued with two year reexamination periods. If there are deficiencies, the proponent must address them and there are commensurate fees for the reexamination.

Above Ground and Ground Contact Issues

The ICC-ES makes clear distinctions in the testing required for above ground versus ground contact uses. It is possible within the ICC-ES framework to obtain an ESR that limits the wood to only above ground applications (AWPA UC3B and less). Since accelerated testing is accepted by ICC-ES for above ground uses, the evaluation process can be expedited and the proponent can enter the market in a timelier manner. This above ground evaluation takes about 12 months from the onset of the testing. Typically, the necessary tests would be underway for ground contact uses and the ESR would be amended later when the supporting ground contact data is available.

Appeal Process

The ICC-ES has an appeal process if say a competitor feels that an issue must not have been correctly addressed during the ESR process. The issue must be a technical one and a response is solicited from the proponent. The complaint and response are reviewed by the staff and the issue is resolved by staff actions to disallow the complaint, amend the ESR or in extreme cases, rescind the ESR.

Recent History

In recent years, most preservative suppliers have first commercialized their systems with an ICC-ES evaluation since it allows for building code uses in a timelier manner. The extent of this approach by preservative suppliers is shown by consideration of the major companies that have or have had wood preservative ESRs as shown in Table 2.

Table 2. ESR Holders

EASTMAN

ES+WOOD (ENVIROSAFE)

LONZA WOOD PROTECTION (ARCH)

OSMOSE

PACIFIC WOOD PRESERVING

PLANETSAVER

RIO TINTO

RUTGERS

VIANCE

Similarly, the wood preservative systems that have been or are covered in ESRs are listed in Table 3. These preservatives are essentially all of the commonly available formulations for residential uses and cover a wide range of applications. Some such as PTI are for above ground uses while others such as barrier wrap are primarily used in ground contact.

Table 3. Preservatives with ESRs

ACQ

ACETYLATED WOOD

BARRIER WRAP

BORATES

CA-B, CA-C

EL2

KDS

MICRONIZED CA (MCA)

MICRONIZED CQ (MCQ)

PTI

Summary

Although somewhat intimidating at first glance, the procedures for obtaining an ICC-ES evaluation report for a new preservative system are reasonably structured and orderly. The critical aspects for success are to carefully plan the testing regime, select appropriate tests, select appropriate testing organizations and to start as soon as reasonable.

The ICC-ES evaluation process requires that testing be done at accredited institutions and there must be an approved quality control manual. The evaluation process is an internal peer review process and competitor input is limited to comments at the AC hearing.

The major preservative suppliers have all used the ICC-ES process to introduce new preservative systems in recent years. The preservatives are thoroughly reviewed in this process and deemed to be the equivalent to those recognized in the codes.

References

1. AC326, Acceptance Criteria for Proprietary Wood Preservative Systems—Common Requirements for Treatment Process, Test Methods and Performance, available upon request to ICC-ES (es@icc-es.org).
2. AC85, Acceptance Criteria for Test Reports, available upon request to ICC-ES (es@icc-es.org).
3. AC10, Acceptance Criteria for Quality Documentation, available upon request to ICC-ES (es@icc-es.org).

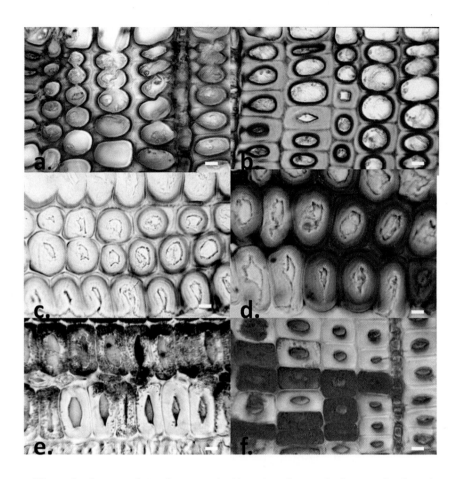

Figure 1. Aspects of simultaneous (a, b) and preferential white rot (c, d) and tunneling (e) and erosion (f) bacteria decay of pine and spruce (e). a) Advanced cell wall thinning of pine leads to early rupture of the tangential cell walls as these are thinner than the radial cell walls; b) Phlebia radiata cell wall thinning with dark brown zones indicating sites of simultaneous attack of all cell wall components and presence of peroxidases; (c, d) Preferential decay by Phlebia radiata Cel 26, after prussian blue staining (c) the rings indicating "in time and space" sites of lignin and hemicellulose attack; d) As for c) but after permanganate staining showing lignin removal; (e) Advanced tunneling bacteria attack developing across all cell wall layers including middle lamellae; f) Advanced bacterial erosion decay of secondary cell walls leaving middle lamellae. Bars: a, b, c, d, e, f, 5.0 μm. (Chapter 2)

Chapter 20

The Development of Consensus-Based Standards for Wood Preservatives/Protectants and Treated Wood Products

Colin A. McCown*

American Wood Protection Association, Post Office Box 361784, Birmingham, Alabama 35236-1784 U.S.A.
*E-mail: mccown@awpa.com.

Consensus-based standards for wood preservatives and treated wood products have been developed by the American Wood Protection Association (AWPA) for over a century. AWPA's standards development process is open to all persons and allows for consideration of all viewpoints. The AWPA Technical Committee Regulations provide the procedures and basic framework for the standards development process, leading to promulgation of credible wood protection standards that are relied upon by engineers, architects, governments, builders, and consumers in the United States and around the world.

Introduction

Standards play a significant role in society, but we are rarely aware of their value. When was the last time you considered the color of traffic signals and signs, or how information is transmitted over the Internet? Standards are very important to society in that they establish minimum levels of quality acceptable to consumers, and often level the playing field between competitors. AWPA's development of standards for treated wood products is subject to an open, consensus-based process which provides due process for all participants. This, in turn, leads to the promulgation of technically competent and reliable standards for use by specifiers of treated wood products.

© 2014 American Chemical Society

AWPA History

The American Wood Protection Association (AWPA) was established in 1904 at the World's Fair in St. Louis, Missouri, by a delegation of individuals engaged in the business of preserving wood, primarily for telegraph poles, railroad crossties, and trestle materials. At their first scheduled meeting, the members set out to create industry standards for preservatives used at that time, which were coal tar creosote and zinc chloride. Since that time over a century ago, AWPA has continued to develop and improve standards for additional preservative systems, methods of analyzing for preservatives, quality control procedures, tests for evaluation of wood preservative performance, and specifications for treated wood products. Another key development was the establishment of the AWPA Use Category System, first promulgated in 1999. AWPA has also expanded its membership to include individuals from diverse interest groups, such as producers of preservatives and components of preservative systems, producers of treated and untreated forest products, end users, engineers, architects, building code officials, government employees, academics, and others interested in wood protection.

Today, AWPA's Use Category System Standards (U1 and T1) are recognized worldwide as credible and reliable standards for preservative treated wood products. AWPA Standard U1 is typically specified by architects and engineers, required by major model building codes, and referenced in material specifications for wood products used in railroad, utility, and other commercial/industrial construction.

Current AWPA Standards

AWPA publishes an annual Book of Standards which contains all standards currently maintained by the Association. The general classes of AWPA Standards are as follows:

U:	User Specifications for Treated Wood
T:	Processing and Treatment
P:	Preservatives/Protectants
HS:	Hydrocarbon Solvents
A:	Analytical Methods
M:	Miscellaneous Standards
E:	Evaluation Methods

At present, there are 172 standards, all of which are under "continuous maintenance". This means that proposals to create new standards, revise or reaffirm existing standards, or to withdraw standards may be made at any time by any person in accordance with the AWPA Technical Committee Regulations. In order to ensure that standards remain up-to date, performance data must be submitted every five years to reaffirm each preservative/protectant standard. All other classes of standards must be either revised or reaffirmed every five years.

ANSI Accreditation

AWPA is an American National Standards Institute (ANSI) accredited standards developer, which means that the procedures which govern AWPA's standards development process have been reviewed by ANSI and are found to be in conformance with the ANSI Essential Requirements. These procedures are found within the AWPA Technical Committee Regulations (TCR). TheTCR are maintained by, and are under the jurisdiction of the AWPA Executive Committee, an elected body representing the Association membership and serving as its board of directors. The TCR contains the elements of due process, consensus, public review, consideration of all viewpoints, incorporation of proposed changes to AWPA Standards, and right to appeal by any participant. ANSI itself does not write or review the technical content of any AWPA Standard. ANSI accreditation simply demonstrates that AWPA's standards development process is open, consensus-based, and affords each participant with due process. AWPA's Executive Committee continues to revise the TCR to maintain ANSI accreditation as well as to make the standards development process more efficient and transparent.

AWPA Technical Committees

To develop effective standards, AWPA relies heavily upon the knowledge of the individual members of its Technical Committees. These volunteers donate their time and expertise to the standards development process. Each committee has jurisdiction over specific standards or portions of standards. These members consider and deliberate all standardization proposals submitted to their committee, and ultimately vote to approve or reject each proposal as originally submitted or as modified during the meeting of the committee. The current AWPA Technical Committees are as follows:

P-1:	Preservatives/Protectants Coordination
P-3:	Oilborne and Creosote-Based Preservative Systems
P-4:	Waterborne Preservative Systems
P-5:	Methods for Chemical Analysis of Preservatives
P-6:	Methods for the Evaluation of Wood Preservatives
P-8:	Nonpressure Preservatives
P-9:	Nonbiocidal Wood Protection
T-1:	Treatments Coordination
T-2:	Lumber and Timbers
T-3:	Piles and Ties
T-4:	Poles and Posts
T-7:	Quality Control and Inspection
T-8:	Composites
T-10:	Millwork and Manufactured Wood Products

Membership in a Technical Committee is limited so as to strive for balance and avoid dominance. Except for Committees P-1 and T-1, which are comprised of the officers of all related committees, there can be no majority representation of any single interest group (i.e., User, Producer, or General Interest), nor can there be more than one member representing any single organization. Any individual, whether or not they are a member of AWPA, may apply for Technical Committee membership. Persons with expertise on the subject matter considered by each committee are sought for active participation.

It is also important to note that AWPA has several Special Committees which do not develop standards, but serve important functions within the Association. These are as follows:

    S-2:   Wood Preservation Research
    S-3:   Treated Wood Use, Handling, and Disposal/ Recycle/Reuse
    S-8:   Plant Operations

Data Requirements

It is understood that when evaluating the performance of a wood protection system against biological agents of deterioration, it can be quite challenging to develop a perfect data package to accompany a standardization proposal. With that in mind, AWPA's Technical Committees have published a number of Guidance Documents to help those developing proposals for new wood protection systems and treated wood products. These documents define the Technical Committees' expectations for the type and quality of data necessary for the Committee to evaluate the product's performance. Due to past experience and the anticipated variability in many evaluation test results, the development of static criteria for approval of wood protection systems has proven futile. For this reason, the data requirements set forth in the Guidance Documents may be subject to waiver by the relevant Technical Committee if justification is provided along with the data package. Some of the current AWPA Guidance Documents which provide data requirements or testing protocols are as follows:

A. Data Requirement Guidelines for Listing Wood Preservatives in the AWPA Standards
B. Guidelines for Evaluating New Fire Retardants for Consideration by AWPA
C. Protocol for Standardization of New Millwork Preservative Systems
D. Protocol for Standardization of New Wood Preservative Finishes
E. Recommended Method for Determining the Treatability of a Species for Inclusion in AWPA Use Category System Commodity Specifications for Sawn Material
F. Guidelines for Evaluating Composite Wood Products Preservative Treated Using Nonpressure Processes

G. Surface Applied Wood Preservative Finish Performance Testing Guideline
H. Evaluating Preservatives for Remedial Treatment
K. Data Requirement Guidelines for Solvents Used with Oil-borne Preservatives
L. Data Requirement Guidelines for Listing Chemically Modified Wood with Enhanced Durability in the AWPA Standards
M. Data Requirement Guidelines for Evaluating Performance Enhancing Additives (PEA)

(Note that Guidance Documents I and J are not in this list, as they do not provide specific data requirements or testing protocols.)

The complete text of each Guidance Document is available on the AWPA website (www.awpa.com) and is also published in the AWPA Book of Standards.

General Procedures for Standardization

As mentioned previously, AWPA's standardization procedures are governed by the Technical Committee Regulations (TCR). AWPA Staff and all other persons involved in the standardization process are required to abide by the TCR so as to promote an open, consensus based standards development environment.

Data Development

In some cases, organizations considering standardization of their products will attend the AWPA Technical Committee Meetings to gauge interest. If the organization chooses to engage in the standardization process, they will often request that a Task Group be formed to assist them in developing a proposal for submittal to that Technical Committee. Much of the data required by the appropriate Guidance Document has already been gathered at that point, so the intent of the task group is to look for deficiencies in the data package, and help the organization develop a proposal for standardization. However, most of the organizations familiar with AWPA's procedures may choose to submit a proposal directly to the appropriate Technical Committee without requesting formation of a Task Group.

Preservatives Review Board

The AWPA Preservatives Review Board (PRB) is an optional procedure which was designed to assist manufacturers of new preservative systems in developing a data package and proposal for submission to the appropriate AWPA Technical Committee. It is intended to take the place of a Task Group if the manufacturer prefers expert review of a data package in an expedited manner. The PRB is beneficial to manufacturers who are unfamiliar with AWPA's standardization procedures and are concerned about the relatively slow rate at which Task Groups operate. This procedure is optional, so a preservative manufacturer is still free to request formation of a Task Group if they so choose. The PRB is randomly selected from a pool of qualified reviewers, and is required to generate a report within a specific time frame. While AWPA charges no fees to promulgate standardization of any product, the cost for this optional review is currently $3500 for the review itself, plus a $500 administrative fee. The PRB procedures are covered in AWPA Guidance Document J, which is published in the AWPA Book of Standards and on the AWPA website.

Proposal Submission

The proponent would first obtain a copy of the standardization proposal form from the AWPA website. A proponent can be any person, whether or not they are members of AWPA or a Technical Committee. The proponent would complete the fields for contact information, the type of proposal and a brief description of and rationale for the proposal. In the "Proposal" section, the proponent would show exactly how a current standard is being revised by using strikethrough text for deletions, and underlined text for insertions. For new standards, a complete version of the proposed new standard without any formatting marks is inserted in the form or attached as a separate document. Finally, all relevant data must be included with the form. The proposal form and data package must be submitted to AWPA at least 60 days prior to the commencement of the Technical Committee Meetings, which take place in September of each year, when all AWPA Technical and Special Committees are required to meet. Time sensitive or urgent proposals may be submitted 60 days prior to the AWPA's Annual Meeting, which occurs in the spring of each year. If proposals for the Spring cycle are received on time, the committee to which the proposal is made will meet in conjunction with the Annual Meeting.

Technical Committee and Public Review

All proposals are then posted to the appropriate committee web page in the member-accessible section of the website as soon as possible after they are received, but normally at least 50 days prior to the commencement of the Technical Committee meetings, or AWPA's Annual Meeting if any time-sensitive or urgent proposals have been submitted. Notification is made to all committee members, and a summary of proposals is posted to the AWPA website. All persons are advised to provide any comments and/or questions to the proponent at least 20 days prior to the technical committee meeting.

Technical Committee Meetings

During the Technical Committee meetings, each proponent is provided an opportunity to briefly present their proposal, answering any questions and responding to any comments made during the meeting. If deemed necessary by the committee, modifications to a proposal can be made from the floor. If a member of the committee makes a motion to adopt the proposal, whether as originally proposed or as modified, and at least two-thirds of the committee members present and voting at the meeting approve the motion, the proposal moves forward to letter ballot of the full committee.

Letter Ballot

The letter ballot is developed by AWPA staff for each proposal approved during the committee meetings. The ballot form itself, along with a combined supporting data file for all proposals is posted to a member-accessible page on the AWPA website. The ballot forms are disseminated to all committee members, and the ballot is open for at least 30 days, but will remain open until at least a 60% return rate is achieved. All ballot items with no negative votes or objections move forward to procedural review and final action by the AWPA Executive Committee. Ballot items not receiving at least two-thirds affirmative votes are submitted to the Executive Committee for final action and a hearing of appeals, if any. Should a ballot item receive at least two-thirds affirmative votes, but also has negative votes and/or objections, these items are subject to resolution and disposition procedures.

Resolution and Disposition of Objections

If there are any public review comments objecting to a proposal, or if there are any negative votes on a proposal, an effort is made by the Technical Committee chair to resolve these comments. Since the TCR does not permit substantive changes to proposals after the meeting of the Technical Committee, only those objections which would result in an editorial change tend to cause the negative voter and/or objector to change their vote or withdraw their comment. In many cases, objectors simply decide that their concerns were very minor and change their vote at this time. In both cases, the objections are deemed "resolved". Those persons strongly objecting to a proposal tend to maintain their negative vote or objection, since no substantive changes are permitted. In this case, the objections are considered "unresolved" and the ballot is recirculated to the appropriate Technical Committee.

Recirculation Ballot (Re-Ballot)

Each unresolved objection is attached to a recirculation ballot and sent to all committee members giving them an opportunity to respond, reaffirm, or change their vote on the original letter ballot. At the end of the recirculation period, each proposal on the ballot must be approved by two-thirds of members voting on the ballot item. If the ballot item passes with at least two-thirds in favor of a proposal, all unresolved negative voters and objectors are then notified that an appeals process exists within the TCR. If the ballot item fails, then the proponent is notified that they may resubmit their proposal, with or without modifications, in a future standardization cycle.

Procedural Review and Final Action

The final step in the process is procedural review and final action by the AWPA Executive Committee. Any appeals received by the deadline are considered, and a determination is made as to whether or not all of the procedural aspects of the TCR were followed. If so, the Executive Committee takes final action on the proposals. The approved revisions or additions to the AWPA standards are fully promulgated at this time and are subsequently published in the latest annual edition of the AWPA Book of Standards.

Conclusion

The development of standards for the protection of wood from biological agents of deterioration isn't always easy. Fortunately, AWPA provides well-structured procedures in an open forum where the performance of products can be reviewed by nearly all of the experts in the field of wood protection in North America. There are no financial barriers to participation in the process, and the views of all persons must be considered. This results in stringent but effective standards which have served as the foundation for the treated wood industry since 1904.

Global Trends

Chapter 21

Wood Protection Trends in North America

Tor P. Schultz,*,[1] Darrel D. Nicholas,[2] and Alan F. Preston[3]

[1]Silvaware, Inc. 303 Mangrove Palm, Starkville, Mississippi 39759
[2]Department of Sustainable Bioproducts/FWRC,
Mississippi State University,
Mississippi State, Mississippi 39762
[3]Apterus Consulting, The Sea Ranch, California 95497
*E-mail: tschultz.silvaware@bellsouth.net.

The economic value of wood protection in North America is about 60 percent of the total worldwide market. Major recent and near term developments will likely occur mainly in the large residential market, with totally organic systems receiving more attention along with chemically-modified wood. The smaller industrial market may experience only minor additional restrictions in the near term, but several alternative systems could be developed from existing biocides. Protection of non-woody products will likely remain a relatively small market in the near term in North America. Long term trends are more difficult to predict, but we expect: 1) it is likely that the traditional three major chemicals historically employed for industrial and agricultural applications will at some point face major long-term restrictions, and 2) enacted governmental regulations will have more of an effect in new product development, and override economic considerations, in residential wood protection. The many favorable environmental, economical, and aesthetical benefits of treated wood strongly suggest that the wood protection industry will continue to be economically viable in the coming years.

© 2014 American Chemical Society

Prior Reviews

A recent review and references therein on wood protection in North America (*1*) provides a good review of the regional status in the prior decade. Another recommended review (*2*) discusses weathering of treated lumber in residential outdoor applications and possible technologies to minimize weathering degradation.

Industrial Wood Preservative Systems

The three major current industrial wood preservatives are creosote, pentachlorophenol (penta) formulated with a heavy oil carrier, and waterborne chromated copper arsenate (CCA). As discussed below, these major systems will likely continue to be employed in the near term in North America, but some usage restrictions have already been enacted and further bans will undoubtedly occur. Another chapter in this book fully discusses industrial systems. Thus, possible changes and alternatives for industrial systems are only briefly covered in this review.

Oilborne copper naphthenate is already being used to some extent as a creosote replacement. This system might further replace creosote in non-aquatic uses, such as railroad ties, if creosote restrictions are enacted in some locations. Another relatively recent creosote formulation involves treated railroad times with borate and creosote, either as a single or dual treatment. These systems enhance creosote's efficacy by having a co-biocide that can diffuse into the untreated center of railroad crossties, and the hydrophobic creosote in the outer shell reduces the leaching of the water soluble borate.

There are already some bans in a few states on creosote marine pilings (*e.g.* (*3*, *4*),). While wood preservative alternatives have been sought to replace creosote in marine applications, at the present time we are not aware of any suitable replacement.

A possible penta and/or creosote alternative for non-aquatic applications is employing the isothiazolone 4,5-dichloro-2-n-octyl-4-isothiazolin-3-one (Kathon 930™, DCOI), or the azole tebuconazole with a co-added insecticide and perhaps a second fungicide, with both of these organic systems formulated with a heavy oil carrier. A DCOI/heavy oil system has shown good long-term efficacy against both decay fungi and termites in field tests (*5*), and we expect that tebuconazole formulated with a heavy oil carrier and a co-added organic insecticide for termite protection would also be effective in field stakes. DCOI and tebuconazole are also active components in two American Wood Protection Association (AWPA) standardized above-ground waterborne residential preservative systems discussed below.

Both penta and copper naphthenate have been traditionally formulated with heavy oil petroleum-based carriers. Recently, some formulations have employed biodiesel, which are plant or animal fatty acids modified by forming methyl esters. These formulation changes were apparently promulgated with less study of possible negative effects prior to commercialization than has historically been the practice for industrial treatment products. A recent study (*6*) has suggested that employing biodiesel may result in increased depletion and/or poorer efficacy with pentachlorophenol. At the present time it appears that copper naphthenate is no longer being formulated with biodiesel, and some treaters who are formulating penta with biodiesel may be also adding the antioxidant BHT to address potential problems with the biodiesel carrier.

In a few locations, principally in the northwest, some restrictions have existed for the past few years which ban use of wood treated with all copper systems in sensitive aquatic environments (*7*). More restrictions for copper based preservatives used in aquatic environments may be enacted. At the present time no alternative organic biocide system appears suitable for freshwater aquatic applications, although wood modification may provide a long term solution in this regard.

Residential Wood Preservative Systems

Copper-Based Systems

Copper-based preservatives are currently the major residential systems in North America and employ copper(II), hereafter copper, and smaller amounts of an organic co-biocide(s) to broaden the range of efficacy against other fungi including those tolerant to copper. Another chapter discusses copper-based residential systems, so these systems are only briefly discussed here. The current systems include amine soluble copper quat (ACQ) and amine soluble copper azole (CA), and copper-based systems employing solid copper(II) ground to mostly sub-micron particles which are then dispersed in water, called micronized, dispersed or particulate copper systems (*8*). The initial micronized copper preservative system, micronized copper quat (MCQ), contained the quat didecyldimethyl ammonium carbonate at copper:quat ratios similar to those used in ACQ-D. This was used commercially for about three years and then replaced with micronized copper azole (MCA), with this latter system sometimes formulated employing both micronized and soluble copper.

Other relatively minor residential copper systems include waterborne copper naphthenate (CuN-W), copper HDO [N-cyclohexyldizendiumdioxide] which is also called copper xyligen (CX), and alkaline copper betaine (KDS). We expect copper systems will continue to dominate the residential market in the near term, especially in ground contact applications, unless unexpected governmental policies or regulations restrict their use.

Non Copper-Based Systems

Borates

The borates have long been known to have good anti-fungal and insecticidal efficacies (*9*). Borates are water-soluble but leachable and, thus, are mainly employed in low deterioration exposure Use Class 1 (UC1) and UC2 applications, with the Use Class system fully described in the AWPA Book of Standards (*10*). They have also been used as a co-biocide in other preservative systems such as

CBA-type A [copper borate azole] and CX-type A [copper xyligen]. While a hindrance in some applications, the water mobility properties of borates allow them to be useful in remedial applications, where a surface brush-on borate treatment or insertion of a fused borate rod into a small drilled hole, allows the borate to diffuse into wet lumber to provide anti-fungal protection.

A fixed, non-leachable borate compound or system has been sought by various groups for many years (*9*). However, this goal remains an unfulfilled vision as results have been disappointing in that, paradoxically, a reduction in borate leachability results in reduced fungal efficacy. However, it is expected that borates will continue to be employed for UC1 and UC2 applications and will likely find increasing uses, such as in the southeastern US which has spreading Formosan termite infestations, with some areas requiring that wood products for indoor use be dip-treated with a borate.

Also currently employed to protect wood composites are non-soluble borates, usually powdered zinc borate (*11*). This in-process solid preservative is added to the wood furnish during manufacture of wood composites, such as particle board, oriented strandboard, or wood plastic composites (WPC). As with the soluble borates, powdered borate systems will undoubtedly be increasingly employed as the use of composites grows.

Nonmetallic Preservative Systems

Waterborne non-metallic, organic or carbon-based systems for residential applications have received increasing attention, and two above-ground systems are now standardized by the AWPA (*10*). These include EcoLife2 (EL2) which uses the isothiazolone DCOI along with the insecticide imidacloprid, and the PTI system which employs the azoles propiconazole and tebuconazole as fungicides and imidacloprid as the insecticide. Both of these systems may include water protection agents to provide enhanced decay resistance along with increased dimensional stability to the treated wood in service. In these systems the organic biocides are all formulated as emulsions to provide a waterborne system which is desirable for residential applications. Other non-biocidal additives are also added to increase product appeal and/or provide enhanced properties, such as colorants and water repellents Manufactures will continue to study additives to provide enhanced product performance or increase market share.

The neonicotinoid insecticides, including imidacloprid, are coming under increasing scrutiny for their possible negative effect on bees and have recently been banned in Europe (*12*). If neonicotinoids in treated wood products are restricted in North America they might be replaced by a synthetic pyrethroid.

At the present time no totally organic waterborne preservative system for ground-contact residential applications has been brought forward to the two major standardization (accreditation) organizations in North America. [Chapters in this book describe the process by which these two organizations, the AWPA and International Code Council – Evaluation Service, Inc. (ICC-ES), standardize new preservatives.] While organic preservatives such as penta and DCOI are known to be effective in ground contact when used in heavy oil carriers, oil treatments are unsuitable for residential applications due to the unpleasant petroleum odor and difficulty in painting or staining the wood surface. Developing an effective and economical waterborne ground-contact carbon based system will be challenging as it is apparent that the oil carrier protects the organic fungicides from biodegradation by bacteria and other microorganisms in ground contact, and this protection is difficult to achieve economically when a water carrier system is used. Adding non-biocidal additives, such as water repellents or antioxidants (*e.g.* (*13*),) is one approach, but such concepts currently are uneconomical due to the relatively high levels necessary, especially for ground-contact applications, and will likely remain so. We anticipate that copper will continue to provide the backbone of ground contact preservatives for residential applications unless there is a drastic change in the regulatory environment, something which appears unlikely in the near future.

Naturally Durable Wood

Naturally-durable lumber, such as the heartwood of western red cedar, redwood, and cypress, has long been employed. However, the old-growth trees which provided good protection are increasingly difficult to obtain, due to harvesting restrictions of these species on US Federal Lands and reduced availability of imported durable tropical hardwoods, and the newer fast-grown plantation trees appear to have reduced natural durability (*14*) as well as greater sapwood to heartwood content ratios. In addition, recent efforts within the AWPA to standardize naturally-durable heartwoods were abandoned due to the difficulty in developing test methods to determine if a particular heartwood sample would, or would not, have a recognized performance in service.

Wood Modification

Chemically-modified wood has long been studied but, until recently, has had only minor commercial success (*15*). Methods examined over the past 60 years include resin impregnation, and chemical modification of the lignocellulosic components with reactants such as furfuryl alcohol, acetic and propionic anhydride, and DMDHEU; many of these processes were reviewed previously in several chapters (*16*) with updates in this book. Another commercial modification practiced in Europe involves various thermal treatments where lumber is heated in an inert atmosphere (*17*), called thermal modification.

Depending on the particular process and extent of reaction, by chemically modifying the lignocellulosic structure the fungal enzymatic wood degradation mechanisms are inhibited to give the modified wood good-to-excellent durability. Further, chemically-modified lumber, while relatively expensive, is a premium product which can provide excellent dimensional and wood surface stability and a visually appealing authentic grain for homeowners. Chemically modified wood usually has similar or greater strength properties as the unmodified wood, although embrittlement occurs with thermally modified wood. It is obvious that chemically modified wood offers the potential to address many of the issues facing the treated lumber industry in North America, but this will require a quantum shift in attitudes towards market acceptability and the achievement of real returns for value added properties.

Secondary Protection Systems

Alternative non-biocidal methods to protect wood, besides chemical modification, include plastic wraps for ground-contact timbers which minimize water uptake such as described in AWPA (*10*) Standard P20, Barrier Protection Systems. The tough plastic wrap with a bitumen inner coating securely wraps around a treated wood product in ground contact. By minimizing the water uptake to reduce fungal deterioration, and preventing underground termite attack by the physical barrier, a reduced biocide retention is possible.

Protection of Cellulosic Renewable Materials

The major alternative products to solid lumber or wood composite products used by homeowners are wood-plastic composites (WPCs) (*17*). The market for WPCs will likely continue to increase for decking and other above-ground applications, albeit at a slower rate than the dramatic increases observed in the prior decade. To increase consumer appeal some producers are developing decking with enhanced properties, such as brighter colors and modified surfaces to resist grease or chipping. These product enhancements will undoubtedly continue as producers strive to increase market share and/or offer value-added products. The potential for an increase in ground-contact WPCs is unclear as this product is very expensive and currently has minimal use in this application.

When first sold, WPC decking was assumed to need no biocide to protect the wood fibers/particles. However, it is now clear that decay will occur, although at a greatly reduced rate compared to solid lumber, and some mold will also grow into the surface of WPC decking. Mold growing into the surface of WPC materials can occur when recycled plastics are used that may carry nutrient contaminants which serve as a growth medium for some molds (*18*). At the present time WPC decking are mainly treated with zinc borate as a powder (*11*). Research has shown the organic DCOI biocide, when added to WPCs, provides improved protection against mold (*19*), and this organic biocide may already be employed commercially with WPCs.

WPCs can be manufactured using non-woody lignocellulosic particles or fibers. Other non-wood panel or structural lumber products, besides WPCs, can also be made using non-wood feedstocks, and this is widespread in regions of the world where bamboo or other lignocellulosic fibers are available, such as described in the China chapter in this book. The availability of panel and other products made from non-woody feedstocks North America in the future depends on many factors, including fiber supply, industry innovation, government subsidies for renewable fuels and any resulting by-products. However, we expect these products to have only minimal market share in North America in the near term with the exception of WPCs. This is based on the availability of wood in North America. Additional economic considerations are the lower density and thus higher transportation costs of non-woody materials such as kenaf and corn stover (20), and that these feedstocks which are only harvested in the fall may require storage to supply a manufacturing facility for the remainder of the year. Any non-wood panel or lumber substitute products manufactured would likely be protected with current commercial preservatives.

A non-wood product which we expect to see increased production of is mold control on drywall and other indoor panel products. At the present time some paper coating on drywall is impregnated with chlorothalonil; other drywall products with mold control agents are also likely being developed.

Utilization of Decommissioned/Waste Wood Products

Another non-traditional feedstock for wood composites is the waste wood which is currently disposed of by landfill. This trend, if it occurs, will come about mainly due to limited landfill capacity and high tipping charges, and likely resulting government regulations. Utilization of this decommissioned wood as a fiber source will require that the producer ensures that any prior biocide system be removed to prevent possible health concerns to the potential consumers. This requirement may be difficult to achieve economically, especially with metallic preservatives.

Long-Term Possibilities

Due to rapid changes in technology and governmental policies and regulations it is impossible to accurately forecast near term trends, much less long term possibilities in the development of new biocidal systems to protect wood and other renewable materials from deterioration. However, a few general possibilities can be suggested. Whatever the future holds, it needs to be emphasized that many changes will undoubtedly be due to public perception followed by resulting governmental regulations, with the outcome forcing changes to the current status quo and economic-based considerations.

First, we expect that the use of chemical protective biocides will continue to face public concerns and resulting restrictions. Thus, we anticipate an increasing market share of non-chemical protective processes, such as chemical modification, to continue. Looking further down the road, future processes to preserve wood may involve returning to naturally durable heartwood. For example, it might be possible through various genetic techniques to develop plantation grown softwoods which form heartwood with durability that approaches that of currently treated wood and, further, have the dimensional stability of the heartwood of species such as redwood. Also, genetic modification to raise and modify the lignin content in wood may provide a pathway to form naturally durable sapwood in some plantation species.

The use of solid lumber decking will, we believe, continue to see further replacement by lumber composites which have greater dimensional stability and more uniform strength properties, and which may also be coated to provide UV protection (*2*). Whether these products with real wood surfaces and enhanced performance properties can compete with WPC decking for a substantial portion of the residential market remains to be seen.

While only some restrictions have occurred to date with industrial wood preservatives, and this will likely continue for the near term, the long term outlook appears more clouded. It is likely that the traditional major wood preservative chemicals for industrial and agricultural uses, creosote, penta, and CCA, will at some time in the future face major long-term restrictions. It remains to be seen if alternative chemical preservatives or processes to economically protect wood for the long service life needed by industrial and agricultural users can be developed, or if non-wood alternatives will be employed. Undoubtedly, non-wood alternatives will be examined. However, the question is whether alternatives can be developed which are effective and economical.

Possible Near Term Problems – and Opportunities

As long recognized by industry and researchers, homeowners are increasingly concerned with the visual aspects of treated wood decking; e.g. pleasing color and grain pattern, lack of mold and/or algae growth, and good dimensional and sunlight stability. These sought-after visual aspects for decking properties have been addressed in the past two decades by several approaches, including greatly expanded production of isotropic WPC decking, premium solid lumber decking co-treated with a water repellent for greater dimensional stability, and chemically-modified wood which undergoes minimal dimensional changes upon being subjected to wetting and drying cycles. Other approaches being examined include using more dimensionally-stable quarter-sawn lumber and/or milling a ribbed surface pattern into decking (*21*). Research has also examined the specialized sawing of logs into a "star" pattern and then gluing the sections together to obtain a quarter-sawn wood composite (*22*). Also being examined is the possibility of producing photostable wood surfaces by coatings which contain UV stabilizers such as nano zinc particles which block UV radiation but are invisible to visible light and, thus, is a colorless protectant (*2, 23, 24*).

Until the last decade, no viable alternative products to treated lumber were readily available that sought to provide such properties. The increasing availability and use of both WPC decking and modified wood treated products such as acetylated wood have changed that situation. Starting in the 1970s with the CCA treatment of southern pine, pressure treated lumber has achieved a well-earned reputation for providing reliable long service life in terms of decay and termite protection for both in-ground and above-ground applications. Factors contributing to this include the widespread practice of treating lumber for both in-ground and above-ground applications to ground contact retentions, a rigorously enforced third-party quality control system, and preservative retentions based on many years of on-going field performance tests.

Looking to the future, the expectations are that consumers will expect that treated lumber continues to provide a long and dependable service life, while at the same time offering durable solid wood products that retain the pleasing natural appearance expected by homeowners which competitive materials usually lack. This will require all participants in the supply chain to participate for the mutual benefit of all. Retailers cannot expect producers of treated lumber to provide reliable and desirable products that compete with alternative materials when price differentials between treated lumber and alternative materials remain starkly different. Also, the treated wood industry, that is the chemical suppliers, wood treaters, and third party inspection agencies, must work together to ensure that all lumber is treated to the highest standards in order to maintain treated lumber's well-earned reputation for longevity in protecting against decay and termite attack. This is especially important as the industry has entered an era of rapidly changing residential preservative formulations with retentions being established to minimize chemical loadings in wood, a challenging situation when wood structure can act as a porous ion exchange matrix. While targeted chemical retentions may benefit appearance properties for treated lumber in service, they may also present challenges from specific aggressive decay organisms, should these be encountered in service situations. The industry and scientific community need to work together in order to strike the optimal balance of these factors in setting retentions that provide the desired service life performance, while allowing all industrial participants sustainable returns that ensure the industry's long term viability.

Many opportunities to address any concerns with the future of treated wood exist. In the near-term we believe that the many benefits of treated wood products – economical, manufactured from a sustainable feedstock, visually appealing, and the fact that trees sequester the greenhouse gas carbon dioxide as they grow – mean that the wood protection industry can continue to provide products to benefit mankind. While we expect some changes to occur in the near term and even more dramatic changes in the long term, the wood protection industry will undoubtedly continue to be economically viable.

References

1. Barnes, H. M. In *Development of Commercial Wood Preservatives: Efficacy, Environmental, and Health Issues*; Schultz, T. P., Militz, H., Freeman, M. H., Goodell, B., Nicholas, D. D., Eds.; ACS Symposium Series 982; American Chemical Society: Washington, DC, 2008; pp 583–597.
2. Evans, P. D. In *Development of Commercial Wood Preservatives: Efficacy, Environmental, and Health Issues*; Schultz, T. P., Militz, H., Freeman, M. H., Goodell, B., Nicholas, D. D., Eds.; ACS Symposium Series 982; American Chemical Society: Washington, DC, 2008; pp 69–117.
3. *California Coastal Nonpoint Source Program. Pilings – Treated Wood and Alternatives*; www.coastal.ca.gov/nps/Pilings-Treated_Wood.pdf (accessed January 14, 2014).

4. Abbot, R. *Courthouse News Service*; www.courthousenews.com/2013/03/06/55486.htm (accessed January 14, 2014).
5. Leightely, L. E.; Nicholas, D. D. *Internat. Res. Group/Wood Preservation*, 1990; IRG/WP 3612.
6. Langroddi, S. K.; Borazjani, H.; Nicholas, D.; Prewitt, L.; Diehl, S. V. *Internat. Res. Group/Wood Preservation*, 2012; IRG/WP 12-30584.
7. Abbott, R. *Courthouse News Service*; www.courthousenews.com/2012/08/01/48885.htm (accessed January 14, 2014).
8. McIntyre, C. R.; Freeman, M. H.; Shupe, T. F.; Wu, Q.; Kamdem, D. P. *Internat. Res. Group/Wood*, 2009; IRG/WP 09-30513.
9. Manning, M. J. In *Development of Commercial Wood Preservatives: Efficacy, Environmental, and Health Issues*; Schultz, T. P., Militz, H., Freeman, M. H., Goodell, B., Nicholas, D. D., Eds.; ACS Symposium Series 982; American Chemical Society: Washington, DC, 2008; pp 440–457.
10. *AWPA 2013 Book of Standards*; AWPA, PO Box 361784, Birmingham, AL 35236.
11. Larkin, G. M.; Merrick, P.; Gnatowski, M. J.; Laks, P. E. In *Development of Commercial Wood Preservatives: Efficacy, Environmental, and Health Issues*; Schultz, T. P., Militz, H., Freeman, M. H., Goodell, B., Nicholas, D. D., Eds.; ACS Symposium Series 982; American Chemical Society: Washington, DC, 2008; pp 458–469.
12. Carrington, D. *The Guardian*; http://www.theguardian.com/environment/2013/apr/29/bees-european-neonicotinoids-ban (accessed January 15, 2014).
13. Green F., III; Schultz, T. P. In *Wood Deterioration and Preservation: Advances in Our Changing World*; Goodell, B.; Nicholas, D. D.; Schultz, T. P., Eds.; ACS Symposium Series 845; American Chemical Society: Washington, DC, 2003; pp 378–389.
14. Taylor, A. M.; Gartner, B. L.; Morrell, J. J. *Wood Fiber Sci.* **2002**, *34*, 587–611.
15. Rowell, R. M. *Wood Mater. Sci. Eng.* **2006**, *1*, 29–33 and references therein.
16. *Development of Commercial Wood Preservatives: Efficacy, Environmental, and Health Issues*; Schultz, T. P., Militz, H., Freeman, M. H., Goodell, B., Nicholas, D. D., Eds.; ACS Symposium Series 982; American Chemical Society: Washington, DC, 2008.
17. Militz, H. In *Development of Commercial Wood Preservatives: Efficacy, Environmental, and Health Issues*; Schultz, T. P., Militz, H., Freeman, M. H., Goodell, B., Nicholas, D. D., Eds.; ACS Symposium Series 982; American Chemical Society: Washington, DC, 2008; pp 372–388.
18. Goodell, B. Personal Communication, Virginia Tech, 2013.
19. Shirp, A.; Ibach, R. E.; Pendleton, D. E.; Wolcott, M. P. In *Development of Commercial Wood Preservatives: Efficacy, Environmental, and Health Issues*; Schultz, T. P., Militz, H., Freeman, M. H., Goodell, B., Nicholas, D. D., Eds.; ACS Symposium Series 982; American Chemical Society: Washington, DC, 2008; pp 480–507.
20. Watson, W. Personal Communication, Mississippi State University, 2002.
21. Evans, P. D.; Cullis, I.; Morris, P. I. *For. Prod. J.* **2010**, *60*, 501–507.
22. Sandberg, D.; Sonderstrom, O. *Wood Mater. Sci. Eng.* **2006**, *1*, 12–20.

23. Clausen, C. A.; Green, F.; Kartal, S. M. *Nanoscale Res. Lett.* **2010**, *5*, 1464–1467.
24. Salla, J.; Pandey, K. K.; Srinivas, K *Polym. Degrad. Stab.* **2012**, *97*, 592–596.

Chapter 22

Preservation of Wood and Other Sustainable Biomaterials in China

Jinzhen Cao[1] and Xiao Jiang[*,2]

[1]Department of Wood Science and Technology, Beijing Forestry University, Qinghua East Road 35, Haidian, Beijing, China 100083
[2]New Application Research - Materials Protection, Lonza Inc., 25 Commerce Drive, Allendale, New Jersey 07401 United States
*E-mail: xiao.jiang@lonza.com.

The past, present and future of the preservation of wood and other sustainable biomaterials in China is reviewed and summarized. The utilization of treated wood and methods to protect wood have a lengthy and well documented history in ancient China, while wood preservation in current China is rapidly changing. Chromated copper arsenate (CCA) currently accounts for more than 70% of treated wood market but it is being phased out for more environmentally friendly wood preservatives. Because of the limited natural forest resource in China, presently wood preservation is primarily used for imported and fast grown plantation lumbers. Two unique local sustainable biomaterials, rubberwood and bamboo, are discussed on availability, treatments and applications. The most recent status on wood preservation standardization in China is reported, but currently wood preservation standards in China are only recommended.

Biological Hazard for Wood in China

Climates and Forests in China

China is located in the east of Asia. The land area of China is about 9.6 million square kilometers, ranking the third largest nation in the world. The spans for both latitude and longitude are fairly broad, resulting in various climate conditions. From south to north, there are five temperature bands; tropic, sub-tropic, warm

© 2014 American Chemical Society

temperate, mid temperate and cold temperate (Figure 1). The vertical temperature zone on the Qinghai-Tibet Plateau has a unique climatic zone. From the southeast to the northwest, the annual rainfall tends to decrease with significant differences among various regions. For example, the annual rainfall along the southeast coast may exceed 1500 mm, while that of the northwest inland is less than 200 mm, as shown in Figure 2. According to the annual rainfall, China can be divided into four regions (Table 1).

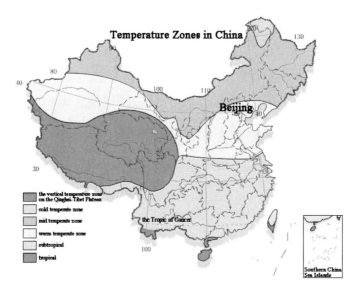

Figure 1. Temperature zones in China.

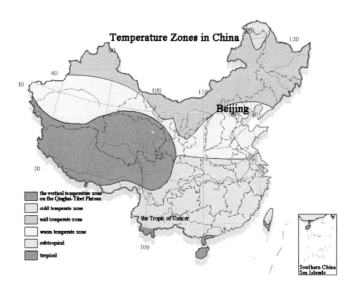

Figure 2. Annual rainfall in China.

Table 1. Classification of Regions in China Based on the Annual Rainfall

Regions	Annual Rainfall (Mm)	Distribution	Vegetation
Humid Region	More Than 800	South China Below The Qinling-The Hwai River, The Eastern Area Of Northeast China, And The Southeast Edge Of The Qinghai-Tibet Plateau	Forest
Semi-Humid Region	400~800	The Northeast Plain, The North China Plain, The Southern Area Of The Loess Plateau, And The Southern Area Of The Qinghai-Tibet Plateau	Forest, Grassland
Semi-Arid Region	200~400	The Eastern Area Of Inner Mongolia Plateau, The Northern Area Of The Loess Plateau, And Most Of The Qinghai-Tibet Plateau	Grassland
Arid Region	Less Than 200	Xinjiang, The Western Area Of Inner Mongolia Plateau, The North-Western Area Of The Qinghai-Tibet Plateau	Dessert

Table 2. Climate Conditions of Typical Cities in China

City	Climate Type	Annual Rainfall	Average Temperature In January	Average Temperature In July
Beijing	Temperate Monsoon Climate	> 400mm	-6°C	28°C
Shanghai	Subtropical Monsoon Climate	> 800mm	2°C	30°C
Haikou, Hainan	Tropical Climate	Around 1600mm	18°C	30°C
Lasa, Tibet	Plateau Climate	> 400mm	-4°C	16°C
Wulumuqi, Xinjiang	Temperate Continental Climate	> 200mm	-14°C	26°C

Considering both characteristics of temperature and rainfall, the climate conditions in China can be classified into five types: plateau climate, temperate continental climate, temperate monsoon climate, subtropical continental climate, and tropical climate (Table 2).

Along with changing climates, the minerals in soil are also different to give various soil types. The climates and soils determine the distributions of forests. The typical forests in China include: a) coniferous forests in the cold temperate zone, e.g., Dahurian larch (*Larix gmelinii*), Mongolia Scots pine (*Pinus sylvestnis* var. *mongolica* Litv.); b) coniferous and broadleaf mixed forests in the mid temperate zone, including softwood species such as Korean pine (*Pinus koraiensis*), Dragon spruce (*Picea asperata* Mast.), fir (*Abies spp.*), and hardwood species such as ribbed birch (*Betula costata*), maple (*Acer spp.*), etc.; c) broadleaf deciduous forests in the warm temperate zone, mainly composed by the species in *Quercus* genus of *Fagaceae* family, e.g., East-Liaoning oak (*Quercus liaotungensis* Koidz.), Sawtooth oak (*Quercus acutissima*), etc.; d) broadleaf evergreen and deciduous mixed forests in the north subtropical zone with a fairly complicated structure of wood species; e) broadleaf evergreen forests in the mid subtropical zone, mainly composed by three families, *Fagaceae*, *Lauraceae*, and *Theaceae*; f) monsoon forests in the south subtropical zone with complicated composition of species and some characteristics of rainforest; and g) rainforests and monsoon forests in the tropical zone.

According to China's 7th survey on forest resources (*1*), the natural forest and plantation forest areas are 119.6925 million and 61.6884 million hectares respectively. The corresponding stocking volumes are 11.402 billion cubic meters and 1.961 billion cubic meters respectively. In October 2000, China formally launched the Natural Forest Protective Project, to be carried out through 2050. The objective of this project is to restore the natural forest resources in China, and require wood industries to mainly consume plantation forests. Wood from the plantation trees is, in general, more susceptible to bio-deterioration than the same species from the natural forests because the chemical compositions are different and less heartwood is formed in plantation wood (*2*). As such, the appropriate preservation treatment of plantation wood is especially important in China. Among the plantation species, pines are the most popular species in coniferous plantations. In southern China, there are mainly Masson pine (*Pinus massoniana*), Slash pine (*Pinus elliottii*), Chinese red pine (*Pinus tabuliformis*), and Latter pine (*Pinus latteri*); In northern China, larch, Mongolian Scots pine, Korean pine, and Masters pine (*Pinus armandii*) are predominant. In plantation broadleaf forests, eucalypts and poplars are predominant.

Biological Hazard for Wood in China

China has a variety of wood-destroying fungi and termites, with white rot more numerous than brown rot fungi. According to Dai et al. and Zeng et al. (*3, 4*), the most common wood decaying species on conifer logs are *Antrodia xantha, Fomitopsis pinicola, Gloeophyllum sepiarium, Laetiporus sulphureus, Trichaptum abietinum*; while the most common decaying fungi on hardwoods are *Abundisporus fuscopurpureus, B jerkandera adusta, Cerrena unicolor, Earliella scabrosa, Funalia trogii, Oxyporus corticola, Phellinus gilvus, Pycnoporus cinnabarinus, Schizopora flavipora, Trametes hirsuta, Trametes ochracea* and *Trametes versicolor*.

Table 3. Types of Termites and Their Distribution in China

Families	Species	Attacked wood type	Distributed area
Rhinotermitidae	Coptotermes formosanus Shiraki	Moist wood, live wood	See Figure 3.
	Reticulitrmes flaviceps (Oshima)	Wood structural elements under the height of 2 meters, such as wood flooring, wood doors, etc.	Mainly in the Eastern China region.
	Reticulitermes chinensis Snyder		Widely distributed in subtropical regions including Sichun, Jiangsu, Henan, Hunan, Shanxi, etc.
	Reticulitermes(Frontotermes) speratus (Kolbe)		Distributed in Northern China, such as Liaoning and Hebei provinces.
Kalotermitidae	Cryptotermes domesticus (Haviland)	Dry wood, such as wood columns, wood flooring, millwork, etc.	Distributed in tropical and subtropical regions, including Hainan, Guangdong, Guangxi, Fujian, and Yunnan provinces.
	Cryptotermes declivis Tsai et Chen		
Termitidae	Odontotermes formosanus	Live tree, buried wood, etc.	Distributed in Hainan, Henan, Jiangsu, Tibet, etc.
	Macrotermes barneyi Light		Distributed in the south area of Changjiang river.

Figure 3. Termite map with northern boundaries for Reticulitermes and Coptotermes.

Figure 4. Biological hazard map in China.

Termites are also a big threat in southern China. Species include *Coptotermes formosanus* Shiraki, *Reticulitermes*, *Cryptotermes*, and *Odontotermes* (Table 3). The termite map and biological map based on the survey of termite activities and Scheffer's Climate Index by Ma et al. (5) shown in Figures 3 and 4. According to the biological map, there are four biological hazard zones: Zone I with low hazard, Zone II with moderate hazard but without termites, Zone III with moderate hazard and termites, and Zone IV with severe hazard.

The biologically hazard conditions for wood products are classified into C1 (interior, dry condition), C2 (interior, moist condition), C3.1 (exterior, above-ground, surface coated), C3.2 (exterior, above-ground, surface uncoated), C4.1 (exterior, ground contact or in contact with water), C4.2 (exterior, ground contact or in contact with water for long-term; critical elements), and C5 (continuous salt water/marine exposure), as regulated in the Chinese National Standard GB/T 27651-2011 (6).

History of Wood Preservation in China

Wood Utilization in Ancient China

Wood was widely utilized in ancient China including construction, transportation, furniture, and musical instruments. The ancient Chinese understood the importance of forest harvest planning. Mencius (372-289 B.C.), an ancient Confucian philosopher, proposed that only the timely harvesting of trees could ensure the continuous supply of wood.

Wood construction achieved magnificent accomplishments. Examples include palaces, temples, pagodas, bridges, mansions, and folk houses. The reasons which Chinese preferred wood for construction can be summarized as follows: a) Wood was easily available and easy to process and transport; b) Wood construction was more efficient and economical than other types of construction based on shorter construction time, and reduced consumption of raw materials and labor; c) Wood construction could adapt to different climate conditions, and had good seismic resistance; d) The practicability of an edifice was the most important consideration for the ancient Chinese as opposed to religious memorability of a structure being most important to western cultures; e) The selection of construction materials was influenced by the theory of Yinyang and five elements in ancient China. According to this theory, wood was a Yang material suitable for living people, while stone belonged to Yin material which was suitable for underground palaces.

Although wood is a vulnerable material to biodegradation and other hazards such as fire, there are ancient wooden constructions over a thousand years old. Typical examples are: the main hall of Nanzenji in Wutai County built in 782 A.D., the Goddess of Mercy Pavilion of Dule Temple in Ji County built in 984 A.D., and the Yingxian Wood Pagoda built in 1056 A.D. The Yingxian Wood Pagoda is the tallest pagoda built entirely from wood, with a height of 67 meters (Figure 5). All of the wood members were connected with mortise and tenon connections, which are thought to be the reason that it has survived seven earthquakes.

Figure 5. Yingxian Wood Pagoda built in 1056 B.C.

Wood Preservation in Ancient China

Selection of Wood

Before Han Dynasty, the Chinese observed that wood harvested during winter tended to be drier, stronger, and more resistant to decay and insects. Lumber cut from dead trees was not recommended for critical frames such as beams and columns, because it might be infected by wood-destroying organisms. The Chinese also recognized that different wood species had different natural durability, with the evidence of different wood species used within one wooden building. For example, in the Forbidden City of Beijing, Nanmu (*Phoebe zhennan*) and Korean larch (*Larix olgensis*) were frequently used for columns, Nanmu, Chinese yew (*Taxus mairei*), Zimu (*Catalpa ovata*) for beams, Chinese fir (*Cunninghamia lanceolata*) for the rafters and sheathings, camphorwood (*Cinnamomum camphora*) for the windowsills, and cypress (*Cupressus funebris*) for members used in moist situations (7). In southern China, Chinese fir was the most widely selected wood species in building materials due to its superior resistance to decay, insects, and deformation. In addition, Nanmu and cypress were used, especially in Sichuan province, because at that time Sichuan had large volumes of these woods (8).

Treatment Methods

There were four main methods used to protect wood in ancient China (*9*). These methods are: surface coating and painting, chemical impregnation, ponding or boiling, and smoking. The first two methods were used most.

Mineral pigments and raw lacquer have long been used in China, and their wide application in wood surface treatment dates back to the Warring States Period. Mineral pigments function as a barrier and fungicide. Raw lacquer can shield wood from air and moisture after forming the film. In case the raw lacquer was not sufficient for protection, toxic mercuric sulfide was usually applied before coating. This technique was also widely used in making wooden utensils, furniture, coffins, etc. According to the investigations on constructions built in the Tang and Song dynasties, exposed wood members were usually coated with both an inorganic pigment, such as ferric oxide, and natural adhesives made from animals or plants. The technique was further developed during the Ming and Qing dynasties with "Dizhang" used as the substrate for the painting. "Dizhang" is composed of difference sizes of brick dust, blood adhesive from processed pig blood, and hemp or cloth. The protection layer on the wood members becomes thick which leads to good water repellency.

In a book written in the 4th century, cupric acetate was suggested as a wood treatment to improve decay resistance. Tung oil was also considered a good preservative, as reported in 6th century. During Ming and Qing dynasties, various chemicals such as ferrous sulfate, boric acid, borax, and sodium chloride were used to impregnate or brush onto wood (*10*).

The ancient folk people in China stored debarked logs in ponds for one to three months, or boiled wood. It was assumed that the wood would become more durable after this process due to the removal of nutrients. In some places, wood was smoked to achieve a better durability.

Design of Wood Structure

The ancient Chinese realized that building design was important to protect wood to prolong the building's longevity. First, broad eaves were employed to protect walls from rain; secondly, wood members were kept from direct contact with the ground; thirdly, air ventilation was designed for wood members embedded within structures. The details of building design for ancient buildings in China can be found in Guo's article (*8*).

Evolution of Wood Preservation

Preservation technologies have evolved over time based on changes in composition of forest resources, application areas, concerns on the toxicity of treating chemicals, and the pursuit of more effective and economical ways to preserve wood. For example, the availability of naturally resistant wood species such as Nanmu has decreased since its uses in ancient China, and the natural durability of the plantation wood is not comparable to that from natural forests.

Before the 1980's, wood preservation in China was only limited to two general products, crossties and utility poles, and rubberwood. The preservatives used for crossties or utility poles were exclusively oil-borne creosote and pentachlorophenol (PCP), although researchers from the Railroad Ministry performed some trials on water-borne wood preservatives, such as chromated copper arsenate (CCA). PCP, alone or mixed with boric acid and borax, was also used for rubberwood treatment (*11*). In November of 2004, PCP was officially prohibited in China, as a result of the global POPS (Persistent Organic Pollutants) of the "Stockholm Convention on Persistent Organic Pollutants", which regulated the phasing out of twelve persistent organic pollutants (*12*). The ban of PCP in China opened an opportunity for more environmentally friendly biocides for wood treatment.

The rapid development of wood preservation in China only began one or two decades ago, with the Chinese government pushing to replace wood crossties with cement to reduce wood consumption (*13*). The crosstie treatment industry switched to small treating plants and started to promote treated wood for residential uses, such as decking and landscaping materials. Recently, there has been a shift to use treated wood in buildings. With the transition of the application fields for treated wood, the preservative systems have also changed accordingly.

Wood Preservation in Modern China

Wood Preservatives Used in China

The wood preservatives currently used in China are: oil-borne preservatives for crossties, copper-containing water-borne preservatives for landscaping and decking materials, and organic preservatives as water-borne emulsions or in organic solvents for rubberwood, bamboo and other woody species. Among these preservatives, copper-containing water-borne preservatives are dominant and employed for the large landscaping and decking market. CCA, alkaline copper quat (ACQ), and copper azole (CuAz) are manufactured in China. Since CCA is not yet prohibited in China, the market share for other water-borne preservatives is very limited due to their high cost compared to CCA. It is estimated that over 70% of landscaping and decking materials in China are treated with CCA, with less than 30% of the market for other water-borne preservatives (*14*).

Organic preservatives are receiving more attention in China. There are three classes of organic preservatives: organic insecticides, organic fungicides, and organic moldicides or anti-sapstain chemicals. If the wood is being claimed as treated with organic preservatives, only the organic preservatives listed in the Chinese National Standard of Wood Preservatives are recommended (*15*). The list of the active organic preservatives is:

- Organic insecticides: deltamethrin, cypermethrin, permethrin, bifenthrin, chlorpyrifos, or imidachloprid;

- Organic fungicides: tebuconazole, propiconazole, chlorothalonil, copper oxine (Cu8), copper naphthenate (CuN), 3-iodo-2-propynyl butyl-carbamate (IPBC), Bis(tri-n-butyltin)oxide (TBTO), and 4,5-dichloro-2-n-octyl-4-isothiazolin-3-one (DCOI).
- Organic moldicides or sapstain inhibitors: chlorothalonil, Cu8, carbendazim, IPBC, didecyldimethylammonium chloride (DDAC), alkyldimethylbenzylammonium chloride (BAC), and propiconazole (PPZ).

The use of auxiliary additives such as water repellents, colorants and photostabilizers are under development. Although there are many well established formulations for colorants developed from advanced wood dyeing technologies in China, their formulations with wood preservatives need further investigation. With the commercialization of organic/metal-free systems, these additives are expected to play a more important future role.

Applications of Preservatives and Preservative-Treated Wood

Treatment Methods

In China, various vacuum-pressure treatments are the most widely used method. With water-borne preservatives the conventional full-cell process is employed. The impregnation is performed at room temperature and elevated pressures in combination with vacuum. The applied pressure varies depending on the treating solution and treatability of the wood species. Non-pressure treatments such as spraying, brushing or dipping are not used frequently and only limited to surface treatment for temporary protection.

In addition to preservative treatment, other types of treatments, such as thermal treatment and phenol-formaldehyde (PF) impregnation, are also used in China. Thermally-treated wood is well recognized in China because of the improved dimensional stability and attractive dark brown color after treatment. The dark brown color, representing elegance in China, is one of the selling points of the thermal wood in the Chinese wood market. At the end of last century, China began to import thermally treated wood from Finland, and thermally treated wood was originally called "Finland Wood" in China. The imported thermally treated wood had only a very small market share in China because of the high cost. In 2003, several wood treaters in China started to thermally-treat wood using equipment originally designed for preservative impregnation. In recent years, thermal treatment technology has been fully developed and commercialized in China (*16*).

PF impregnation is usually used for the treatment of poplar wood, a fast growing species, to improve its durability and, equally importantly, the strength of the wood. Like the conventional process for PF impregnation, low molecular weight PF is impregnated into wood and then cured at elevated temperatures. The PF impregnation method is mainly used in northern China where poplar wood is readily available.

Wood Species Used for Treatment

As mentioned earlier, forest coverage in China is small and most of the natural forests are protected under the Natural Forests Protective Project. Although China has significant stocking volume of plantation forests, the utilization of these wood resources is still under development. As such, the wood supply for preservative treatment largely depends on imported wood. This includes Mongolian scots pine from Russia, southern pine and Spruce-Pine-Fir (SPF) from North America, scots pine from Europe, and radiata pine from New Zealand. China also imports other wood species with good natural durability, for example, western red cedar (*Thuja plicata* Donn) from Canada, Merbau (*Intsia biujga*) and Bangkirai (*Shorea laevis*) from Southeast Asian countries. Suitable treatments are occasionally applied to the sapwood of these species.

The domestic wood species used for preservative treatment include Chinese fir, Masson pine, poplar, and rubberwood. Chinese fir, a special wood species widely planted in Southeast China, belongs to the genus of *Cunninghamia* in the family of *Taxodiaceae*. It grows fast and straight, and has a high ratio of strength to weight. The average specific gravity is about 0.39. The heartwood of Chinese fir is resistant to decay and insects. According to the Chinese National Standard GB/T 13942.1-2009 (*17*), the heartwood of Chinese fir is rated as durable among highly durable, durable, slightly durable, and nondurable heartwoods (*18*). That is why Chinese fir has been used in ancient Chinese constructions in the history. Aspirated bordered pits are very common in the heartwood. Therefore, treatability of Chinese fir heartwood is very poor.

Masson pine belongs to the genus of *Pinaceae*. It has straight grain and a coarse structure with an average specific gravity between 0.39-0.49. Masson pine has long wood fibers, making it a good raw material for pulping and paper industry. It has numerous big resin canals, which often result in "bleeding" and which directly influences the coating and adhesion properties. Masson pine is generally classified as a typical "nondurable" wood species due to its large portion of the nondurable sapwood and also its vulnerability to sapstain fungi, although the heartwood of Masson pine is rated as durable (*19*).

Poplar covers all the wood species in the genus of *Populus*. In China, there are more than fifty poplar wood species including *Populus ussuriensis* Kom., *Populus suaveolens* Fisch., *Populus koreana*, *Populus davidiana* Dode, *Populus alba* var. *pyramidalis* Bunge, *Populus tomentosa Carr.*, *Populus euphratica*, and also a series of natural or hybrid *Populus nigra* Linn., which is being introduced from Europe and North America. The plantation poplars grow rapidly in most regions of China, but have many drawbacks such as a poor visual figure or appearance, low density giving inferior hardness and strength, an inclination to deformation, and poor decay resistance. Because of these defects, its utilization has usually been limited to raw materials for wood-based composites, pulp and paper, and packaging materials. With the recent development of poplar processing technologies, these deficiencies might be overcome. In various processes poplar lumber is compressed to increase its density, impregnated with resins to improve its density and durability, treated with dyes to optimize the color, or heat treated to improve the decay resistance and dimensional stability. Among all these treatment methods, the strengthening treatments in combination with biocides for durability enhancement are the focus and have good potential.

Figure 6. Rubberwood treated inefficiently with resultant blue stains.

Rubber trees are replanted every 25-30 years when they become uneconomical for latex production. Traditionally, felled rubber trees were mainly used as fuel wood with low commercial value. Since the 1980's rubberwood has become popular as a source of timber, particularly for the furniture industry. In 1951, China began to plant rubber trees to satisfy the demand for latex supply, most particularly in southern China including Hainan, Yunnan, Guangdong, and Guangxi provinces. Two rubberwood species were introduced, namely, *Hevea brasiliensis* and *Ficus elastica*. According to statistics in 2011 (*20*), the plantation area of rubberwood in China was 1.08 million hectares. Up until 2008, lumbers made from rubberwood exceeded 5 million m^3 in total. In recent years, there have been about 0.6 million m^3 of lumber that have been produced from rubberwood every year. Rubberwood is a low density hardwood and easy to treat. The density of rubberwood ranges from 480 to 650 kg/m^3 depending on their age and clonal variation (*21*). Rubberwood is white to light yellow in color with a visually appealing grain that is desired by the furniture industry. The continuous supply of the rubberwood lumbers makes it an important raw material for furniture manufacturing. It is well known, however, that rubberwood sapwood is extremely susceptible to sapstain, mold fungi and insect attack because of a high level of starch, sugar, proteins and other nutrients. In the past, rubberwood was usually treated with PCP to improve durability. Since the ban of PCP, industries have used alternative biocides such as boric acid and borax (*11*). Although boron is effective in inhibiting insects and decay fungi, the mold issue is still a problem, especially during the rainy season (Figure 6). Other commercial moldicides, such as chlorothalonil and carbendazim, have been used in combination with the boron compounds to improve the efficacy against molds, in spite of the recognized compatibility issues with the boron compounds. Finding effective and suitable moldicides for rubberwood treatment is an ongoing research effort.

The treatment of bamboo and its utilization is unique in China. Bamboo is a monocotyledonous plant belonging to the subfamily of *Bambusoideae* in the family of *Graminales*. There are more than 1,200 bamboo species all over the world. China has approximately five hundred bamboo species. The area covered by bamboo is around 22 million hectares in the world, and about 20% is in China. Some bamboo species are of great economic value. The species of *Phyllostachys pubescens*, also known as Mao bamboo, is one of the dominant species in China and occupies about 3 million hectares. Bamboo grows very fast. It becomes mature in six months and is usually harvested after four years. For instance, Mao bamboo can grow about two to three feet in the first 24 hours when spouting from ground in spring. In 2012, the production of bamboo for industrial utilization was 1.644 billion sticks, among which 1.115 billion were *P. pubescens*. Because of the distinctive anatomical structure, bamboo possesses a fairly high tensile strength but low shear strength (*22*). As a typical graminaceous plant, bamboo contains higher percentage of hydroxy-phenyl lignin than wood. A relatively high amount of extractives, ranging from 2.5% to 12.5%, has been found in bamboo depending on the extraction agents (*23*). The extractives of bamboo comprise water-soluble sugars, fats, and proteins, a group of ideal food for the growth of the microorganism. These rich nutritious extractives in bamboo make it highly susceptible to wood decaying fungi, mold fungi and insect attack. Therefore, appropriate chemical or physical treatments are required to improve the biological resistance of bamboo to ensure its utilization in different applications. Physical treatment methods, such as heating, solarizing, boiling, steaming and water soaking, are currently used for bamboo treatment in China. These methods aim to remove the nutritious extractives from bamboo and reduce mold growth. After physical treatments, bamboo needs to be stored in dry conditions to avoid the reinfection of molds. Chemical treatment methods include brushing, dipping and pressure treatment with different fungicides and moldicides. Brushing treatment of the bamboo surface can only provide minimal protection against decay fungi and molds because of the poor penetration. Dip treatment, when applied under elevated temperatures, can improve biocide penetration. Pressure treatments are also used. For chemicals used in bamboo protection, biocides such as CCA, ACQ, CuAz, CuN, borates, chlorothalonil have been recently tested (*24, 25*).

Applications of Treated Wood

The application of preservative-treated wood was almost entirely limited to railroad ties before the 1980's. With competition from concrete crossties, preservative-treated wood began to change to other applications. At present, treated wood used for landscaping and construction is dominant, with a demand of about 1.2-1.3 million m^3 per year, accounting for more than 70% of the total wood treatment market. On the other hand, treated crossties are now less than 10% of the total treated lumbers. Rubberwood treated with moldicides and insecticides are primarily used as raw materials, such as finger joint boards, for furniture industry (Figure 7). Treated bamboos are largely used for flooring and furniture materials (Figure 8). There is a small portion of treated eucalypt lumbers and bamboos that are used in agricultural as support for banana trees and other vegetable greenhouses. A small amount of treated lumbers is also used in the restoration of ancient wood constructions. In recent years, many historical wood constructions were restored with support from the Chinese government, such as the Potala Palace in Lhasa, Tibet, and the Tiananmen Gate in Beijing. Because of a low forest resource and a growing concern for the environment, the use of treated wood as a renewable material will certainly receive more and more attention in China.

Figure 7. Finger joint boards made from treated rubberwood.

Figure 8. A living room decorated with treated bamboo.

Standardization of Wood Preservation in China

The standardization of wood preservation in China is a fairly recent development, using similar standards from the USA, Canada and Europe as references. There are three main types of standards regarding wood preservation in China, namely, GB standards representing the Chinese national standards, LY standards issued by the Ministry of Forestry, China, and SB standards issued by the Ministry of Commerce, China. There is also another YB standard concerning creosote (YB/T 5168-2000) (26), which was issued by the Ministry of Metallurgical Industry, China. Standard with "/T" means the standard is recommended, not mandatory. The GB and LY standards regulate mostly fundamental, evaluative, and analytical standards. For example, GB/T 14019-2009 "Glossary of Terms Used in Wood Preservation", GB/T 27651-2011 "Use Category and Specification for Preservative-Treated Wood", GB/T 27654-2011 "Wood Preservatives", GB/T 23229-2009 "Methods for Analysis of Waterborne Wood Preservatives", GB/T 13942.1-2011 "Durability of Wood - Part 1: Method for Laboratory Test of Natural Decay Resistance", GB/T 13942.2-2011 "Durability of Wood - Part 2: Method for Field Test for Natural Durability", and GB/T 27655-2011 "Method of Evaluating Wood Preservatives by Field Tests with Stakes". The evaluation for the laboratory and field leaching test methods have been well established in the standard. In GB/T 27651-2011, the penetration depth was regulated only for sapwood, but not for heartwood due to lack of data. In addition, the GB/T 23229-2009 only presents the analytical methods for the preservatives that are widely used in China. The standards for analyzing other preservatives, especially the organic wood preservatives, have not yet been well established.

Tables 4 and 5 are examples of the national standards regarding the water-borne wood preservatives and their required retention levels corresponding to different use categories.

Table 4. Water-Borne Wood Preservatives Listed in GB/T 27654-2011

Preservative	Type	Effective ingredients
Chromated copper arsenate (CCA)	CCA-C	CrO_3: 47.5%; CuO: 18.5%; As_2O_5: 34%
Alkyl ammonium compounds (AAC)	AAC-1 AAC-2	DDAC≥90%, dodecyl ammonium chloride or dimethyldioctyl ammonium chloride≤10%; BAC≥90%, other AAC≤10%.
Boron compounds (BX)		DOT, boric acid etc. and their compounds
Alkaline copper quat (ACQ)	ACQ-2 ACQ-3 ACQ-4	CuO: 66.7%, DDAC: 33.3% (using aqueous ammonia as solvent) CuO: 66.7%, BAC: 33.3% (using aqueous ammonia or amine as solvent) CuO: 66.7%, DDAC: 33.3% (using amine as solvent)
Micronized copper quat (MCQ)		CuO: 66.7%, $DDACO_3$: 33.3%
Copper azole (CuAz)	CuAz-1 CuAz-2 CuAz-3 CuAz-4 CuAz-5	Cu: 49%, H_3BO_3: 49%, Tebuconazole: 2% Cu: 96.1%, Tebuconazole: 3.9% Cu: 96.1%, Propiconazole: 3.9% Cu: 96.1%, Tebuconazole: 1.95%, Propiconazole: 1.95% Cu: 98.6%, Cyproconazole: 1.4%
Copper citrate (CC)		CuO: 62.3%, citric acid: 37.7%
N′-hydroxy -N-cyclohexyl-diazenium oxide (CuHDO)		CuO: 61.5%, HDO: 14%, H_3BO_3: 24.5%
Propiconazole Tebuconazole Imidacloprid (PTI)		Tebuconazole: 47.6%, Propiconazole: 47.6%, Imidacloprid: 4.8% (can be omitted in termite-free situations or be replaced by dimethrins)

$DDACO_3$: didecyldimethylammonium bicarbonate/carbonate; DOT: Disodium octaborate tetrahydrate.

Table 5. Required Minimal Retentions of the Effective Ingredients for Different Wood Preservatives while Used in Different Use Categories (Reproduced from GB/T 27651-2011)

Use categories	Minimal retentions of effective ingredients (unit in kg/m^3)														
	CC-A-C	BX	MCQ	ACQ-				CuAz-				CC	CuHDO	PTI	TEB **	Cu8 ***	CuN ***
				2	3	4	1	2	3	4						
C1	NR	2.8	4.0	4.0	4.0	4.0	3.3	1.7	1.7	1.0	4.0	2.4	0.21	0.24	0.32	NR
C2	NR	4.5	4.0	4.0	4.0	4.0	3.3	1.7	1.7	1.0	4.0	2.4	0.21	0.24	0.32	NR
C3.1	4.0	NR	4.0	4.0	4.0	4.0	3.3	1.7	1.7	1.0	4.0	2.4	0.21	0.24	0.32	0.64
C3.2	4.0	NR	4.0	4.0	4.0	4.0	3.3	1.7	1.7	1.0	4.0	2.4	0.29*	0.24	0.32	0.64
C4.1	6.4	NR	6.4	6.4	6.4	6.4	6.5	3.3	3.3	2.4	6.4	3.6	NR	NR	NR	NR
C4.2	9.6	NR	9.6	9.6	9.6	9.6	9.8	5.0	5.0	4.0	NR	4.8	NR	NR	NR	NR
C5	24.0	NR	NR	24.0	NR	NR	NR	NR	NR	NR	NR	NR	NR	NR	NR	NR

Note: NR: not recommended. *: replaceable by 0.21+water repellent. **: retention based on tebuconazole. ***: retention based on copper. The effective ingredients of all other preservatives are listed in Table 4.

Future Expectations

The demand for treated wood has been increasing. It is anticipated that the treated wood market in China will become significant within Asia and perhaps eventually as a global influence (*14*). At present, CCA is still the dominant wood preservatives in China. However, some measures have already restricted CCA use. A national standard titled "Code for use of CCA-treated wood" has been drafted and discussed in 2012, though it has not been formally accepted. In this code, CCA is proposed to be prohibited for not only C1 and C2 use categories, but also for above-ground applications, namely, C3.1 and C3.2 use categories. In addition, all residential uses which have direct human contact have been proposed to be banned. It should be noted, however, that the proposal is only a recommended standard, not a mandatory one. It is believed with the limitation of CCA use, more opportunities will be available for developing alternative wood preservatives.

The development of economic, effective, and environmentally friendly wood preservatives is a world-wide pursuit in wood preservation, but the research focuses depend on the particular country due to the variability in local climate, type of wood species, and the region's culture. The following aspects are believed to be the focus for the wood preservation research in China: a) organic biocides in both oil-borne and water-borne formulations; b) cost effective moldicides for bamboo, rubberwood and thermally-treated wood; c) multi-functional wood protection system in wood construction; d) restoring technologies for traditional wood constructions; and e) wood preservative technologies for wood-based or other bio-based composites.

In 1970's, China proposed to replace some wood utilization with bamboo due to limited wood resources. With the build-up of the first production line of sliver plybamboo in Sichuan province in 1975 (*27*), numerous products based on bamboo were developed or created. These products were widely used in both outdoor and indoor environments. As a result, bamboo industries have developed rapidly. As mentioned in the previous section, bamboo preservative treatment is especially challenging. Although a variety of biocides have been examined for bamboo treatment, the bamboo preservation technology is still under development.

Thermally treated wood was successfully introduced in China because of the enhanced durability and the attractive darkened color. However, mold growth as well as the termite attack on thermally treated wood, especially in southern China, posed obstacles for its application. In addition, color stabilization and prevention of color fading for thermally treated wood for outdoor applications is another area that needs to be studied.

Wood protection technologies for both ancient and modern wood constructions are required. For ancient wood constructions, new restoration technologies are needed urgently. For example, for those columns in a historical construction, an in-situ remediation is needed. Wood constructions seemed to phase out in China during an extensive period due to the limited availability of wood resources. With the development of wood preservation technologies and advanced wood structure design, the wood construction has been returning to the market recently in China. Therefore, it is necessary to develop technologies that can provide multiple protections of decay, mold and insect attack. Wood fire protection treated with the fire retardants is equally important.

Wood-based composites have high production in China, and their applications are mainly for indoor decoration or furniture materials. The structural or exterior applications for wood-based composites or other bio-fiber based composites have relatively short history in China, but the market for structural or exterior use is increasing quickly with more use of renewable resources including wood and bamboo. For example, oriented strand boards (OSBs) based on wood or bamboo manufactured in China are used as building materials. Therefore, the methods to improve the bio-efficacy against wood destroying fungi, molds and insects for wood-based composites and other bio-fiber based composites, are currently being studied.

References

1. China Ministry of Forestry. *Report of the 7th survey on forest resources (2004-2008)*; 2012.
2. Liu, L.; Shu, H.; Liang, L.; Zhang, Y. *Guangdong For. Technol.* **2004**, *20* (4), 17–20.
3. Dai, Y.; Xu, M.; Yang, Z.; Jiang, M. *For. Res.* **2008**, *21* (1), 49–54.
4. Zeng, X.; Cui, B.; Xu, M.; Piao, C.; Liu, H. *For. Res.* **2008**, *21* (6), 783–791.
5. Ma, X.; Wang, J.; Jiang, M.; Li, X. *Sci. Silvae Sin.* **2011**, *47* (12), 129–135.
6. Chinese National Standards Management Association; *Chinese National Standard GB/T 27651-2011*; 2011.
7. Team of Maintenance of Wood Structural Members in Ancient Constructions. *Sichuan Building Sci.* **1994** 1, 11−14.
8. Guo, H. *Ancient Wood Constructions*; China Building Industry Press: Beijing, China, 2004; p 20.
9. Deng, Q. *Architecture Tech.* **1979**, *10*, 62–65.
10. Chen, Y.; Li, W. *Restoration of Ancient Wood Constructions and Wood Cultural Relics*; China Forestry Press: Beijing, China, 1995; pp 5−7.
11. Li, J.; Lin, W.; Xie, G.; An, F. *Chin. J. Trop. Agric.* **2010**, *30* (3), 67–70.
12. Jiang, M. *China Wood Ind.* **2006**, *20* (2), 23–25.
13. Li, Y.; Cao, J.; Tian, Z. *A Guide to the Uses of Treated wood*; Chinese Building Industry Press: Beijing, China, 2006; pp 86−89.
14. Preston, A.; Jin, L. In *Development of Commercial wood Preservatives, Efficacy, Environmental and Health Issues*; Schultz, T., Militz, H., Freeman, M. H., Goodell, B., Nicholas, D. D., Eds.; ACS Symposium Series 982; American Chemical Society: Washington DC, 2008; pp 599−607.
15. Chinese National Standards Management Association; *Chinese National Standard GB/T 27654-2011*; 2011.
16. Gu, L.; Ding, T. *Wood-Based Panels in China* **2008**, *9*, 14–18.
17. Chinese National Standards Management Association; *Chinese National Standard GB/T 13942.1-2011*; 2011.
18. Ma, X.; Yang, Z.; Jiang, M.; Liu, L.; Su, H. *China Wood Ind.* **2009**, *123* (13), 34–36.

19. Zhou, M. *Sci. Silvae Sin.* **1981**, *2*, 145–154.
20. China Ministry of Forestry. *Annual Statistic Report of Forests in China*; 2011.
21. Lim, S. C.; Gan, K. S.; Choo, K. T. *Timber Tech. Centre* **2003**, *26*, 1–11.
22. Pan, B. In *Wood Science*, 2nd ed.; Liu, Y., Zhao, G., Ed.; China Forestry Press: Beijing, China, 2012, p 273.
23. Xu, Y.; Hao, P.; Liu, Q. *J. Northeast For. Univ.* **2003**, *31* (5), 73–76.
24. Sun, F.; Duan, X. *World Bamboo Rattan* **2004**, *2* (4), 1–4.
25. Li, N.; Chen, Y.; Bao, Y.; Wu, Z. *J. Central South Univ. For. Technol.* **2012**, *32* (6), 172–176.
26. Ma, X.; Jiang, M.; Duan, X.; Yu, H. *China For. Prod. Ind.* **2011**, *38* (4), 12–15.
27. Xu, J.; Zhao, R.; Fei, B. *Wood Process. Mach.* **2007**, *3*, 39–42.

Editors' Biographies

Tor P. Schultz

Tor P. Schultz obtained a B.S. in wood science from University of Florida and an M.S. in wood technology, followed by a Ph.D. in wood chemistry from North Carolina State University. He was employed 32 years at Mississippi State University and now works for Silvaware, Inc., as a consultant in wood preservation, wood chemistry, and wood science. He is a Fellow in both the ACS/Cellulose Division and the International Academy of Wood Science, has previously co-edited four books in biomass, wood chemistry, and wood deterioration and preservation, is co-inventor of seven patents with one new submission, and has published about 200 articles.

Darrel D. Nicholas

Darrel D. Nicholas received B.S. and M.S. degrees in forest products at Oregon State University, followed by a Ph.D. degree in wood science and technology at North Carolina State University. He has extensive experience in both industry and academia in the areas of wood science and wood preservation and is currently a professor at Mississippi State University. He is a Fellow in the International Academy of Wood Science, has over 100 publications, has served as editor and co-editor for five books on Wood Biodeterioration and Preservation, and has been awarded nine U.S. patents.

Barry Goodell

Barry Goodell has over 29 years of experience in the sustainable biomaterials and wood science and engineering fields, including work in bioconversion and bioenergy, structural biocomposites, and sustainable nanomaterials fields. He holds a Doctorate from Oregon State University and was previously a professor and program leader at the University of Maine in the U.S.A. He also previously served as head of the Department of Sustainable Biomaterials, which was restructured with that name, with two new degrees developed, under Dr. Goodell's leadership. He currently serves as a professor in the department at Virginia Polytechnic Institute and State University (Virginia Tech). Dr. Goodell has published over 100 articles on wood and biomaterials degradation and protection, biochemical mechanisms related to free radical bioconversion processes, engineered wood composites as related to FRPs and PMCs, and the development of novel products including advanced hybrid biocomposites. He also holds four patents and two provisional patents, with other patents pending. He has co-edited three ACS books, and publishes in journals ranging from Science to the Journal of Nanoscience and Nanotechnology to Applied and Environmental Microbiology.

© 2014 American Chemical Society

Indexes

Author Index

Altgen, M., 269
Arantes, V., 3
Averdunk, H., 227
Cao, J., 363
Cookson, L., 159
Daniel, G., 23
Dickerson, J., 301
Diehl, S., 59
Eastwood, D., 93
Evans, P., 227
Ewart, D., 159
Goodell, B., xi, 3, 147
Jiang, X., 363
Kataoka, Y., 227
Kiguchi, M., 227
Kirker, G., 81, 113
Krause, A., 287
Lebow, P., 255
Lebow, S., 239
Limaye, A., 227
Matsunaga, H., 227
McCown, C., 341

McIntyre, C., 331
Militz, H., 269, 287
Morrell, J., 203
Morris, P., 131
Nicholas, D., xi, 255, 351
Preston, A., 351
Rowell, R., 301
Schmitt, S., 217
Schultz, T., xi, 217, 255, 351
Senden, T., 227
Shields, S., 217
Stirling, R., 185
Tang, J., 59
Taylor, A., 203
Temiz, A., 185
Turner, M., 227
Wakeling, R., 131
Winandy, J., 113
Xie, X., 147
Xie, Y., 287
Zhang, J., 217

Subject Index

A

Above ground performance of treated wood
  service life prediction, North American research needs, 127
  standardized test methodologies
    above ground tests, 126
    use class system, 127
  standards for wood preservation, international governing bodies, 127t
Acetylation of wood, 301, 303
  change in volume in wood, 305t
  chemistry, 304
  fiber saturation point depression, 307f
  Southern Yellow Pine, distribution of acetyl group, 306t
  stability of acetyl groups, 306t
Assessing leaching, test methods
  biocide leaching, models, 248
  non-standard test methods, 247
  standardized laboratory test methods, 246
AWPA's standardization, general procedures
  data development, 345
  letter ballot, 347
  objections, resolution and disposition, 347
  Preservatives Review Board, 345
  procedural review and final action, 348
  proposal submission, 346
  recirculation ballot (re-ballot), 347
  technical committee and public review, 346
  technical committee meetings, 346

B

Biological hazard for wood in China, 366
  annual rainfall, classification of regions, 365t
  annual rainfall in China, 364f
  biological hazard map, 368f
  climates and forests, 363
  temperature zones in China, 364f
  termite map with northern boundaries, 368f
  types of termites and their distribution, 367t
  typical cities, climate conditions, 365t
Biotic factors, 120
  insects
    carpenter ants, 125
    carpenter bees, 126
    dampwood termites, 125
    drywood termites, 124
    subterranean termites, 124
    termites, 123
  wood decay fungi
    brown-rot fungi, 121
    mold fungi, 121
    sapstains, 121
    soft-rot fungi, 121
    white-rot fungi, 121
    yeasts, 122
Brown rot fungi, biology, 5
Brown-rot fungal biodegradation
  mechanisms, 3
  fenton-based free radicals, 7
  hydrogen peroxide, 9
  iron-reducing agents, 10
  lignin repolymerization, 15
  lignin side-chain linkages, oxidative alteration, 15
  non-enzymatic brown-rot pathways, 8
Building code, 333
  ICC organization, 334
  International Code Council-evaluation Service (ICC-ES), 334
  recognition
    executive summary, 331
    ICC-ES procedure, 333f

C

CCA. *See* Chromated copper arsenate (CCA)
China, preservation of wood and other sustainable biomaterials, 363
Chromated copper arsenate (CCA), 217
Colony control
  baits, 167
  dusts, 168
  other colony controls, 168
  risk reduction strategies, 168
Commercialization of acetylated wood, 321
  application expertise, 323
  application opportunities and market development, 324

391

byproduct acid, removal and use, 322
challenges, 322
expectations and market risk, 324
wood as raw material, 322
Consensus-based standards
American National Standards Institute (ANSI) accreditation, 343
American Wood Protection Association (AWPA) history, 342
AWPA technical committees, 343
current AWPA standards, 342
data requirements, 344
Copper-based wood preservative systems
Europe, waterborne copper-based systems, 223
carbon-based co-biocides, 224
copper-based wood preservatives, 224
micronized copper systems, 224
preservative penetration, 224
North America, trends, 222

D

Degraded wood, physical properties, 5
Dimethyloldihydroxyethyleneurea (DMDHEU) and its derivatives, wood protection, 287
chemical agents, 288
reaction mechanism and wood treatment, 289
Diversification in fungal kingdom, 101

E

Effect of treatment parameters on leaching
post-treatment conditioning, 242
retention of biocide, 241
Environmentally benign organic biocides, 16
Enzymatic/non-enzymatic white rot decay, 35
Erosion and tunneling decay, features, 49
Erosion bacteria, 49
Evolution of fungal wood decay, 93
agaricomycete nutritional modes, 104
brown rots, 108
Ceriporiopsis subvermispora, 107
comparative analysis, 109
hydroxyl radicals, 109
manganese and lignin peroxidise genes, 107
putative lignocellulose, 106t
transcriptomic and proteomic comparison, 109
white rots, 105
agaricomycota saprotrophic and ectomycorrhizal species genome sequences, 96t
anoxic swamps, 111
comparative genomics, 99
functional genomics, 100
genome size and gene complement, 95t
promoter analysis, 111
review and future work, 110
secondary metabolites, 111
species adaptation, 111
whole genome sequencing, 94
Exposure factors affecting leaching, 242
construction and site parameters, 244
effect of soil properties, 245
finishes and wraps, application, 244
water characteristics, 245
wood used above-ground or above water
effect of rainfall pattern, 243
other climatic factors, 244

F

Fungal and bacterial biodegradation, 23
advanced simultaneous white rot, 27f
aspects of tunneling bacterial attack of wood, 46f
bacteria erosion decay of wood cells, TEM and Cryo-FE-SEM micrographs, 48f
bacterial decay leading to increased permeability, 43
cryo-FE-SEM X-ray analysis, 38
enzymes system involved, 31
erosion bacteria attack, 47f
FEM-SEM X-ray microanalysis, 38
in-vitro culturing of fungi, 30
laccases, 33
lignified tissues, true bacterial decay, 44
lignin degrading enzymes, 32
localization of extracellular lignin- (Lp) and manganese peroxidases (Mn) and laccase (La), 37f
morphological decay types, 25
morphological white rot decay, other types, 40
non-enzymatic processes, 34
other bacteria colonizing wood, 50
oxidoreductase enzymes, 38
pine and birch secondary cell walls, preferential white rot decay, 29f

reactive oxygen species, 34
simultaneous and preferential white rot and tunneling and erosion bacteria decay, 26*f*
soft rot decay, 41
striped decay pattern, 48
tunneling bacteria, 44
white rot decay, importance of slime, 38
white rot enzymes in situ in wood, detection, 36
wood and lignocellulose
  bacterial decay, 43
  brown rot decay, 41
  white rot decay, 24
wood polysaccharides and lignin, biochemical aspects, 30
Fungi, genetic identification. *See* Genetic methods
cultural, 82
genetic methods
immunological assays, 83
methods based on DNA sequence information
phospholipid analysis, 82
traditional methods for identification, morphological, 82
Fungi and plants, early evolution, 100
Fungicides, 186
Fungicides and insecticides used in wood preservation
future outlook, 194
inorganic actives in wood preservation, 190*t*
inorganic biocides, 188
  copper, 189
  other metals, 190
  zinc, 189
modified wood, 195
nanotechnologies, 195
natural biocides, 195
oilborne preservatives, 187
organic actives in wood preservation, 192*t*
organic biocides, 191
  Carbamate fungicides, 193
  Chlorothalonil, 193
  Dichlofluanid, 193
  Isothiazolones, 193
  quaternary ammonium compounds, 193

G

Gene ontology (GO), 65

Genetic methods
general considerations
  central dogma, 83
  PCR primer selection, 84
  polymerase chain reaction, 84
Genome annotation, 63
  BLAST or basic local alignment search tool, 65
  functional annotation, 64
  gene structure from genome, summary, 64*t*
  GO annotations, 66
  protein analysis, 65
  protein sequence homology, 65
  protein signature, 65
  structural annotation, 63
Genomes of fungi, wood decay, 61
Gloeophyllum trabeum, 11
GO. *See* Gene ontology (GO)

H

History of wood preservation in China
  wood preservation, evolution, 371
  wood utilization, ancient China, 369
Holocellulose loss, 11
Hyphae of brown-rot fungi, 6
Hyphal sheath, 6

I

ICC-ES preservative evaluation procedures
  above ground and ground contact issues, 337
  acceptance criteria, 335
  accredited testing and inspection organizations, 336*t*
  appeal process, 338
  ESR holders, 338*t*
  evaluation service report, 337
  preservatives with ESRs, 339*t*
  quality control and inspection, 343
  recent history, 338
  reviews, 336
  testing and accreditation, 335
Industrial wood preservative systems, 352

L

Land colonisation, 101
Leaching in service, evaluations, 248

Leaching of biocides, 239
  dip-immersion methods, 250
  factors affecting leaching
    wood anatomy and chemistry, 241
    wood dimensions and proportion of end-grain, 240
Lignin demethylation, 11
Lignin evolution, 101
Lignin modification, 14

M

Methods based on DNA sequence information
  amplified fragment length polymorphisms (AFLP), 87
  cloning and sequencing, 85
  gradient gel electrophoresis, 88
  multiplex PCR methods, 86
  quantitative PCR (Q-PCR), 85
  restriction fragment length polymorphisms (RFLP), 86
  species-specific probes, 85
  terminal restriction fragment length polymorphisms (T-RFLP), 88
Micro-distribution of preservatives in wood, 227
  particulate wood preservatives, 228
Microscopical events of wood decay, 28

N

Nature of termites, 169
N-methylol compounds treated wood, properties
  durability against biological decay
    brown and white rot decay resistance, 291
    mass loss of mDMDHEU/DEG-treated wood, 292f
    soft-rot decay resistance, 291
  mechanical properties, 293
  moisture content and dimensional stability, 290
  surface properties, 293
    coating performance, 296
    effect of modification of beech wood with DMDHEU/MgCl$_2$.6H$_2$O, 294f
    Scots pine latewood after impact fracture, 295f
    weathering resistance, 296
    wet adhesion of coatings on wood, pull-off method, 295f

North America, wood preservation systems employed
  dispersed particulate copper systems
    dispersed copper azole (DCA), 220
    dispersed or micronized copper quat (MCQ), 220
  dissolved copper systems, 218
    alkaline copper quat (ACQ), 219
    copper azole (CA), 219
    non-biocidal additives, 222
    other standardized copper-based waterborne systems, 221
North America, wood protection trends, 351
  future of treated wood, 359
  long-term possibilities, 357
  possible near term problems (and opportunities), 358
  prior reviews, 352
  treated lumber, durability, 359

P

Pesticides, 186
Plant biomass
  carbonization, carbon structure evolution, 149
  carbonized plant material, homogeneous morphology, 155f
  cellulose and lignin carbon
    apparent kinetic parameters, 156t
    pore volume and surface area, 157t
  graphene sheets, orientation, 151
  large monolithic carbon block, 153f
  large monolithic carbon panels, 153f
  mass yield, 149t
  shrinkage, 150t
  step-wise oxidative carbonization process, 155
  thermal conversion, 154
  thermal decomposition, 148
  thermal degradation and conversion, 147
  Young's Modulus of samples, 152t
Polysaccharide and lignin biodegradation, chemistry, 13
Polysaccharide biodegradation, 13
Prior commercial wood preservation systems, 255
  effect of field ground-contact exposure time, preliminary study, 261
  copper-based residential system, 262
  positive control, 262
  PXTS, 262

prior commercial systems with efficacy concerns
  Australia and Asia, quaternary ammonium compounds, 258
  commercial-sized poles, 256
  common factors, 260
  eucalypts utility poles, CCA treatment, 258
  penta leaching, 257
  tributytin oxide, 260
  volatile solvent pentachlorophenol treatment, 256
  waterborne pentachlorophenol, 259
Progenitor, 101
Properties of acetylated wood
  acetylated pine and aspen, equilibrium moisture content, 309t
  acetylated solid pine, repeated antishrink efficiency (ASE), 310t
  acetylated southern pine exposed to marine environment, 319t
  control and acetylated flakeboards, deflection-time curve, 313f
  control and acetylated pine fiber, thermal properties, 319t
  dimensional stability of solid wood, 310t
  equilibrium moisture content and antishrink efficiency, 311t
  liquid water of pine fiberboards, 309t
  mechanical properties, 320
  moisture and water sorption, 308
  resistance to biological attack
    fungi, fungal cellar test, 315
    fungi, in-ground tests, 315
    fungi, lab tests, 311
    marine organisms, 317
    termites, in-ground tests, 317
    termites, lab tests, 316
  SEM of brown-rot fungal attack on wood, 312f
  sorption/desorption isotherms, 308f
  sugar analysis on pine sample, 314t
  thermal properties, 318
  weathering, 320
PXTS (polymeric xylenol polysulfide) system, 261

R

Repellents, 186
Residential wood preservative systems
  cellulosic renewable materials, protection, 356
  copper-based systems, 353
  decommissioned/waste wood products, utilization, 357
  non copper-based systems
    borates, 353
    naturally durable wood, 355
    nonmetallic preservative systems, 354
    secondary protection systems, 356
    wood modification, 355

S

Service life of wood and wood-based materials above ground, 113. *See* Biotic factors
  abiotic degradation, modes
    anatomical/species related issues, 119
    earlywood and latewood, differential weathering, 115f
    mechanical damage, 117
    surface changes, 114
    weathering process, progression, 116f
    wood-plastic composite deck boards, 118f
  biotic factors
Sustainable biomaterials. *See* Genome Annotation
  accelerating discovery, 71
    comparative genomics, 72
    wood attacking insects, functional genomics, 74
    wood decay, functional genomics, 73
    wood preservative tolerance, functional genomics, 74
  basidiomycota, 61
  brown and white rot fungi, sequenced genomes, 62t
  future perspectives, 75
  genome annotation
  genomics era, 60
  integrating omics data, 70
  metabolomics, 69
  metagenomics, 70
  NGS and genomics, 67t
  omics and future, 59
  other omes, 67
  proteomics, 68
  sequencing technology history, 60
  transcriptomics, 67

T

Termite detection, 167
Termite risk management, 169

Termites and timber. *See* Colony control
colony control
  managing subterranean termites, 160
  resistant framing
    composites, 164
    modified wood, 162
    natural durability, 161
    preservatives, 163
  whole-of-structure approaches
    particle barriers, 165
    pipe collars, 166
    planar physical barriers, 165
    reticulation systems, 166
    termiticide applied to soil, 166
    termiticide impregnated plastic sheet, 166
Thermally modified wood manufactured
  anatomical and chemical changes, 272
  defects in thermally modified spruce, 273f
  coating performance, 280
  gluability, 280
  heat-treated wood, products and production, 272
  properties
    colour and odor, 280
    equilibrium moisture content, 275f
    fracture of heat treated poplar specimen, 279f
    mass loss of untreated and heat treated scots pine, 277f
    maximal volumetric swelling, 276f
    mechanical properties, 278
    resistance against fungi and insects, 276
    sorption and dimensional stability, 274
  quality assessment, 281
  treatment processes, 270
    PLATO-process, 271
    thermally modified wood, 271
    vacuum press dewatering method, 271
    WTT (wood treatment technology), 271
Trametes versicolor and *Phanerochaete chrysosporium*, 38
Treated wood, microdistribution of particulate copper, 232
  location of copper in block of southern pine wood, 236f
  secondary cell wall and middle lamella in latewood tracheids, 234f
  southern pine wood treated with particulate preservative
    X-ray micro-CT images, 235f
    radial longitudinal surface, 233f
  summary and concluding remarks, 236
Treated wood, microdistribution of preservatives
  electron microscopy and energy dispersive analysis of X-rays, 229
  X-ray fluorescence microscopy, 230
  X-ray microcomputed tomography, 231
Treatment technologies
  fixed biocides, 207
  flow equation, 205
  full cell process, 211
  future wood treatments, 213
    composites, treatment, 214
    refractory woods, resist impregnation, 214
    supercritical fluid (SCF) processes, 215
    vapour phase treatments, 215
  pressure treatments, 210
  relative difficulty of impregnating heartwoods, 206t
  sapwood and heartwood, moisture contents, 209t
  treatment process, 208
  treatment results, 212
  viscosity of the fluid, 205
  wood moisture content, 207
  wood substrate, 204
  wood treating vessel or retort, 211f

U

Unicellular wood degrading bacteria, 137

W

White rot decay, type, 28
Wood decay
  early evolution, 102
    Basidiomycota, 103
    carboniferous period, 103
    coastal swamp forests, 103
    Fungal class II, 104
    oxygen-requiring peroxidation mechanism, 103
    versatile peroxidases, 104
  fungi, genetic identification, 81
Wood deterioration
  detoxifying organisms, 141
  predicting durability of wood in soil, methods, 142
  wood-rotting basidiomycetes, 132
    brown rot fungi, 134

396

copper-tolerant brown rot fungi, 134
mycelium, 133
soft rot-fungi, 135
white-rot inoculum potential, 133
wood degrading bacteria, 136
Wood preservation, ancient China
  selection of wood, 370
  treatment methods, 371
  wood structure, design, 371
Wood preservation in modern China
  effective ingredients for different wood preservatives, minimal retentions, 380t
  future expectations, 381
  preservatives and preservative-treated wood
    rubberwood treated inefficiently with resultant blue stains, 375f
  treatment methods, 373
  treatment of bamboo, 376
  wood species used for treatment, 374
  standardization, 378
  treated wood, applications, 377
  water-borne wood preservatives, 379t
  wood preservatives used, 372
Wood product service life in ground contact
  ecology of wood decay and other factors, 138
  copper-based preservative systems, 139
  copper-tolerant decay fungi, 139
  degrees of field study, 140
  specific soil factors, 140
  wood substrate, physico-chemical properties, 140
Woody materials, biodegradation, 4